C-Type Lectin Receptors in Immunity

Sho Yamasaki
Editor

C-Type Lectin Receptors in Immunity

Springer

Editor
Sho Yamasaki
Medical Institute of Bioregulation
Kyushu University
Fukuoka
Japan

ISBN 978-4-431-56013-5 ISBN 978-4-431-56015-9 (eBook)
DOI 10.1007/978-4-431-56015-9

Library of Congress Control Number: 2016933249

Printed on acid-free paper

This Springer imprint is published by Springer Nature
The registered company is Springer Japan KK

Contents

Part I
ITAM-Coupled Activating Receptors

Chapter 1
Dectin-2 in Antimicrobial Immunity and Homeostasis

Rikio Yabe and Shinobu Saijo

Abstract Dendritic cell-associated lectin-2 (Dectin-2) is one of the most well-characterized members of the C-type lectin family. Recent studies have revealed its indispensable functions as a pattern recognition receptor (PRR) for a wide variety of pathogens, including fungi, bacteria, and viruses. This receptor recognizes microbial carbohydrates as a pathogen-associated molecular pattern (PAMP). Upon ligand ligation, Dectin-2 induces secretion of the pro-inflammatory cytokines such as interleukin (IL)-1β, IL-6, and TNF, as well as the inhibitory cytokine IL-10. These cytokines differentiate T cells into IL-17-producing Th17 cells to eliminate pathogens. In addition to microbes, Dectin-2 also binds to allergens such as those of house dust mites and helminths to activate the NLRP3 inflammasome. In vivo, Dectin-2 plays a key role in antimicrobial infection, especially antifungal infections. Owing to these abilities, Dectin-2 agonists could be promising adjuvants in vaccinations. In this section, we summarize the current knowledge of Dectin-2 in detail, describing its structure, ligand recognition, signaling, and associated human diseases.

Keywords Dectin-2 • Fungal infection • Bacterial infection • Innate immunity • Inflammation • Carbohydrate • High mannose

1.1 Introduction

Dendritic cell-associated lectin-2 (Dectin-2, gene symbol: *Mus musculus Clec4n*, *Homo sapiens CLEC6A* or *CLEC4N*) was originally found as a Langerhans cell-specific C-type lectin that recognizes self-ligands expressed in $CD4^{+}CD25^{+}$ T cells (Ariizumi et al. 2000). Subsequently, Dectin-2 expression was found in myeloid cells, including monocytes, tissue macrophages, neutrophils, several dendritic cell (DC) subsets, and B lymphoid cells (McDonald et al. 2012; Robinson et al. 2009; Seeds et al. 2009; Taylor et al. 2005). The expression of the gene is rather low in

R. Yabe • S. Saijo (✉)
Division of Molecular Immunology, Medical Mycology Research Center, Chiba University, 1-8-1 Inohana Chuo-ku, Chiba City, Chiba 260-8673, Japan
e-mail: saijo@faculty.chiba-u.jp

S. Yamasaki (ed.), *C-Type Lectin Receptors in Immunity*,
DOI 10.1007/978-4-431-56015-9_1

these cells, although it is greatly enhanced with the inflammatory stimuli. Nevertheless, the molecular mechanisms underlying the regulation of Dectin-2 expression still remain to be elucidated.

Dectin-2 is encoded in the mouse chromosome 6F and in the syntenic region on human chromosome 12q13, where several C-type lectin genes are located and form clusters, including Dectin-1 and Dectin-2 clusters (Fig. 1.1a) (Balch et al. 2002; Fujikado et al. 2006). In the mouse, nine members, DCIR4, DCIR3, DCIR2, DCAR2, DCIR1, DCAR-1, Dectin-2, MCL, and Mincle (gene symbols: *Clec4a1*, *Clec4a3*, *Clec4a4*, *Clec4b1*, *Clec4a2*, *Clec4b2*, *Clec4n*, *Clec4d*, and *Clec4e*), are mapped in close vicinity on the Dectin-2 cluster of chromosome 6, sharing a common structure (see next section). By contrast, fewer molecules are found in the Dectin-2 cluster in humans, i.e., BDCA-2, DCIR, DECTIN-2, MCL, and MINCLE (gene symbols: *CLEC4C*, *CLEC4A*, *CLEC6A*, *CLECSF8*, and

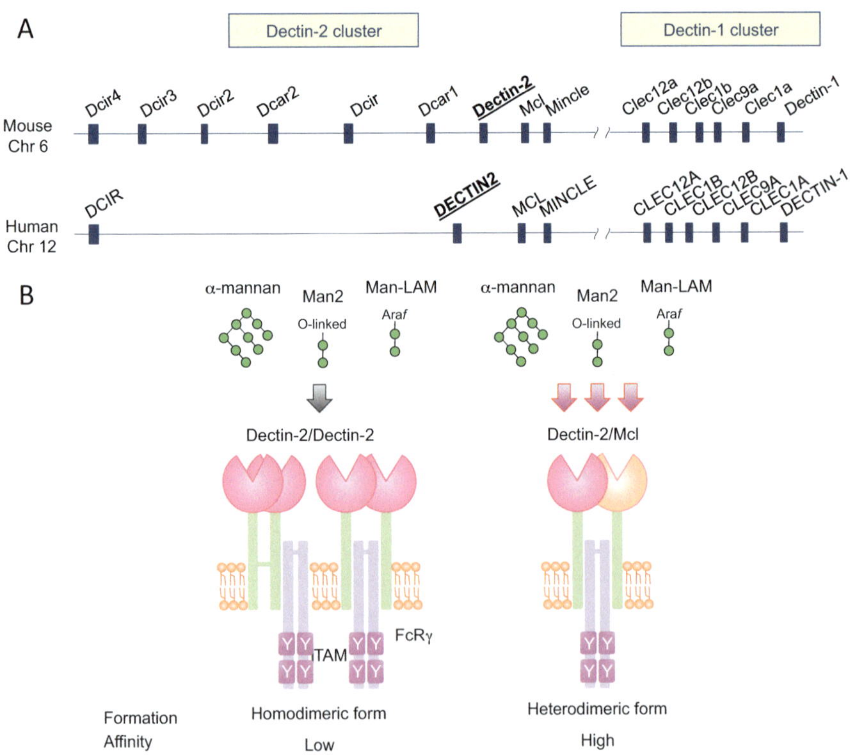

Fig. 1.1 **a** Organization of mouse and human Dectin-1 and Dectin-2 clusters. **b** Dectin-2 senses α-mannosylated chains and initiates cellular responses through association with the Fc receptor γ chain (FcRγ), which contains immunoreceptor tyrosine-based activation motif (ITAMs). Dectin-2 is presented as a homodimer through a disulfide bond or in association with FcRγ, whereas it forms a heterodimer with MCL linked by FcRγ. The heterodimeric complex has relatively strong affinity to pathogen-associated molecular patterns (PAMPs) in comparison with the homodimeric complex

CLEC4E). Due to the common gene structure shared by the receptors and the variation of the family members between species, it is speculated that this gene cluster may have been established by gene duplication. Interestingly, this region of the chromosome harboring the Dectin-2 and Dectin-1 clusters has been implicated in several autoimmune diseases by linkage studies (Fujikado et al. 2006; Wandstrat and Wakeland 2001). Therefore, it is possible that C-type lectins could be one of the susceptibility genes for these diseases (Caliz et al. 2013).

Of note, Dectin-2 may have an immune inhibition role because of its unique ability to produce anti-inflammatory cytokine, interleukin (IL)-10. However, compared to its immune-activating or antimicrobial functions, the physiological role of the immune-inhibiting activity of Dectin-2 remains largely unknown. In this section, we update paradigm and summary findings on Dectin-2 and its functions that have been made in recent years.

1.2 Structure of Dectin-2

Dectin-2 is a glycosylated type II transmembrane protein, whose C-terminal portion encodes the extracellular region and the N-terminal portion encodes the cytoplasmic region of the receptor (Ariizumi et al. 2000). This protein is encoded by six exons and has a single carbohydrate recognition domain (CRD) in the extracellular region, a stalk region, a transmembrane region, and a short cytoplasmic domain with no known signaling motif. These structures are shared among other C-type lectins in the Dectin-2 cluster with the exception of DCIR and DCIR2, which have an inhibitory motif, immunoreceptor tyrosine-based inhibitory motif (ITIM), in their cytoplasmic region. Dectin-2 associates with an adapter molecule, Fc receptor γ chain (FcRγ, gene symbol: *Fcer1g*) to transduce its signaling. Although Dectin-2 contains an arginine residue, which often mediates associations with immunoreceptor tyrosine-based activation motif (ITAM)-containing adaptor molecules in the transmembrane region, the interaction between these two molecules is not dependent on the arginine residue unlike other FcRγ-coupled receptors, but instead requires the cytoplasmic tail of Dectin-2 (Sato et al. 2006). As Dectin-2 has a conserved cysteine residue in its stalk region, which can form disulfide-linked homodimers, it was implicated that this receptor forms homodimers by ligand recognition. Recently, Zhu et al. demonstrated that Dectin-2 forms heterodimers with another C-type lectin, MCL, as well as homodimers, and the heterodimers showed stronger affinity to the ligand than the homodimers (Fig. 1.1b). Requirement of the cysteine residue of Dectin-2 for the heterodimerization and the molecular mechanism for this dimer formation are still unknown.

1.3 Ligands of Dectin-2

Dectin-2 contains an acid–proline–asparagine (EPN) amino acid triplet in its extracellular CRD, a common feature that is known to facilitate binding to mannose, glucose/*N*-acetylglucosamine, and fucose in a Ca^{2+}-dependent manner and has attracted much attention with respect to its role as a pattern recognition receptor (PRR) (Fernandes et al. 1999). Indeed, a study using a glycan array revealed that Dectin-2 binds to high-mannose structures that are distributed in a wide range of species, including fungi, parasites, bacteria, and mammals (McGreal et al. 2006). Consistent with this, using a recombinant soluble form of Dectin-2 CRD as a probe, or Dectin-2-expressing reporter cells, it was shown that Dectin-2 binds to a variety of pathogenic microorganisms, including *Candida* spp., *Saccharomyces cerevisiae*, *Cryptococcus neoformans*, *Trichophyton rubrum*, *Paracoccidioides brasiliensis*, *Malassezia* spp., and *Aspergillus fumigatus*, *Mycobacterium* spp., *Streptococcus pneumoniae*, and *Ebola* virus, whose cell wall polysaccharides contain a high-mannose structure (Table 1.1) (Barrett et al. 2009; Brudner et al. 2013; McGreal et al. 2006; Miyasaka et al. 2013; Ritter et al. 2010; Sato et al. 2006).

Among these pathogens, the best-studied interactions are with *Candida albicans* (*C. albicans*), *Malassezia*, and *Mycobacterium*. The *C. albicans* cell wall is composed of multiple layers, including the outermost mannans (polymers of mannose), middle-layer β-glucans (polymers of D-glucose linked by β-glycosidic bonds), and the inner-layer chitins (polymers of *N*-acetylglucosamine) (Odds 1988). Mannans from *C. albicans* have β-1,2-linked mannose residues attached to their α-1,2-linked oligomannose side chains with α-1,6-linked mannose backbone (Fig. 1.2a) (Cutler 2001; Shibata et al. 2003). Since the β-1,2-linked mannose residues are not synthesized when this fungus is cultured in a carbon-limited (C-limited), low-pH

Table 1.1 Dectin-2 sensing microbial pathogens

Category	Microbial pathogens	Ligands
Bacteria	*Mycobacterium* spp.	Man-LAM
	S. pneumonia	
Fungi	*Candida* spp.	α-mannans, high-mannose-type *N*-glycans
	Malassezia spp.	*O*-linked mannobioses
	A. fumigates	
	C. neoformans (non-capsulated)	
	Saccharomyces cerevisiae	
	Coccidioides spp.	
	B. dermatitidis	
	H. capsulatum	
	T. rubrum	
	P. brasiliensis	
Virus	Ebola	
Others	*S. mansoni*	
	House dust mite allergen	

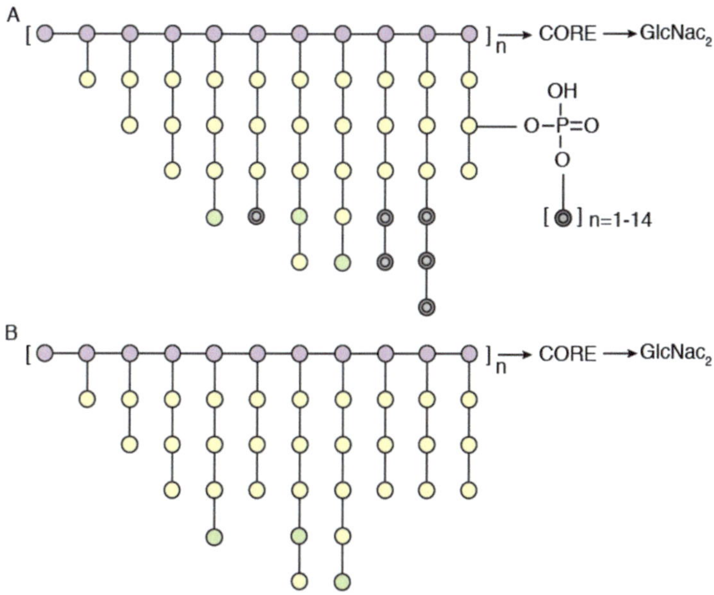

Fig. 1.2 The structure of the mannosyl residues of an *N*-linked mannoprotein. **a** *C. albicans* cultured at 37 °C with normal medium. **b** *C. albicans* cultured at 27 °C with C-limiting medium. α-1,6-linked mannose (*pink circles*), α-1,2-linked mannose (*yellow circles*), β-1,2-linked mannose (*gray circles*), α-1,3-linked mannose (*green*) (Adapted from Hobson et al. 2004)

medium at a low temperature, only α-1,2-mannose residues are synthesized and secreted into the medium as a water-soluble fraction (Shinohara et al. 2006) (Fig. 1.2b). When Dectin-2-deficient bone marrow-derived DCs (BMDCs) are stimulated with these α-mannans, the production of cytokines such as interleukin (IL)-6 and TNF was abolished, indicating that Dectin-2 is a receptor for *C. albicans* α-mannans. *C. albicans* is a dimorphic fungus that can exist in yeast or hyphal form depending on its growth environment (Saijo et al. 2010). A previous study revealed that Dectin-2 selectively binds to the hyphal form of *C. albicans* (Sato et al. 2006). Indeed, significant, but not complete, reduction in cytokine secretion was found in Dectin-2-deficient BMDCs upon stimulation with the hyphal form of *C. albicans*. Notably, cytokine production of Dectin-2-deficient BMDCs was eradicated in response to the yeast form of *C. albicans*, indicating that Dectin-2 is the only receptor to produce cytokines for the yeast form of the fungus.

Malassezia, which is found in the normal flora of the human skin, is another opportunistic fungal pathogen. Ishikawa et al. identified the *O*-linked mannoprotein fractionated from the *Malassezia* cell wall as a distinct Dectin-2 ligand. By using Dectin-2-expressing reporter cells, they also showed that α-1,2-mannosyl residues were necessary and sufficient for recognition by the receptor (Table 1.1) (Ishikawa et al. 2013). Recently, Dectin-2 was found to recognize a cell wall component of *Mycobacterium*, mannose-capped lipoarabinomannan (Man-LAM), which consists of a mannosyl-phosphatidyl-myo-inositol anchor, a mannose backbone, and an arabinan domain with mannose capping (Yonekawa et al. 2014). Interestingly, a distinct mycobacterial cell wall component, trehalose-6,6′-dimycolate (TDM), is recognized by MINCLE and MCL, whose genes are closely located to the Dectin-2 locus (Ishikawa et al. 2009; Miyake et al. 2013).

1.4 Dectin-2 Signaling

Upon recognition of the carbohydrate structures in pathogens, Dectin-2 initiates a series of cellular responses beginning with the association of the ITAM-containing adaptor molecule FcRγ, followed by recruitment of phosphorylated Syk (Fig. 1.3). Subsequently, phosphorylated Syk activates a CARD9–BCL10–MALT1 (CBM) complex, resulting in activation of nuclear factor (NF)-κB (Hara et al. 2007; Saijo et al. 2010). The activated NF-κB induces expression of inflammatory cytokines such as IL-23 and pro-IL-1β. At the same time, Syk activation induces reactive oxygen species (ROS) production. ROS are important for the direct killing of pathogens and activation of the NLRP3 inflammasome that enhances processing of pro-IL-1β into mature IL-1β (Ritter et al. 2010). The mitogen-activated protein kinase (MAPK) pathway is also activated simultaneously, although the biological significance of this pathway in the host defense mechanism is not known (Fig. 1.3).

It is noteworthy that Dectin-2 signaling also induces the production of an inhibitory cytokine, IL-10, by the stimulation of α-mannans of the *C. albicans* cell wall as well as Man-LAM of the *Mycobacterium tuberculosis* cell wall (Saijo et al. 2010; Yonekawa et al. 2014). Since the capping moieties at the terminal extremity of the arabinan domain in LAM differ among *Mycobacterium* species, Man-LAM is rather unique to pathogenic mycobacteria such as *M. tuberculosis*. In this context, it has been shown that *M. tuberculosis* suppresses the host immune system and phagosome–lysosome fusion, suggesting that IL-10 production via the Man-LAM–Dectin-2 interaction might be an explanation for the severe pathogenicity of *M. tuberculosis*.

1.5 Role of Dectin-2 in Diseases

The importance of Dectin-2 in antifungal immunity has been clearly shown in mouse models (Robinson et al. 2009). Dectin-2-deficient mice showed decreased survival against *C. albicans* infection, because of increased fungal growth in the kidney (Saijo et al. 2010). This demonstrated the important role of Dectin-2 in pathogen elimination; however, the in vivo mechanism remains unclear. One possibility is that the pro-inflammatory cytokines secreted from Dectin-2-expressing myeloid cells enhance IL-17 expression in lymphocytes. Indeed, IL-17A-deficient and IL-23-p19-deficient mice are more vulnerable to *C. albicans* infection, suggesting that IL-17A is crucial for the host protection

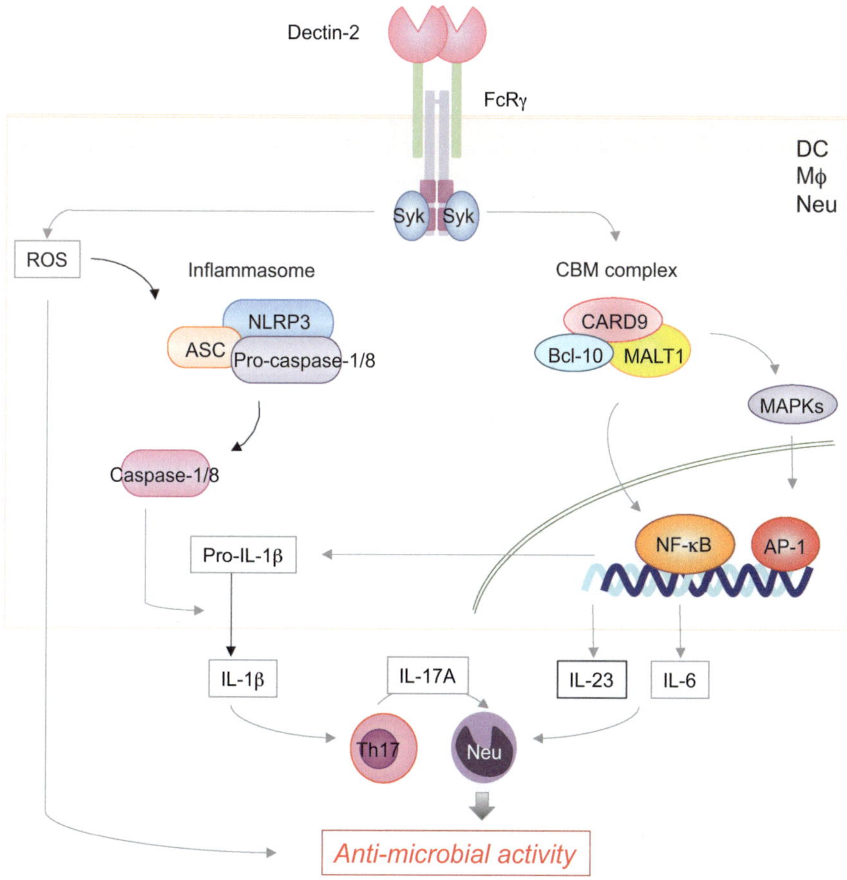

Fig. 1.3 Upon ligand binding, Dectin-2 recruits phosphorylated Syk to ITAM of the FcRγ, leading to activation of the CARD9–BCL10–MALT1 (CBM) complex. At the same time, reactive oxygen species (ROS) production is induced in a Syk-dependent manner, resulting in the direct killing of pathogens and activation of the NLRP3 inflammasome. The CBM complex activates NF-κB, which induces the production of cytokines such as pro-IL-1β, IL-6, and IL-23. In contrast, the NLRP3 inflammasome activates caspase 1 and/or caspase 8 to process pro-IL-1β into mature IL-1β. IL-1β, IL-6, and IL-23 preferentially induce the differentiation of Th17 cells, which play an important role in the host defense against microbes by recruiting neutrophils

against this fungus (Kagami et al. 2010; Saijo et al. 2010). Although single nucleotide polymorphisms (SNPs) or inborn errors in the human Dectin-2 gene have not been reported yet, mutations that affect human IL-17F and IL-17 receptor A (IL-17RA) functions cause the development of chronic mucoepithelial fungal infection (CMC), which is due to impaired IL-17-mediated immunity (Puel et al. 2011). Consistent with these observations, Dectin-2-induced cytokines preferentially promote the differentiation of Th17 cells in vitro, although the specific IL-17A- and IL-17F-producing cells upon fungal infection remain to be elucidated.

On the other hand, it was recently reported that Dectin-2-induced IL-6 and IL-23 enhanced IL-17A production in neutrophils that constitutively express the transcription factor RORγt, upon infection with *A. fumigatus*. IL-6 and IL-23 also induce the expression of IL-17RC and Dectin-2 in an autocrine manner, resulting in the production of ROS and increased fungal killing (Taylor et al. 2014) (Fig. 1.3).

Regarding Th17 differentiation caused by Dectin-2-mediated signaling, as described above, Dectin-2 recognizes *Mycobacterium* Man-LAM, which has been known to have both inhibitory and stimulatory effects on host immunity (Briken et al. 2004; Chan et al. 2001; Gringhuis et al. 2009; Mazurek et al. 2012). Man-LAM stimulation of BMDCs cocultured with T cells led to increased production of IL-17 in T cells. More importantly, Man-LAM stimulation was sufficient to promote host immunity for the development of experimental autoimmune encephalomyelitis (EAE), a mouse model of multiple sclerosis. Given that Th17 cells play a pivotal role in the pathogenesis in these mice, this *Mycobacterium* component or other Dectin-2 agonists might serve as a beneficial adjuvant (Yonekawa et al. 2014).

Dectin-2 also senses mannan-containing parasites, including house dust mite and *Schistosoma mansoni* (Barrett et al. 2011; Norimoto et al. 2014; Ritter et al. 2010). In these cases, Dectin-2 seems to induce Th1, Th2, as well as Th17 differentiation, suggesting a broader swath of its immunological roles.

1.6 Concluding Remarks

The significant functions of Dectin-2 in the host defense against pathogen infection have been studied and are described in this review. However, the in vivo roles of Dectin-2 remain to be unveiled. Interestingly, some other C-type lectins have redundant biological functions to Dectin-2. For example, MCL senses the same ligand, *C. albicans* α-mannan, as Dectin-2, and, more importantly, MCL and Dectin-2 form heterodimers as well as homodimers (Zhu et al. 2013). On the other hand, MCL also stabilizes Mincle expression, and both receptors sense mycobacterial TDM (Miyake et al. 2013, 2015). These findings suggest that the C-type lectin system provides a sophisticated and fail-safe method to respond to pathogens using a limited number of molecules. Determining the collaboration between C-type lectins and other innate immune receptors such as toll-like receptors (TLRs), retinoic acid-inducible gene-I-like receptors (RLRs), or nucleotide oligomerization domain-like receptors (NLRs) could be another interesting area of research. For instance, Dectin-1 and TLR-2 collaboratively enhance IL-12 production to induce Th1 differentiation, which is important for fungal and bacterial protection (Dennehy et al. 2009; Gerosa et al. 2008). These studies could help to elucidate the whole picture of the role of the C-type lectin system in host defense, with valuable implications for the development of new therapeutic and vaccine strategies.

References

Ariizumi K, Shen GL, Shikano S, Ritter R 3rd, Zukas P, Edelbaum D, Morita A, Takashima A (2000) Cloning of a second dendritic cell-associated C-type lectin (dectin-2) and its alternatively spliced isoforms. J Biol Chem 275:11957–11963

Balch SG, Greaves DR, Gordon S, McKnight AJ (2002) Organization of the mouse macrophage C-type lectin (Mcl) gene and identification of a subgroup of related lectin molecules. Eur J Immunogenet: Off J Br Soc Histocompatibility Immunogenet 29:61–64

Barrett NA, Maekawa A, Rahman OM, Austen KF, Kanaoka Y (2009) Dectin-2 recognition of house dust mite triggers cysteinyl leukotriene generation by dendritic cells. J Immunol 182:1119–1128

Barrett NA, Rahman OM, Fernandez JM, Parsons MW, Xing W, Austen KF, Kanaoka Y (2011) Dectin-2 mediates Th2 immunity through the generation of cysteinyl leukotrienes. J Exp Med 208:593–604

Briken V, Porcelli SA, Besra GS, Kremer L (2004) Mycobacterial lipoarabinomannan and related lipoglycans: from biogenesis to modulation of the immune response. Mol Microbiol 53:391–403

Brudner M, Karpel M, Lear C, Chen L, Yantosca LM, Scully C, Sarraju A, Sokolovska A, Zariffard MR, Eisen DP et al (2013) Lectin-dependent enhancement of Ebola virus infection via soluble and transmembrane C-type lectin receptors. PLoS One 8:e60838

Caliz R, Canet LM, Lupianez CB, Canhao H, Escudero A, Filipescu I, Segura-Catena J, Soto-Pino MJ, Exposito-Ruiz M, Ferrer MA et al (2013) Gender-specific effects of genetic variants within Th1 and Th17 cell-mediated immune response genes on the risk of developing rheumatoid arthritis. PLoS One 8:e72732

Chan ED, Morris KR, Belisle JT, Hill P, Remigio LK, Brennan PJ, Riches DW (2001) Induction of inducible nitric oxide synthase-NO* by lipoarabinomannan of Mycobacterium tuberculosis is mediated by MEK1-ERK, MKK7-JNK, and NF-kappaB signaling pathways. Infect Immun 69:2001–2010

Cutler JE (2001) N-glycosylation of yeast, with emphasis on Candida albicans. Med Mycol 39 (Suppl 1):75–86

Dennehy KM, Willment JA, Williams DL, Brown GD (2009) Reciprocal regulation of IL-23 and IL-12 following co-activation of Dectin-1 and TLR signaling pathways. Eur J Immunol 39:1379–1386

Fernandes MJ, Finnegan AA, Siracusa LD, Brenner C, Iscove NN, Calabretta B (1999) Characterization of a novel receptor that maps near the natural killer gene complex: demonstration of carbohydrate binding and expression in hematopoietic cells. Cancer Res 59:2709–2717

Fujikado N, Saijo S, Iwakura Y (2006) Identification of arthritis-related gene clusters by microarray analysis of two independent mouse models for rheumatoid arthritis. Arthritis Res Ther 8: R100

Gerosa F, Baldani-Guerra B, Lyakh LA, Batoni G, Esin S, Winkler-Pickett RT, Consolaro MR, De Marchi M, Giachino D, Robbiano A et al (2008) Differential regulation of interleukin 12 and interleukin 23 production in human dendritic cells. J Exp Med 205:1447–1461

Gringhuis SI, den Dunnen J, Litjens M, van der Vlist M, Geijtenbeek TB (2009) Carbohydrate-specific signaling through the DC-SIGN signalosome tailors immunity to Mycobacterium tuberculosis, HIV-1 and Helicobacter pylori. Nat Immunol 10:1081–1088

Hara H, Ishihara C, Takeuchi A, Imanishi T, Xue L, Morris SW, Inui M, Takai T, Shibuya A, Saijo S et al (2007) The adaptor protein CARD9 is essential for the activation of myeloid cells through ITAM-associated and Toll-like receptors. Nat Immunol 8:619–629

Hobson RP, Munro CA, Bates S, MacCallum DM, Cutler JE, Heinsbroek SE, Brown GD, Odds FC, Gow NA (2004) Loss of cell wall mannosylphosphate in Candida albicans does not influence macrophage recognition. J Biol Chem 279:39628–39635

Ishikawa E, Ishikawa T, Morita YS, Toyonaga K, Yamada H, Takeuchi O, Kinoshita T, Akira S, Yoshikai Y, Yamasaki S (2009) Direct recognition of the mycobacterial glycolipid, trehalose dimycolate, by C-type lectin Mincle. J Exp Med 206:2879–2888

Ishikawa T, Itoh F, Yoshida S, Saijo S, Matsuzawa T, Gonoi T, Saito T, Okawa Y, Shibata N, Miyamoto T et al (2013) Identification of distinct ligands for the C-type lectin receptors Mincle and Dectin-2 in the pathogenic fungus Malassezia. Cell Host Microbe 13:477–488

Kagami S, Rizzo HL, Kurtz SE, Miller LS, Blauvelt A (2010) IL-23 and IL-17A, but not IL-12 and IL-22, are required for optimal skin host defense against Candida albicans. J Immunol 185:5453–5462

Mazurek J, Ignatowicz L, Kallenius G, Svenson SB, Pawlowski A, Hamasur B (2012) Divergent effects of mycobacterial cell wall glycolipids on maturation and function of human monocyte-derived dendritic cells. PLoS One 7:e42515

McDonald JU, Rosas M, Brown GD, Jones SA, Taylor PR (2012) Differential dependencies of monocytes and neutrophils on dectin-1, dectin-2 and complement for the recognition of fungal particles in inflammation. PLoS One 7:e45781

McGreal EP, Rosas M, Brown GD, Zamze S, Wong SY, Gordon S, Martinez-Pomares L, Taylor PR (2006) The carbohydrate-recognition domain of Dectin-2 is a C-type lectin with specificity for high mannose. Glycobiology 16:422–430

Miyake Y, Toyonaga K, Mori D, Kakuta S, Hoshino Y, Oyamada A, Yamada H, Ono K, Suyama M, Iwakura Y et al (2013) C-type lectin MCL is an FcRgamma-coupled receptor that mediates the adjuvanticity of mycobacterial cord factor. Immunity 38:1050–1062

Miyake Y, Oh-Hora M, Yamasaki S (2015) C-type lectin receptor MCL facilitates Mincle expression and signaling through complex formation. J Immunol 194:5366–5374

Miyasaka T, Akahori Y, Toyama M, Miyamura N, Ishii K, Saijo S, Iwakura Y, Kinjo Y, Miyazaki Y, Oishi K et al (2013) Dectin-2-dependent NKT cell activation and serotype-specific antibody production in mice immunized with pneumococcal polysaccharide vaccine. PLoS One 8:e78611

Norimoto A, Hirose K, Iwata A, Tamachi T, Yokota M, Takahashi K, Saijo S, Iwakura Y, Nakajima H (2014) Dectin-2 promotes house dust mite-induced T helper type 2 and type 17 cell differentiation and allergic airway inflammation in mice. Am J Respir Cell Mol Biol 51:201–209

Odds FC (1988) Candida and candidosis, 2nd edn. Baillere-Tindall, London

Puel A, Cypowyj S, Bustamante J, Wright JF, Liu L, Lim HK, Migaud M, Israel L, Chrabieh M, Audry M et al (2011) Chronic mucocutaneous candidiasis in humans with inborn errors of interleukin-17 immunity. Science 332:65–68

Ritter M, Gross O, Kays S, Ruland J, Nimmerjahn F, Saijo S, Tschopp J, Layland LE, Prazeres da Costa C (2010) Schistosoma mansoni triggers Dectin-2, which activates the Nlrp3 inflammasome and alters adaptive immune responses. Proc Natl Acad Sci U S A 107:20459–20464

Robinson MJ, Osorio F, Rosas M, Freitas RP, Schweighoffer E, Gross O, Verbeek JS, Ruland J, Tybulewicz V, Brown GD et al (2009) Dectin-2 is a Syk-coupled pattern recognition receptor crucial for Th17 responses to fungal infection. J Exp Med 206:2037–2051

Saijo S, Ikeda S, Yamabe K, Kakuta S, Ishigame H, Akitsu A, Fujikado N, Kusaka T, Kubo S, Chung SH et al (2010) Dectin-2 recognition of alpha-mannans and induction of Th17 cell differentiation is essential for host defense against Candida albicans. Immunity 32:681–691

Sato K, Yang XL, Yudate T, Chung JS, Wu J, Luby-Phelps K, Kimberly RP, Underhill D, Cruz PD Jr, Ariizumi K (2006) Dectin-2 is a pattern recognition receptor for fungi that couples with the Fc receptor gamma chain to induce innate immune responses. J Biol Chem 281:38854–38866

Seeds RE, Gordon S, Miller JL (2009) Characterisation of myeloid receptor expression and interferon alpha/beta production in murine plasmacytoid dendritic cells by flow cytomtery. J Immunol Methods 350:106–117

Shibata N, Kobayashi H, Okawa Y, Suzuki S (2003) Existence of novel beta-1,2 linkage-containing side chain in the mannan of Candida lusitaniae, antigenically related to Candida albicans serotype A. Eur J Biochem 270:2565–2575

Shinohara H, Nagi-Miura N, Ishibashi K, Adachi Y, Ishida-Okawara A, Oharaseki T, Takahashi K, Naoe S, Suzuki K, Ohno N (2006) Beta-mannosyl linkages negatively regulate anaphylaxis and vasculitis in mice, induced by CAWS, fungal PAMPS composed of mannoprotein-beta-glucan complex secreted by Candida albicans. Biol Pharm Bull 29:1854–1861

Taylor PR, Reid DM, Heinsbroek SE, Brown GD, Gordon S, Wong SY (2005) Dectin-2 is predominantly myeloid restricted and exhibits unique activation-dependent expression on maturing inflammatory monocytes elicited in vivo. Eur J Immunol 35:2163–2174

Taylor PR, Roy S, Leal SM Jr, Sun Y, Howell SJ, Cobb BA, Li X, Pearlman E (2014) Activation of neutrophils by autocrine IL-17A-IL-17RC interactions during fungal infection is regulated by IL-6, IL-23, RORgammat and dectin-2. Nat Immunol 15:143–151

Wandstrat A, Wakeland E (2001) The genetics of complex autoimmune diseases: non-MHC susceptibility genes. Nat Immunol 2:802–809

Yonekawa A, Saijo S, Hoshino Y, Miyake Y, Ishikawa E, Suzukawa M, Inoue H, Tanaka M, Yoneyama M, Oh-Hora M et al (2014) Dectin-2 is a direct receptor for mannose-capped lipoarabinomannan of mycobacteria. Immunity 41:402–413

Zhu LL, Zhao XQ, Jiang C, You Y, Chen XP, Jiang YY, Jia XM, Lin X (2013) C-type lectin receptors Dectin-3 and Dectin-2 form a heterodimeric pattern-recognition receptor for host defense against fungal infection. Immunity 39:324–334

Chapter 2
Pathogen-Sensing by Mincle: Function and Molecular Aspects

Masahiro Nagata, Zakaria Omahdi, and Sho Yamasaki

Abstract The C-type lectin receptor called Mincle (macrophage-inducible C-type lectin) is a pattern recognition receptor (PRR) mainly expressed by myeloid cells. Over the years, Mincle has been reported to recognize pathogen-associated molecular patterns (PAMPs) from several microorganisms. Among these PAMPs, the most studied is trehalose-6,6-dimycolate (TDM), which is also the most abundant glycolipid present in the cell wall of *Mycobacterium tuberculosis*. Moreover, it has also been demonstrated that Mincle is involved in fungi recognition, and a growing number of reports show that this PRR may recognize other pathogens. However, in some cases the ligands are still unknown, or the exact role of Mincle in the immune response against these pathogens is unclear. In this chapter, we will begin by presenting the pathogens recognized by Mincle. Then, the Mincle-glycolipid interaction will be described at the molecular level. And last but not least, we will discuss the immune response triggered through Mincle.

Keywords Mincle • C-type lectin • Glycolipid recognition • Adjuvant receptor

2.1 Introduction

Prior to the discovery of pattern recognition receptors (PRRs) in the 1990s, the innate immune system was thought to be nonspecific and solely implicated in the clearance of pathogens. The PRRs expressed by antigen-presenting cells (APCs) in fact allow innate immune cells to recognize and discriminate pathogens, which is of tremendous importance for the establishment of a robust and adequate acquired response (Kawai and Akira 2010). By recognizing PAMPs (pathogen-associated molecular pattern), molecular signatures from essential components shared by entire groups of microorganisms, PRRs allow the innate immune system to orient the acquired response efficiently (Geijtenbeek and Gringhuis 2009).

M. Nagata • Z. Omahdi • S. Yamasaki (✉)
Division of Molecular Immunology, Medical Institute of Bioregulation, Kyushu University, Fukuoka, Japan
e-mail: yamasaki@bioreg.kyushu-u.ac.jp

S. Yamasaki (ed.), *C-Type Lectin Receptors in Immunity*,
DOI 10.1007/978-4-431-56015-9_2

One of the most important and studied PRR families is the C-type lectin family of receptors (CLRs). Lectins are proteins characterized by their capacity to bind carbohydrates (Teuschl et al. 2014). C-type lectins are a particular subfamily of lectins that require Ca^{2+} to bind sugar moieties (Zelensky and Gready 2005; Richardson and Williams 2014). They share homology in their compact globular module responsible for the carbohydrate-binding activity called CRD (carbohydrate recognition domain) (Zelensky and Gready 2005; Cummings and McEver 2009). Due to this ability, CLRs can elicit the immune response through the recognition of various glycosylated or even non-glycosylated targets. Macrophage-inducible C-type lectin (Mincle), also called Clec4e or Clecsf9, is a type II transmembrane CLR, genetically mapped to mouse chromosome 6F2 and human chromosome 12p31 (Matsumoto et al. 1999; Flornes et al. 2004). Mincle was at first characterized as one of the transcriptional targets of NF-IL6 (nuclear factor for interleukin-6, formerly known as C/EBPβ), and it was shown that although its expression in resting cells is low, it can be induced by various cellular stresses and stimuli in professional APCs (dendritic cells, macrophages, neutrophils, and B cells). However, its function remained elusive for about a decade (Richardson and Williams 2014; Matsumoto et al. 1999). It is in 2008 that the first reports, revealing the function and the ligands recognized by Mincle, were published. On one hand, Wells et al. showed that Mincle is an essential actor in the immune response against *Candida albicans* by inducing inflammatory signaling (Wells et al. 2008). On the other hand, it was demonstrated that Mincle is also involved in sensing danger signals through the recognition of SAP130 (one of the nuclear proteins forming the spliceosome complex) released by dead cells. In addition, the signaling pathway initiated by Mincle was also revealed in the same report (Yamasaki et al. 2008). As many other activating receptors, Mincle is incapable of transducing activating signals by itself. A positively charged arginine present in the transmembrane region of Mincle allows its interaction with the adaptor protein containing an immunoreceptor tyrosine-based activation motif (ITAM), called Fc receptor γ-chain (FcRγ). When Mincle recognizes a PAMP, its activation induces a dual phosphorylation of the ITAM of FcRγ that becomes a high-affinity docking site for Syk (spleen tyrosine kinase) family proteins through their SH2 (Src homology 2) domain. The signal is then transmitted to NF-κB by the CARD9-BCL10-MALT1 pathway that leads to the production of pro-inflammatory cytokines that orient the maturation of naïve T cells (Richardson and Williams 2014; Yamasaki et al. 2008; Mócsai et al. 2010).

Shortly after, Mincle was also shown to be involved in mycobacterial recognition (Ishikawa et al. 2009). It is now well established that most ligands recognized by Mincle are glycolipids mainly found in mycobacteria and fungi (Ishikawa et al. 2013; Richardson and Williams 2014). However, recent articles show evidence about the recognition of other pathogens by Mincle (Fig. 2.1). Consequently, we wish to firstly describe the PAMPs recognized by Mincle. We will then present the molecular aspects of the ligand recognition by this CLR before describing the establishment of the acquired immune response through Mincle.

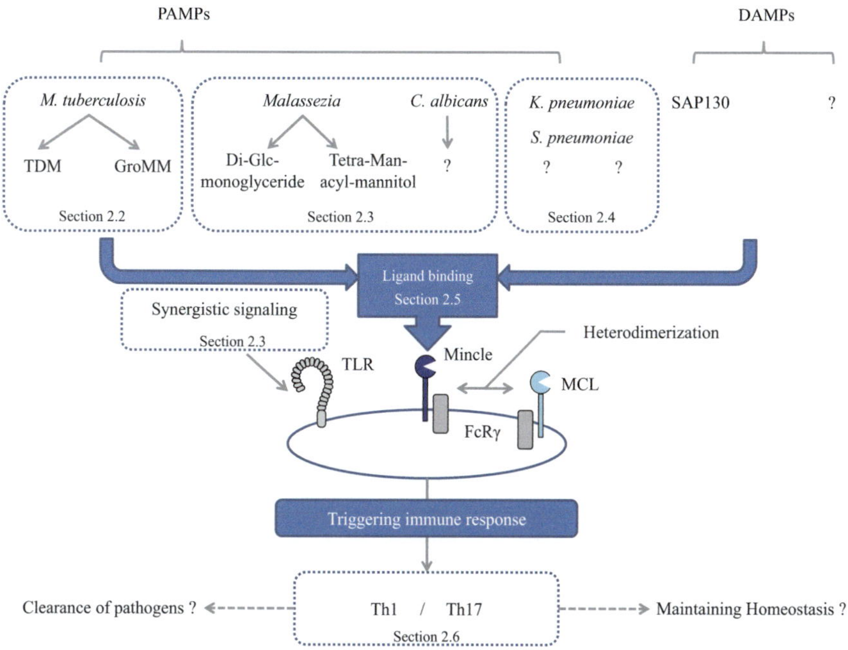

Fig. 2.1 Danger sensing by the C-type lectin receptor Mincle. Mincle is a pleiotropic pattern recognition receptor that has been reported to recognize several danger signals. Mincle binds to ligands from mycobacteria (Sect. 2.2) and the fungi family *Malassezia* (Sect. 2.3). However, many others remain to be determined (Sects. 2.3 and 2.4). Mincle interacts with FcRγ to trigger downstream signaling pathway, but there is also evidence about a Mincle/TLR synergistic signaling (Sect. 2.3) or a possible interaction with MCL. Mincle-dependent signaling is essential to the Th1/Th17 polarization in response to TDM (Sect. 2.6). Yet, whether Mincle is crucial or not for the clearance of pathogens or homeostasis maintenance is still unclear

2.2 Mincle Recognizes Mycobacterial PAMPs

On March, on the occasion of the World Tuberculosis Day, the World Health Organization called on the countries to continue the efforts against this disease. Although remarkable progress has been made in the past few years, the mortality is still alarming with 1.5 million people dying of tuberculosis each year (World Health Organization Global tuberculosis report 2015). *Mycobacterium tuberculosis*, the bacilli that causes this infectious disease, possesses various cell wall components that affect host immune response, and some of them are Mincle ligands (Fig. 2.1; Sect. 2.2).

2.2.1 Mincle Recognizes TDM

In the beginning of the 1950s, a toxic glycolipid was isolated from *M. tuberculosis* cell wall by Bloch. As it was related to the capacity of forming cord, a characteristic feature of the virulence of the bacilli, Bloch named it cord factor (Bloch et al. 1953). Shortly after, its final structure was determined and shown to be trehalose-6,6′-dimycolate (TDM) (Noll et al. 1956; Julián et al. 2010). TDM is presently known as one of the most bioactive components of *M. tuberculosis*, triggering pro-inflammatory cascades and granuloma formation as well as having adjuvant activity (Geisel et al. 2005; Hunter et al. 2006). However, despite all the studies on this cell wall component, the actors implicated in its recognition stayed indefinable until a few years ago.

Our team showed that the receptor at the origin of TDM-dependent inflammatory responses is Mincle. In this study, by using nuclear factor of activated T-cell (NFAT)-driven green fluorescent protein (GFP) reporter cells, Mincle was first shown to recognize mycobacteria. Afterwards, lipid extraction and separation strategy with high-performance thin-layer chromatography (HPTLC) revealed that TDM is a direct ligand of Mincle from *M. tuberculosis*. Moreover, TDM-induced granuloma formation in the lung was abrogated, and pro-inflammatory cytokine production was highly reduced in Mincle-deficient mice (Ishikawa et al. 2009).

In a recent report, we also studied the recognition of TDM in guinea pig (Toyonaga et al. 2014). It is a highly relevant animal model since the clinical and morphological features of guinea pig tuberculosis are very similar to those of human tuberculosis. We showed that in guinea pig, TDM binds to and induces activation signal through Mincle. Group 1 CD1 proteins (involved in the presentation of mycobacterial glycolipid antigens to T-cells) are present in human and guinea pig but are not expressed in mice and rats (Moody et al. 1997; Layre et al. 2009). Besides, it has been reported that TDM orients the maturation of T-helper (Th) cells toward Th1 and Th17 subsets (Schoenen et al. 2010). Thus, this animal model would be appropriated to investigate the role of Mincle in the entire T-cell response against *M. tuberculosis*. Taken together, these two reports demonstrated that the TDM-induced immune response relies on Mincle.

The function of Mincle in macrophages and dendritic cells is well studied. However, during mycobacterial- or TDM-induced inflammation in the lung, other immune cells are recruited to the inflammation site. Berry et al. found that interferon (IFN)-inducible transcripts are overexpressed in purified blood neutrophils from tuberculosis patients (Berry et al. 2010). This result hinted that neutrophils are probably involved in the acute phase of mycobacterial infection. Furthermore, it has been shown that Mincle in neutrophils may be important for the lung inflammation triggered by TDM (Lee et al. 2012). Interestingly, by inoculating TDM-coated beads as well as activated bone marrow macrophages (BMMs) subcutaneously, Lee et al. could demonstrate that neutrophils are recruited to the inoculation site in wild-type (WT) mice but not in Mincle-deficient mice. The presence of BMMs, that should play the role of resident macrophages, did not rescue the impaired recruitment observed in knockout mice. Although this result is likely to indicate that neutrophil recruitment in the lung depends on direct recognition of TDM by Mincle on neutrophil, further experiments should be done to confirm this conclusion. A first possibility would be that either BMMs did not mimic correctly the resident macrophages or that another cell type, such as endothelial cells, is implicated in the recruitment of neutrophils. Also, to strongly confirm the role of Mincle in this cell type, a conditional knockout mice line, in which Mincle is not expressed in neutrophils exclusively, should be used.

Nevertheless, this study reported an important and currently poorly understood function of Mincle in neutrophil migration and is among the growing number of articles showing the implication of Mincle in neutrophil immune response. This report also demonstrated that TDM is recognized by Mincle in neutrophils as well. However, TDM is not the unique mycobacterial PAMP recognized by Mincle.

2.2.2 Human Mincle Recognizes Glycerol Monomycolate

As we mentioned in the previous section, it has been shown that CD1b presents a mycobacterial cell wall component to T cells. This glycolipid, glycerol monomycolate (GroMM), is a mycoloyl-based lipid that has a single mycolate linked to a glycerol unit (Layre et al. 2009). The presence of GroMM in the cell wall of *Mycobacterium* strains was assessed in the 1950s (Tsumita 1956; Bloch et al. 1957; Noll 1957; Noll and Jackim 1958). Andersen et al. reported the immunostimulatory activity of GroMM and its capacity to directly activate human dendritic cells, but the recognition mechanism has been uncovered only recently (Andersen et al. 2009). By using NFAT-driven GFP reporter cells expressing either human or mouse Mincle, Hattori et al. reported that human Mincle, contrarily to its mouse homologue, recognizes GroMM (Hattori et al. 2014), a result that was further confirmed by challenging WT and human Mincle transgenic mice with GroMM liposomes. Unlike WT mice, a marked accumulation of macrophages and eosinophils was observed in transgenic mice expressing human Mincle. The authors also identified two stretches of amino acids (positions 174–176 and 195–196) that are implicated in the human Mincle-GroMM interaction. In fact, replacing these two stretches in mouse Mincle by their human equivalent induced GFP expression by reporter cells in response to GroMM.

These results show that human Mincle recognizes GroMM and illustrate that Mincle ligand binding can differ from mouse to human. It also raises the question as to how the recognition of GroMM contributes to the control of mycobacterial infection. However, to push the investigation further, using the same human Mincle transgenic mice or humanized mice and examine the immune response after GroMM stimulation should improve our understanding on this subject. In the case of a possible use for GroMM as an adjuvant for vaccination, this data should also be confirmed in an animal model that displays this recognition. The short amino acid stretches reported by Hattori et al. are likely to be of importance for the interaction between GroMM and human Mincle. These motifs are conserved in chimpanzee and differ in one amino acid in the case of rhesus monkey. Analysis in these model organisms would therefore generate valuable information. Also, TDM and GroMM both induce Mincle-dependent inflammatory response. Yet the differences between these two responses are still unclear and need further investigations.

2.2.3 Role of Mincle in the Clearance of Mycobacteria

With the emergence of resistant mycobacteria, new therapeutic strategies against tuberculosis are needed. Mincle, with its role in mycobacterial recognition, is therefore considered to be a potential target. Hence, Behler et al. first investigated the role of Mincle by using intratracheal mycobacterial infection. Tuberculosis is initiated by the inhalation of aerosol droplets containing the bacilli. As a result, the first immune cells to encounter the pathogen are generally lung alveolar macrophages and newly immigrating macrophages. The authors showed in a first report that in Mincle-deficient mice, the cytokine and chemokine production as well as the leukocyte recruitment after BCG (bacillus Calmette-Guérin) infection were severely impaired (Behler et al. 2012). Plus, they detected a delayed expression of Mincle following primary infection with BCG which suggests that Mincle participate in the late antimycobacterial response in mice lung. In a subsequent study, they challenged mice intravenously with *M. bovis* to mimic a systemic infection (Behler et al. 2015). They could detect higher mycobacterial loads in the spleen and liver of Mincle-deficient mice compared to WT mice. An interesting result is that granuloma formation was attenuated in the spleen but not in the liver of Mincle-deficient mice. These preliminary data indicate that Mincle may contribute to granuloma formation in an organ-specific manner and show that another PRR(s) could be implicated in this phenomenon. The results presented in these two articles show a probable contribution of Mincle in the control of mycobacterial infection in mice.

Conversely, another study reported that Mincle is not required to efficiently control *M. tuberculosis* infection (Heitmann et al. 2013). In their study, Heitmann et al. used *M. tuberculosis* (H37Rv strain) to examine the consequences of Mincle deficiency in mice after aerosol infection. They could show that mycobacterial burden in the lung of Mincle-deficient mice and WT mice was comparable at early or late stage of infection. Then, they measured the production of pro-inflammatory cytokines from lung homogenates at the transcriptional and protein levels. In both cases, no significant differences were detected. TDM strictly depends on Mincle to induce the development of Th1 and Th17 (Sect. 2.6.1). Therefore, the authors assessed the activity of these two subsets. Again, the number of IFN- and IL-17-producing cells as well as gene expression of these cytokines in the lung was similar in WT and Mincle-deficient mice. Lastly, they performed aerosol infection with a high dose of *M. tuberculosis* and evaluated the mycobacterial loads in the lung, spleen, and liver. Consistent with their previous results, bacterial burden was once more not influenced by Mincle deficiency, in contrast with the data showed by Behler et al. with the *M. bovis* strain BCG (Behler et al. 2015). However, Lee et al. demonstrated that after aerosol infection with the Erdmann *M. tuberculosis* strain cytokine levels and mycobacterial burden were higher in the lung of Mincle-deficient mice (Lee et al. 2012). As the major difference between those three studies is the strain used for the infection, Heitmann et al. consequently proposed that the abundance of cell wall components might vary from one strain to another and generate these different results. In sum, the exact role of Mincle in the immune response against tuberculosis is still unsolved and remains controversial. Additional studies are necessary in order to clarify this subject and to determine whether or not Mincle could be a therapeutic target for treating tuberculosis.

Nevertheless, it has been clearly demonstrated that the cytosolic adaptor caspase recruitment domain family member 9 (CARD9), which is involved in transducing signals from several ITAM-coupled receptors (including Mincle), is essential to control pulmonary *M. tuberculosis* infection (Dorhoi et al. 2010). The authors showed that CARD9 deficiency resulted in accelerated death and granulocyte recruitment, higher mycobacterial burden as well as elevated production of pro-inflammatory cytokines. Since Mincle is one of the PRRs upstream of CARD9 signaling pathway, it remains a relevant candidate for the establishment of a new treatment for tuberculosis.

2.3 Mincle Recognizes Fungal PAMPs

Fungi are part of the normal human flora, mainly found in the gastrointestinal tract (Bernhardt and Knoke 1997) but also on the skin and oral mucosa (Roth and James 1988). While most fungal microorganisms are usually harmless, some species can have high pathogenicity if host immune defense weakens. It results in several diseases called mycoses, ranging from superficial infection to fatal systemic invasion (Liu 2011). PRRs are also critically involved in the recognition of this family of pathogens. For instance, TLR2 is a key receptor for pathogenic fungi due to its recognition of zymosan. In the CLR family of receptors, Dectin-1, which recognizes β-glucan, also has the ability to trigger immune response after binding this fungal PAMP (Goodridge and Underhill 2008). In this section, we will thus discuss the contribution of Mincle in the immune response against fungi (Fig. 2.1; Sect. 2.3).

2.3.1 Mincle Recognizes Candida albicans

As we mentioned in the introduction, the implication of Mincle in fungal recognition was first illustrated by Wells et al. (2008). They studied the role of Mincle in the macrophage-dependent immune response to *Candida albicans*, a fungus responsible for invasive candidiasis particularly in hospital environments. The authors showed that the production of the pro-inflammatory cytokine TNF (tumor necrosis factor) was reduced in Mincle-deficient macrophages. In addition, the absence of Mincle enhanced mice susceptibility to systemic candidiasis, and exposure to live *C. albicans* induced the upregulation of Mincle contrarily to the other CLRs tested. BMMs produced TNF when stimulated by *C. albicans*, but in Mincle-deficient BMMs, or when an anti-Mincle antibody was used, this response was lost. On the other hand, they showed that Mincle was not required for phagocytosis of this pathogen by macrophages. Wells et al. also examined the implication of Mincle in the case of a systemic infection. Intravenous injection of the fungi in Mincle-deficient mice resulted in a higher colony number of *C. albicans* in the kidney. This study suggested that the major role of Mincle in the response against *C. albicans* is the induction of pro-inflammatory cytokines. In another report published the same year, Bugarcic et al. investigated the interaction between *C. albicans* and Mincle (Bugarcic et al. 2008). They generated refolded recombinant Mincle CRD (from mouse and human) and showed the recognition of soluble extracts as well as whole yeast cell by ELISA. Using Mincle-expressing NFAT-GFP reporter cells, our team screened 50 fungi species for a possible recognition (Yamasaki et al. 2009). Interestingly, the three *Candida albicans* species screened in this study did not induce the expression of GFP. Although the species were different from the one used by Wells et al., this indicated that Mincle alone is not sufficient but may be implicated in the immune response against *Candida albicans*. Furthermore, Mincle ligand(s) in *C. albicans* has yet to be determined, and identifying it (or them) may provide new insights and details about Mincle ligand recognition.

2.3.2 *Mincle Recognizes* Malassezia

Among the 50 different fungal species (*Aspergillus*, *Microsporum*,...) tested in the screen referred to in the previous section, we could determine that only *Malassezia* species elicited Mincle signaling, and the use of anti-Mincle antibody completely inhibited *Malassezia*-induced GFP expression (Yamasaki et al. 2009). Similar results were observed in vivo with intraperitoneal injection of *Malassezia* to WT and Mincle-deficient mice. The production of pro-inflammatory cytokines detected in WT mice after infection was lost in the absence of Mincle. These results showed the involvement of Mincle in the recognition of *Malassezia*, but the ligand was determined a few years later (Ishikawa et al. 2013). Our team extracted compounds from *Malassezia* using different solvents. We showed that only the chloroform/methanol extract induced GFP expression in reporter cells. This indicated that the Mincle ligands from *Malassezia* were most likely lipids. This extract was then separated with silica gel column chromatography, followed by HPTLC, and the obtained fractions were used to stimulate NFAT-GFP reporter cells. This allowed us to discover two activating ligands for Mincle, and their chemical structures were identified by mass spectrometry and nuclear magnetic resonance analysis. Both ligands are glycolipids and induced the production of pro-inflammatory cytokines in a Mincle-dependent manner. Moreover, we could also demonstrate that the intraperitoneal injection of one of the two ligands has the ability to induce neutrophil infiltration in vivo. These two studies showed strong evidences of the implication of Mincle in *Malassezia*-induced immune response. This family of fungi is thought to be involved in atopic dermatitis since it was reported that patients suffering from this disease develop enhanced allergic reaction to *Malassezia* (Young et al. 1989). Furthermore, *Malassezia*-specific IgE was detected in the serum of atopic dermatitis patients (Devos and van der Valk 2000). It was also shown that the expression of Mincle was upregulated in mast cells when incubated with extracts from *Malassezia sympodialis* (Ribbing et al. 2011). Moreover, some reports showed the improvement of symptoms of atopic dermatitis patients after antifungal treatment (Kobayashi et al. 2006). Hence, it suggests that Mincle activation may contribute to atopic dermatitis, but it warrants further investigation.

2.3.3 *Mincle/TLR Synergistic Response to* Fonsecaea pedrosoi

The chronic fungal subcutaneous disease called chromoblastomycosis is caused by the infection of the skin and subcutaneous tissue, generally after saprophagous dematiaceous fungi enters through an open wound (Kim et al. 2011). In most cases, chromoblastomycosis originates from the infection by the melanin-producing pathogen, *Fonsecaea pedrosoi* (Alviano et al. 2004). Information about this fungus was scarce but Sousa Mda et al. investigated the chronicity of chromoblastomycosis and showed interesting results about *F. pedrosoi* recognition (Sousa Mda et al. 2011). The authors used a mouse model that displays, after intraperitoneal infection, a similar cytokine profile to the one assessed in chromoblastomycosis patients. They could show that *F. pedrosoi* binds to macrophages but does not trigger inflammatory response. The use of heat-killed fungi had the same outcome. Fungi cell wall components are thought to be shielded, and it has been suggested that heat treatment lay bare underlying PAMPs (Robinson et al. 2009). Therefore, these results showed that, even if *F. pedrosoi* binds to macrophages, the activation is not successful which results in a lack in inflammatory response. In a subsequent experiment, Sousa et al. co-stimulated macrophages and bone marrow-derived dendritic cells (BMDCs) with *F. pedrosoi* and TLR agonists or ligands, which elicited robust inflammatory response. This response was severely impaired in MyD88 (the essential adaptor protein for TLR signaling)-deficient and Mincle-deficient BMDCs.

This result shows that Mincle and TLR signaling are required to trigger efficient inflammatory response against *F. pedrosoi*. However, the PAMP(s) from *F. pedrosoi* that binds to Mincle is also still unknown. In addition, Mincle was recently shown to recognize another member of this fungi family, called *Fonsecaea monophora*, and also responsible for chromoblastomycosis (Wevers et al. 2014). But in this case, it interferes with the Dectin-1-induced immune response. Wevers et al. showed that *F. monophora* induced Dectin-1 signaling that triggers interleukin (IL)-12 production through IRF-1 (interferon regulatory factor 1). *F. monophora* also simultaneously activated Mincle signaling which leads to proteasomal degradation of IRF-1 and consequently prevents IL-12 transcription.

Although chromoblastomycosis is nonfatal, treatment usually involves heavy procedures such as chemotherapy, surgical excisions... Understanding the recognition mechanisms might lead to the establishment of new and less abrasive therapeutic solutions. But the studies discussed above illustrate that many aspects of fungal recognition by Mincle remain obscure.

2.4 A Role for Mincle in Pneumonia?

Recent reports showed evidence about the Mincle-dependent recognition of *Klebsiella pneumoniae* and *Streptoccocus pneumoniae*, two bacteria responsible for the development of pneumonia (Fig. 2.1; Sect. 2.4). *K. pneumoniae* belongs to the gram-negative family of bacteria and colonizes the nasopharynx as well as the intestinal tract (Podschun and Ullmann 1998), while *S. pneumoniae*, also called pneumococcus, are gram-positive commensal bacteria that colonize the nasopharynx (Chaguza et al. 2015).

It has been shown that neutrophils have a protective role in *K. pneumoniae* infection through phagocytosis but also by extracellular traps (NETs) formation (Papayannopoulos et al. 2010). As mentioned in Sects. 2.2.1 and 2.3.2, Mincle is expressed by this cell subset and contributes to immune responses against *Candida* and mycobacteria. Thus, Sharma et al. investigated the involvement of Mincle signaling in the infection by *K. pneumoniae* (Sharma et al. 2014). WT and Mincle-deficient mice were infected with a sublethal dose of *K. pneumoniae*. Every Mincle-deficient mice died by day 6 after intranasal infection, whereas more than 70 % of WT mice survived until at least day 14. Mincle deficiency also resulted in higher bacterial systemic dissemination. Subsequently, to explain the high susceptibility of Mincle-deficient mice to *K. pneumoniae*, the authors examined the neutrophil-dependent immune response in the absence of Mincle. They could show that the two protective mechanisms involved in neutrophil clearance of bacteria were impaired. Indeed, it resulted in a significant reduction of phagocytosis of nonopsonized *K. pneumoniae*. It was shown recently that, in macrophages, Mincle may not be required for the uptake of TDM (Kodar et al. 2015). These results show that Mincle could possibly be implicated in phagocytosis by neutrophils but not by macrophages. The lack of Mincle caused NET formation impairment upon infection as well. Hence, it links this CLR to a new biological function, and this report demonstrates that Mincle plays an important protective role in *K. pneumoniae* infection.

Rabes et al. recently examined if CLRs are implicated in the recognition of Streptococcus pneumoniae (Rabes et al. 2015). Using a CLR-Fc fusion protein library, they discovered that Mincle binds heat-killed *S. pneumoniae*. The interaction between Mincle-Fc and live *S. pneumoniae* was confirmed afterwards by flow cytometry. However, Mincle signaling seems to be redundant in the infection by the invasive serotype 2 and 3 pneumococcal strains. The production of inflammatory cytokines by alveolar and bone marrow macrophages isolated from mice and exposed to *S. pneumoniae* did not vary in the absence of Mincle and FcRγ. The same observation was made for phagocytosis. To confirm these results, the authors evaluated the innate immune response of Mincle- and FcRγ-deficient mice during *S. pneumoniae* infection. Consistent with the previous results, survival, bacterial loads in blood and bronchoalveolar fluid, as well as macrophage and neutrophil recruitment in the lung were similar in knockout and WT mice. This study, in contrast to the one published by Sharma et al., suggests that Mincle is not involved in the immune response against invasive *S. pneumoniae*. Nevertheless, Mincle specifically binds to this pathogen, and its expression is upregulated in the lung after pneumococcal exposure in vivo. Therefore, complementary studies may unravel a possible role for Mincle in *S. pneumoniae*-induced immune response (Rabes et al. 2015).

These two reports show how Mincle-induced immune response can differ drastically depending on the pathogen recognized. Although in both cases the ligands have yet to be described, the results discussed in this section are innovative and promising. They provide new insights in the role of Mincle in neutrophils and NET formation, opening new fields of research for this C-type lectin receptor.

2.5 Glycolipid Recognition by Mincle

We showed that a significant number of pathogens and ligands have been reported to bind to Mincle. An important topic to explore is the nature of the receptor-ligand interaction at the molecular level. Mincle, like the other members of the CLRs family, has a CRD in its extracellular region that allows carbohydrate recognition (Zelensky and Gready 2005). This compact globular domain contains a three amino acid motif implicated in the binding specificity. In the case of Mincle, a glutamic acid and an asparagine separated by a proline in a cis conformation, called the EPN motif (residues 169–171 in human), coordinate a Ca^{2+} ion that interacts with monosaccharides through hydrogen bonds (Furukawa et al. 2013). This motif is usually associated with glucose/mannose-binding ability while the QPD motif (glutamine-proline-aspartic acid), another typical motif found in CLRs, generally implies galactose-binding activity. Feinberg et al. generated crystals of a minimal bovine CRD and firstly demonstrated that the overall fold was comparable to other CLRs and that the Ca^{2+} ion was found at the cation-binding site (Feinberg et al. 2013). Then, the authors crystallized the bovine Mincle CRD bound to trehalose, the disaccharide of TDM. Interestingly, in addition to the canonical

EPN-Ca^{2+} site, two complementary sites were shown to be important for the interaction with TDM. Another sugar-binding site interacts with the second glucose residue of TDM and a hydrophobic groove that binds its diacyl group. Moreover, mutations in these two nonconventional sites resulted in a loss of binding affinity.

In a collaborative study with our team, Furukawa et al. further confirmed the results presented by Feinberg et al. (Furukawa et al. 2013). The authors crystallized the extracellular domain of human Mincle and found that its CRD structure was highly similar to the CRD of DC-SIGNR, except for the regions in the vicinity of Ca^{2+} binding site. To assess the implication of the EPN motif in Mincle-TDM interaction, the authors mutated it into the galactose-recognition sequence, QPD. As expected, it induced the loss of interaction between soluble Mincle and TDM-coated plates. Then, the role of the unusual region surrounding the sugar-binding site was investigated. This region contains hydrophobic residues that, when mutated, induced a diminished NFAT activity of reporter cells expressing Mincle after TDM stimulation. Furthermore, replacing the hydrophobic region by the corresponding amino acids of the CLR Dectin-2, that does not contain this unusual groove, resulted in a severe impairment of NFAT-GFP activity. However, the secondary sugar-binding site in human Mincle CRD described by Feinberg et al. has not been reported by Furukawa et al. But in a recent paper, Jegouzo et al. demonstrated that this secondary binding site was conserved in human Mincle (Jégouzo et al. 2014). Another interesting result was the comparison between monoacyl and diacyl binding affinity. Between monoacyl and diacyl derivatives with the same acyl length, human Mincle CRD bound diacyl derivatives with much higher affinity.

In sum, the reports discussed above demonstrate that TDM recognition by Mincle displays singular aspects compared to other CLRs. In addition to the Ca^{2+} -dependent carbohydrate-binding site, from which originates the name of C-type lectins, a hydrophobic region and a secondary carbohydrate-binding site participate in TDM recognition, by interacting with the acyl chains and the second sugar moiety, respectively. Moreover, in the report by Hattori et al. showing the recognition of GroMM by human Mincle (Sect. 2.2.2), the authors also demonstrated that two short stretches of amino acids were important for the interaction (Hattori et al. 2014). One of the two stretches, Val-Thr at positions 195–196, is part of the hydrophobic region discussed above. Thus, this entire hydrophobic groove may also be important for the recognition of the acyl chain of GroMM.

Many predicted mannose- or galactose-specific lectins do not actually bind to these carbohydrates (Richardson and Williams 2014). The discovery of additional binding region in the CRD of Mincle implies that the carbohydrate-binding motif may not be the unique major motif implicated in PAMP recognition by CLRs (Kiyotake et al. 2015).

2.6 Mincle-Induced Acquired Immunity

Professional APCs are the pivots of the immune system for the reason that they assure the link between innate and acquired immunity. The PRRs expressed by APCs allow them to recognize pathogens and then to orchestrate the acquired immune response accordingly. Consequently, PRRs raised increasing interest, particularly for the establishment of new adjuvants. In this section, we will present the immune response triggered by Mincle and its probable role as an adjuvant receptor (Fig. 2.1; Sect. 2.6).

2.6.1 Mincle Elicits Th1/Th17 Immune Response

At first, three complementary studies allowed to improve our understanding of the Mincle-induced acquired immune response. Werninghaus et al. investigated the signaling pathway triggered by TDM and TDB (trehalose-6,6-dibehenate), the latter being a synthetic analogue of TDM (Werninghaus et al. 2009). Using immunization strategy in several knockout mice, the authors could demonstrate that the adaptor protein FcRγ is required for the engagement of the Syk-CARD9-BCL10-MALT1 pathway by both glycolipids. This activation was shown to promote T-cell maturation towards Th1 and Th17 subsets that produce IFNγ and IL-17, respectively. As discussed in Sect. 2.2.1, our team then revealed that TDM is a direct ligand of Mincle (Ishikawa et al. 2009). And shortly after, Schoenen et al. confirmed the previous results and linked the activation of Th1/Th17 immunity to Mincle in vivo (Schoenen et al. 2010). In Mincle-deficient mice, when TDB was used as an adjuvant to the H1 subunit vaccination protocol, the Th1 response as well as IL-17 production were highly weakened. Altogether, these reports put forward evidence suggesting that Mincle is essential for the adjuvant activity of TDM and TDB. However, it was proposed not long ago that another pathway may be involved in this mechanism (Desel et al. 2013). Werninghaus et al. demonstrated that TDM and TDB induce APC activation in absence of MyD88 in vitro (Werninghaus et al. 2009). But it was previously reported that in MyD88-deficient mice, the Th1 immune response after TDB immunization is significantly decreased (Agger et al. 2008). Thus, Desel et al. investigated the implication of MyD88 in TDB adjuvanticity in vivo (Desel et al. 2013). The authors first showed that footpad swelling after subcutaneous TDB immunization was reduced in MyD88-deficient mice, confirming that this adaptor protein is involved in this process. Interestingly, cells from footpads secreted IL-1β, and blockade of its receptor IL-1R diminished IFNγ and IL-17 secretion. As IL-1β is firstly produced as an immature cytokine (pro-IL-1β) and necessitates inflammasome activation to be functional, Desel et al. observed the consequences of the loss of an essential adaptor protein for this process called ASC (apoptosis-associated speck-like protein containing a caspase recruitment domain). The IFNγ secretion was not affected while footpad swelling and IL-17 production was reduced. This could mean that the inflammasome activation is not the unique mechanism implicated in T-cell maturation in response to TDB. It is therefore plausible that Mincle recognizes TDB and induces the production of IL-1β, which in turn activates IFNγ and IL-17 secretion through IL-1R-MyD88 signaling. These results were corroborated by the study by Shenderov et al. in which they observed the Th17 response induced by complete Freund's adjuvant (CFA) (Shenderov et al. 2013). This adjuvant, a water-in-oil emulsion containing heat-killed *M. tuberculosis*, was used to immunize mice and examine the intracellular production of IL-17. In accordance with the results presented by Desel et al., the authors showed that MyD88, IL-1β, and IL-1R were needed for the CFA-induced Th17 maturation, and the role of the inflammasome was confirmed as well. Interestingly, by fractionating heat-killed H37Ra strain, they could also observe that peptidoglycan alone activated the inflammasome. Moreover, CARD9 deficiency and, to a lesser extent, Mincle deficiency resulted in impaired IL-1β production and Th17 polarization. This result hints that Mincle, although important, is not the unique PRR to signal through CARD9 in response to CFA. Therefore, it would appear that Mincle, after recognizing TDM, induces CARD9-dependent signaling and pro-IL-1β production. Peptidoglycan, by an unknown mechanism in this case, triggers inflammasome activation and cleavage of pro-IL-1β into its mature form which in turn elicits Th17 maturation through the IL-1R-MyD88 pathway.

Taken together, these studies showed evidence that improved our understanding of how Mincle stimulates T-helper maturation. To fully understand the mechanism by which Mincle induces Th1 and Th17 response might facilitate the establishment of adjuvants that precisely modulate the immune response.

2.6.2 Mincle Regulates the Activation of B Cells

As stated earlier, Mincle is mainly studied in macrophages and dendritic cells. However, throughout this chapter we could see that other cell types, and especially neutrophils, might utilize Mincle to induce important immune responses. B cells express Mincle and also have the ability to present antigens (Matsumoto et al. 1999). Therefore, Kawata et al. studied the role of Mincle in human B cells by analyzing its expression upon different stimuli (Kawata et al. 2012). They showed that the number of Mincle-expressing B cells increased significantly when stimulated with the TLR ligand CpG-B, while no variation of total B-cell number was detectable. When peripheral blood mononuclear cells (PBMC) are cultured in presence of TDB, the number of Mincle-expressing B cells raised as well, although not as strongly as with CpG-B alone. The authors subsequently measured the secretion of the immuglobulins IgG, IgA, and IgM by PBMCs and Mincle-positive B cells when stimulated by TDB, CpG-B, or both. TDB seems to inhibit IgG and IgA, but not IgM secretion. On the other hand, TDB had no impact on the secretion of inflammatory cytokines, and IL-1β was not detected in supernatant fluid, which may suggests that B cells do not strongly participate in helper T cell differentiation. These results put forward the possibility that Mincle-dependent signaling in B cells might promote IgM production. Still, the data on this particular subject is scarce and needs further study.

2.7 Discussion and Conclusion

Throughout this chapter, we described the role and characteristics of Mincle in the immune response to pathogens. We could see that this CLR recognizes PAMPs from various microorganisms, resulting in pro-inflammatory signals for the most part. Depending on the cell type, tissue, or the pathogen recognized, the Mincle-dependent immune response ranges from redundant to essential. However, as we could emphasize numerous times, many aspects concerning Mincle-dependent immune response require further investigation. For instance, we can cite the conditions of up- and downregulation of this CLR. As the expression level of Mincle in resting cells is low, understanding these conditions would be valuable in order to be able to target Mincle efficiently in therapies.

Another topic that we did not discuss in this chapter, but of high significance nonetheless, is the recognition of endogenous ligands by Mincle. As pointed out in the introduction, one of the first Mincle ligand discovered was SAP130 (Yamasaki et al. 2008). The recognition of this protein released by necrotic cells triggers danger signaling through Mincle. Moreover, the interaction between SAP130 does not require the CRD, which suggests that Mincle possesses a distinct binding motif that has yet to be determined. Vaccari et al. also showed the implication of this recognition in the inflammatory response following traumatic brain injury (de Rivero Vaccari et al. 2015). Likewise, this recognition also contributes to the inflammation subsequent to ischemia (Suzuki et al. 2013). Moreover, Mincle is thought to participate in the chronic inflammation of obese adipose tissue through the recognition of endogenous ligand by macrophages (Tanaka et al. 2014). We also recently reported that human Mincle binds to cholesterol crystals (Kiyotake et al. 2015). These studies suggest a pathophysiological role for the recognition of danger signal by Mincle.

Recent reports showed interesting results concerning the functional conformation of Mincle. It has been demonstrated that Mincle forms a heteromeric complex with FcRγ and MCL (macrophage C-type lectin), also member of the CLR family of receptor. Lobato-Pascual et al. showed the interaction between these three proteins and that in the presence of MCL, Mincle expression is increased in rats (Lobato-Pascual et al. 2013). These results were confirmed in mice and human (Miyake et al. 2015; Kerscher et al. 2015). These studies reported that two CLRs could interact and influence the expression or signaling of one another, an unusual and novel function in this family of receptor. Further investigations may uncover the role of this function in the recognition of pathogens by CLRs. Lobato-Pacual et al. also hypothesized that this heteromeric complex may possibly recognize ligands that Mincle alone cannot (Lobato-Pascual et al. 2013). This might explain why Mincle-expressing reporter cells and APCs or in vivo experiments occasionally result in opposite outcomes. However, it may not explain the results observed for *Candida Albicans* (Sect. 2.3.1). Indeed, it was shown that MCL deficiency does not result in a lack of immune response to this fungus in mice (Graham et al. 2012).

To finish, we saw that Mincle-induced immune response orient T-cell maturation toward Th1/Th17 subsets. Mincle was therefore considered as a highly relevant adjuvant receptor candidate. At present, TDB has made its way to phase I of clinical trials in humans and shows promising results. The adjuvant called CAF01 is the association of TDB and a cationic liposome vehicle. van Dissel et al. recently showed that this novel adjuvant was safe and induced long-lasting *Mycobacterium tuberculosis*-specific cellular immune response in humans (van Dissel et al. 2014). This result is encouraging for the establishment of a vaccine against mycobacteria but also other pathogens that require a robust cellular immune response.

References

Agger EM, Rosenkrands I, Hansen J, Brahimi K, Vandahl BS, Aagaard C, Werninghaus K, Kirschning C, Lang R, Christensen D, Theisen M, Follmann F, Andersen P (2008) Cationic liposomes formulated with synthetic mycobacterial cordfactor (CAF01): a versatile adjuvant for vaccines with different immunological requirements. PLoS One 3(9):e3116. doi:10.1371/journal.pone.0003116

Alviano DS, Franzen AJ, Travassos LR, Holandino C, Rozental S, Ejzemberg R, Alviano CS, Rodrigues ML (2004) Melanin from Fonsecaea pedrosoi induces production of human antifungal antibodies and enhances the antimicrobial efficacy of phagocytes. Infect Immun 72 (1):229–237

Andersen CS, Agger EM, Rosenkrands I, Gomes JM, Bhowruth V, Gibson KJ, Petersen RV, Minnikin DE, Besra GS, Andersen P (2009) A simple mycobacterial monomycolated glycerol lipid has potent immunostimulatory activity. J Immunol 182(1):424–432

Behler F, Steinwede K, Balboa L, Ueberberg B, Maus R, Kirchhof G, Yamasaki S, Welte T, Maus UA (2012) Role of Mincle in alveolar macrophage-dependent innate immunity against mycobacterial infections in mice. J Immunol 189(6):3121–3129. doi:10.4049/jimmunol.1201399

Behler F, Maus R, Bohling J, Knippenberg S, Kirchhof G, Nagata M, Jonigk D, Izykowski N, Mägel L, Welte T, Yamasaki S, Maus UA (2015) Macrophage-inducible C-type lectin Mincle-expressing dendritic cells contribute to control of splenic Mycobacterium bovis BCG infection in mice. Infect Immun 83(1):184–196. doi:10.1128/IAI.02500-14

Bernhardt H, Knoke M (1997) Mycological aspects of gastrointestinal microflora. Scand J Gastroenterol Suppl 222:102–106

Berry MP, Graham CM, McNab FW, Xu Z, Bloch SA, Oni T, Wilkinson KA, Banchereau R, Skinner J, Wilkinson RJ, Quinn C, Blankenship D, Dhawan R, Cush JJ, Mejias A, Ramilo O, Kon OM, Pascual V, Banchereau J, Chaussabel D, O'Garra A (2010) An interferon-inducible neutrophil-driven blood transcriptional signature in human tuberculosis. Nature 466 (7309):973–977. doi:10.1038/nature09247

Bloch H, Sorkin E, Erlenmeyer H (1953) A toxic lipid component of the tubercle bacillus (cord factor). I. Isolation from petroleum ether extracts of young bacterial cultures. Am Rev Tuberc 67(5):629–643

Bloch H, Defaye J, Lederer E, Noll H (1957) Constituents of a toxic-lipid obtained from Mycobacterium tuberculosis. Biochim Biophys Acta 23(2):312–321

Bugarcic A, Hitchens K, Beckhouse AG, Wells CA, Ashman RB, Blanchard H (2008) Human and mouse macrophage-inducible C-type lectin (Mincle) bind Candida albicans. Glycobiology 18 (9):679–685. doi:10.1093/glycob/cwn046

Chaguza C, Cornick JE, Everett DB (2015) Mechanisms and impact of genetic recombination in the evolution of Streptococcus pneumoniae. Comput Struct Biotechnol J 13:241–247. doi:10.1016/j.csbj.2015.03.007

Cummings RD, McEver RP (2009) Essentials of glycobiology, 2nd edn. Cold Spring Harbor Laboratory Press, Cold Spring Harbor, Chapter 31

de Rivero Vaccari JC, Brand FJ 3rd, Berti AF, Alonso OF, Bullock MR, de Rivero Vaccari JP (2015) Mincle signaling in the innate immune response after traumatic brain injury. J Neurotrauma 32(4):228–236. doi:10.1089/neu.2014.3436

Desel C, Werninghaus K, Ritter M, Jozefowski K, Wenzel J, Russkamp N, Schleicher U, Christensen D, Wirtz S, Kirschning C, Agger EM, Prazeres da Costa C, Lang R (2013) The Mincle-activating adjuvant TDB induces MyD88-dependent Th1 and Th17 responses through IL-1R signaling. PLoS One 8(1):e53531. doi:10.1371/journal.pone.0053531

Devos SA, van der Valk PG (2000) The relevance of skin prick tests for Pityrosporum ovale in patients with head and neck dermatitis. Allergy 55(11):1056–1058

Dorhoi A, Desel C, Yeremeev V, Pradl L, Brinkmann V, Mollenkopf HJ, Hanke K, Gross O, Ruland J, Kaufmann SH (2010) The adaptor molecule CARD9 is essential for tuberculosis control. J Exp Med 207(4):777–792. doi:10.1084/jem.20090067

Feinberg H, Jégouzo SA, Rowntree TJ, Guan Y, Brash MA, Taylor ME, Weis WI, Drickamer K (2013) Mechanism for recognition of an unusual mycobacterial glycolipid by the macrophage receptor mincle. J Biol Chem 288(40):28457–28465. doi:10.1074/jbc.M113.497149

Flornes LM, Bryceson YT, Spurkland A, Lorentzen JC, Dissen E, Fossum S (2004) Identification of lectin-like receptors expressed by antigen presenting cells and neutrophils and their mapping to a novel gene complex. Immunogenetics 56(7):506–517

Furukawa A, Kamishikiryo J, Mori D, Toyonaga K, Okabe Y, Toji A, Kanda R, Miyake Y, Ose T, Yamasaki S, Maenaka K (2013) Structural analysis for glycolipid recognition by the C-type lectins Mincle and MCL. Proc Natl Acad Sci U S A 110(43):17438–17443. doi:10.1073/pnas.1312649110

Geijtenbeek TB, Gringhuis SI (2009) Signaling through C-type lectin receptors: shaping immune responses. Nat Rev Immunol 9(7):465–479. doi:10.1038/nri2569

Geisel RE, Sakamoto K, Russell DG, Rhoades ER (2005) In vivo activity of released cell wall lipids of Mycobacterium bovis bacillus Calmette-Guérin is due principally to trehalose mycolates. J Immunol 174(8):5007–5015

Goodridge HS, Underhill DM (2008) Fungal recognition by TLR2 and Dectin-1. Handb Exp Pharmacol 183:87–109

Graham LM, Gupta V, Schafer G, Reid DM, Kimberg M, Dennehy KM, Hornsell WG, Guler R, Campanero-Rhodes MA, Palma AS, Feizi T, Kim SK, Sobieszczuk P, Willment JA, Brown GD (2012) The C-type lectin receptor CLECSF8 (CLEC4D) is expressed by myeloid cells and triggers cellular activation through Syk kinase. J Biol Chem 287(31):25964–25974. doi:10.1074/jbc.M112.384164

Hattori Y, Morita D, Fujiwara N, Mori D, Nakamura T, Harashima H, Yamasaki S, Sugita M (2014) Glycerol monomycolate is a novel ligand for the human, but not mouse macrophage inducible C-type lectin, Mincle. J Biol Chem 289(22):15405–15412. doi:10.1074/jbc.M114.566489

Heitmann L, Schoenen H, Ehlers S, Lang R, Hölscher C (2013) Mincle is not essential for controlling Mycobacterium tuberculosis infection. Immunobiology 218(4):506–516. doi:10.1016/j.imbio.2012.06.005

Hunter RL, Olsen MR, Jagannath C, Actor JK (2006) Multiple roles of cord factor in the pathogenesis of primary, secondary, and cavitary tuberculosis, including a revised description of the pathology of secondary disease. Ann Clin Lab Sci 36(4):371–386

Ishikawa E, Ishikawa T, Morita YS, Toyonaga K, Yamada H, Takeuchi O, Kinoshita T, Akira S, Yoshikai Y, Yamasaki S (2009) Direct recognition of the mycobacterial glycolipid, trehalose dimycolate, by C-type lectin Mincle. J Exp Med 206(13):2879–2888. doi:10.1084/jem.20091750

Ishikawa T, Itoh F, Yoshida S, Saijo S, Matsuzawa T, Gonoi T, Saito T, Okawa Y, Shibata N, Miyamoto T, Yamasaki S (2013) Identification of distinct ligands for the C-type lectin receptors Mincle and Dectin-2 in the pathogenic fungus Malassezia. Cell Host Microbe 13 (4):477–488. doi:10.1016/j.chom.2013.03.008

Jégouzo SA, Harding EC, Acton O, Rex MJ, Fadden AJ, Taylor ME, Drickamer K (2014) Defining the conformation of human mincle that interacts with mycobacterial trehalose dimycolate. Glycobiology 24(12):1291–1300. doi:10.1093/glycob/cwu072

Julián E, Roldán M, Sánchez-Chardi A, Astola O, Agustí G, Luquin M (2010) Microscopic cords, a virulence-related characteristic of mycobacterium tuberculosis, are also present in nonpathogenic mycobacteria. J Bacteriol 192(7):1751–1760. doi:10.1128/JB.01485-09

Kawai T, Akira S (2010) The role of pattern-recognition receptors in innate immunity: update on Toll-like receptors. Nat Immunol 11(5):373–384. doi:10.1038/ni.1863

Kawata K, Illarionov P, Yang GX, Kenny TP, Zhang W, Tsuda M, Ando Y, Leung PS, Ansari AA, Eric Gershwin M (2012) Mincle and human B cell function. J Autoimmun 39(4):315–322. doi:10.1016/j.jaut.2012.04.004

Kerscher B, Wilson GJ, Reid DM, Mori D, Taylor JA, Besra GS, Yamasaki S, Willment JA, Brown GD (2015) The mycobacterial receptor, Clec4d (CLECSF8, MCL) is co-regulated with Mincle and upregulated on mouse myeloid cells following microbial challenge. Eur J Immunol. doi:10.1002/eji.201545858

Kim DM, Hwang SM, Suh MK, Ha GY, Choi GS, Shin J, Han SH (2011) Chromoblastomycosis caused by Fonsecaea pedrosoi. Ann Dermatol 23(3):369–374. doi:10.5021/ad.2011.23.3.369

Kiyotake R, Oh-Hora M, Ishikawa E, Miyamoto T, Ishibashi T, Yamasaki S (2015) Human Mincle binds to cholesterol crystals and triggers innate immune responses. J Biol Chem 290 (42):25322–25332. doi:10.1074/jbc.M115.645234

Kobayashi T, Yamada M, Aihara M, Ikezawa Z (2006) Immedeate and delayed-type reactivity to fungi and effects of antifungal drugs on atopic dermatitis. Arerugi 55(2):126–133

Kodar K, Eising S, Khan AA, Steiger S, Harper JL, Timmer MS, Stocker BL (2015) The uptake of trehalose glycolipids by macrophages is independent of Mincle. Chembiochem 16(4):683–693. doi:10.1002/cbic.201402506

Layre E, Collmann A, Bastian M, Mariotti S, Czaplicki J, Prandi J, Mori L, Stenger S, De Libero G, Puzo G, Gilleron M (2009) Mycolic acids constitute a scaffold for mycobacterial lipid antigens stimulating CD1-restricted T cells. Chem Biol 16(1):82–92. doi:10.1016/j.chembiol.2008.11.008

Lee WB, Kang JS, Yan JJ, Lee MS, Jeon BY, Cho SN, Kim YJ (2012) Neutrophils promote mycobacterial trehalose dimycolate-induced lung inflammation via the mincle pathway. PLoS Pathog 8(4):e1002614. doi:10.1371/journal.ppat.1002614

Liu D (ed) (2011) Molecular detection of human fungal pathogens. CRC Press/Taylor & Francis, Boca Raton, p 5

Lobato-Pascual A, Saether PC, Fossum S, Dissen E, Daws MR (2013) Mincle, the receptor for mycobacterial cord factor, forms a functional receptor complex with MCL and FcεRI-γ. Eur J Immunol 43(12):3167–3174. doi:10.1002/eji.201343752

Matsumoto M, Tanaka T, Kaisho T, Sanjo H, Copeland NG, Gilbert DJ, Jenkins NA, Akira S (1999) A novel LPS-inducible C-type lectin is a transcriptional target of NF-IL6 in macrophages. J Immunol 163(9):5039–5048

Miyake Y, Oh-Hora M, Yamasaki S (2015) C-type lectin receptor MCL facilitates mincle expression and signaling through complex formation. J Immunol 194(11):5366–5374. doi:10.4049/jimmunol.1402429

Mócsai A, Ruland J, Tybulewicz VL (2010) The SYK tyrosine kinase: a crucial player in diverse biological functions. Nat Rev Immunol 10(6):387–402. doi:10.1038/nri2765

Moody DB, Reinhold BB, Guy MR, Beckman EM, Frederique DE, Furlong ST, Ye S, Reinhold VN, Sieling PA, Modlin RL, Besra GS, Porcelli SA (1997) Structural requirements for glycolipid antigen recognition by CD1b-restricted T cells. Science 278(5336):283–286

Noll H (1957) The chemistry of some native constituents of the purified wax of Mycobacterium tuberculosis. J Biol Chem 224(1):149–164

Noll H, Jackim E (1958) The chemistry of the native constituents of the acetone-soluble fat of Mycobacterium tuberculosis (Brevannes). I. Glycerides and phosphoglycolipides. J Biol Chem 232(2):903–917

Noll H, Bloch H, Asselineau J, Lederer E (1956) The chemical structure of the cord factor of Mycobacterium tuberculosis. Biochim Biophys Acta 20(2):299–309

Papayannopoulos V, Metzler KD, Hakkim A, Zychlinsky A (2010) Neutrophil elastase and myeloperoxidase regulate the formation of neutrophil extracellular traps. J Cell Biol 191 (3):677–691. doi:10.1083/jcb.201006052

Podschun R, Ullmann U (1998) Klebsiella spp. as nosocomial pathogens: epidemiology, taxonomy, typing methods, and pathogenicity factors. Clin Microbiol Rev 11(4):589–603

Rabes A, Zimmermann S, Reppe K, Lang R, Seeberger PH, Suttorp N, Witzenrath M, Lepenies B, Opitz B (2015) The C-type lectin receptor Mincle binds to Streptococcus pneumoniae but plays a limited role in the anti-pneumococcal innate immune response. PLoS One 10(2):e0117022. doi:10.1371/journal.pone.0117022, eCollection 2015

Ribbing C, Engblom C, Lappalainen J, Lindstedt K, Kovanen PT, Karlsson MA, Lundeberg L, Johansson C, Nilsson G, Lunderius-Andersson C, Scheynius A (2011) Mast cells generated from patients with atopic eczema have enhanced levels of granule mediators and an impaired Dectin-1 expression. Allergy 66(1):110–119. doi:10.1111/j.1398-9995.2010.02437.x

Richardson MB, Williams SJ (2014) MCL and Mincle: C-type lectin receptors that sense damaged self and pathogen-associated molecular patterns. Front Immunol 5:288. doi:10.3389/fimmu.2014.00288

Robinson MJ, Osorio F, Rosas M, Freitas RP, Schweighoffer E, Gross O, Verbeek JS, Ruland J, Tybulewicz V, Brown GD, Moita LF, Taylor PR, Reis e Sousa C (2009) Dectin-2 is a Syk-coupled pattern recognition receptor crucial for Th17 responses to fungal infection. J Exp Med 206(9):2037–2051. doi:10.1084/jem.20082818

Roth RR, James WD (1988) Microbial ecology of the skin. Annu Rev Microbiol 42:441–464

Schoenen H, Bodendorfer B, Hitchens K, Manzanero S, Werninghaus K, Nimmerjahn F, Agger EM, Stenger S, Andersen P, Ruland J, Brown GD, Wells C, Lang R (2010) Cutting edge: Mincle is essential for recognition and adjuvanticity of the mycobacterial cord factor and its synthetic analog trehalose-dibehenate. J Immunol 184(6):2756–2760. doi:10.4049/jimmunol.0904013

Sharma A, Steichen AL, Jondle CN, Mishra BB, Sharma J (2014) Protective role of Mincle in bacterial pneumonia by regulation of neutrophil mediated phagocytosis and extracellular trap formation. J Infect Dis 209(11):1837–1846. doi:10.1093/infdis/jit820

Shenderov K, Barber DL, Mayer-Barber KD, Gurcha SS, Jankovic D, Feng CG, Oland S, Hieny S, Caspar P, Yamasaki S, Lin X, Ting JP, Trinchieri G, Besra GS, Cerundolo V, Sher A (2013) Cord factor and peptidoglycan recapitulate the Th17-promoting adjuvant activity of mycobacteria through mincle/CARD9 signaling and the inflammasome. J Immunol 190 (11):5722–5730. doi:10.4049/jimmunol.1203343

Sousa Mda G, Reid DM, Schweighoffer E, Tybulewicz V, Ruland J, Langhorne J, Yamasaki S, Taylor PR, Almeida SR, Brown GD (2011) Restoration of pattern recognition receptor costimulation to treat chromoblastomycosis, a chronic fungal infection of the skin. Cell Host Microbe 9(5):436–443. doi:10.1016/j.chom.2011.04.005

Suzuki Y, Nakano Y, Mishiro K, Takagi T, Tsuruma K, Nakamura M, Yoshimura S, Shimazawa M, Hara H (2013) Involvement of Mincle and Syk in the changes to innate immunity after ischemic stroke. Sci Rep 3:3177. doi:10.1038/srep03177

Tanaka M, Ikeda K, Suganami T, Komiya C, Ochi K, Shirakawa I, Hamaguchi M, Nishimura S, Manabe I, Matsuda T, Kimura K, Inoue H, Inagaki Y, Aoe S, Yamasaki S, Ogawa Y (2014)

Macrophage-inducible C-type lectin underlies obesity-induced adipose tissue fibrosis. Nat Commun 5:4982. doi:10.1038/ncomms5982

Teuschl AH, Neutsch L, Monforte X, Rünzler D, van Griensven M, Gabor F, Redl H (2014) Enhanced cell adhesion on silk fibroin via lectin surface modification. Acta Biomater 10 (6):2506–2517. doi:10.1016/j.actbio.2014.02.012

Toyonaga K, Miyake Y, Yamasaki S (2014) Characterization of the receptors for mycobacterial cord factor in Guinea pig. PLoS One 9(2):e88747. doi:10.1371/journal.pone.0088747

Tsumita T (1956) Studies on the lipid of BCG. II. On the mycolic acids of wax A fraction of BCG. Jpn J Med Sci Biol 9(4–5):217–222

van Dissel JT, Joosten SA, Hoff ST, Soonawala D, Prins C, Hokey DA, O'Dee DM, Graves A, Thierry-Carstensen B, Andreasen LV, Ruhwald M, de Visser AW, Agger EM, Ottenhoff TH, Kromann I, Andersen P (2014) A novel liposomal adjuvant system, CAF01, promotes long-lived Mycobacterium tuberculosis-specific T-cell responses in human. Vaccine 32 (52):7098–7107. doi:10.1016/j.vaccine.2014.10.036

Wells CA, Salvage-Jones JA, Li X, Hitchens K, Butcher S, Murray RZ, Beckhouse AG, Lo YL, Manzanero S, Cobbold C, Schroder K, Ma B, Orr S, Stewart L, Lebus D, Sobieszczuk P, Hume DA, Stow J, Blanchard H, Ashman RB (2008) The macrophage-inducible C-type lectin, mincle, is an essential component of the innate immune response to Candida albicans. J Immunol 180(11):7404–7413

Werninghaus K, Babiak A, Gross O, Hölscher C, Dietrich H, Agger EM, Mages J, Mocsai A, Schoenen H, Finger K, Nimmerjahn F, Brown GD, Kirschning C, Heit A, Andersen P, Wagner H, Ruland J, Lang R (2009) Adjuvanticity of a synthetic cord factor analogue for subunit Mycobacterium tuberculosis vaccination requires FcRgamma-Syk-Card9-dependent innate immune activation. J Exp Med 206(1):89–97. doi:10.1084/jem.20081445

Wevers BA, Kaptein TM, Zijlstra-Willems EM, Theelen B, Boekhout T, Geijtenbeek TB, Gringhuis SI (2014) Fungal engagement of the C-type lectin mincle suppresses dectin-1-induced antifungal immunity. Cell Host Microbe 15(4):494–505. doi:10.1016/j.chom.2014.03.008

World Health Organization Global tuberculosis report 2015 (2015)

Yamasaki S, Ishikawa E, Sakuma M, Hara H, Ogata K, Saito T (2008) Mincle is an ITAM-coupled activating receptor that senses damaged cells. Nat Immunol 9(10):1179–1188. doi:10.1038/ni.1651

Yamasaki S, Matsumoto M, Takeuchi O, Matsuzawa T, Ishikawa E, Sakuma M, Tateno H, Uno J, Hirabayashi J, Mikami Y, Takeda K, Akira S, Saito T (2009) C-type lectin Mincle is an activating receptor for pathogenic fungus, Malassezia. Proc Natl Acad Sci U S A 106 (6):1897–1902. doi:10.1073/pnas.0805177106

Young E, Koers WJ, Berrens L (1989) Intracutaneous tests with pityrosporon extract in atopic dermatitis. Acta Derm Venereol Suppl (Stockh) 144:122–124

Zelensky AN, Gready JE (2005) The C-type lectin-like domain superfamily. FEBS J 272 (24):6179–6217

Chapter 3
The DAP12-Associated Myeloid C-Type Lectin 5A (CLEC5A)

Shie-Liang Hsieh

Abstract Innate immunity is the first line of host defense mechanism against pathogen invasion. In order to recognize various pathogen-associated molecular patterns, myeloid cells express abundant innate immunity receptors on cell surface to recognize diverse pathogen-associated molecular patterns (PAMPs). Toll-like receptors (TLRs) are the most well-characterized innate immunity receptors for pattern recognition, and activation of TLRs triggers the MyD88- and TRIF-dependent pathways to induce the secretion of pro-inflammatory cytokines and interferons (Akira and Takeda 2004; Athman and Philpott 2004). In addition to TLRs, natural killer cells and myeloid cells, which are the key players in innate immunity, recognize glycan and non-glycan structures on pathogen surface via the C-type lectin receptors (denoted as CLRs), which are the most abundant lectins in human genome. Among the myeloid CLRs, the spleen tyrosine kinase (Syk)-coupled C-type lectin receptors (Syk-CLRs) have been shown to play critical roles in host defense against pathogen invasion (Osorio et al. 2011; Sancho et al. 2012). Here, we discuss the potential roles of CLEC5A, also known as "myeloid DAP12-associating lectin-1 (MDL-1)," in host defense and autoimmunity. We would also discuss the impact of dual recognition by CLEC5A and TLRs in future study of host-pathogen interaction.

Keywords TLR • Syk • CLR • Lectins • Innate immunity • PAMPs • ITAM

S.-L. Hsieh (✉)
Genomics Research Center, Academia Sinica, 128 Academia Road, Section 2, Nankang, Taipei 115, Taiwan

Institute of Clinical Medicine, National Yang-Ming University School of Medicine, Taipei, Taiwan

Department of Medical Research and Education, Taipei Veterans General Hospital, Taipei, Taiwan
e-mail: slhsieh@gate.sinica.edu.tw; slhsieh@ym.edu.tw

S. Yamasaki (ed.), *C-Type Lectin Receptors in Immunity*,
DOI 10.1007/978-4-431-56015-9_3

3.1 The Syk-Coupled C-Type Lectins (Syk-CLRs) in Human Genome

There are approximately 150 proteins with “C-type lectin-like domain” (denoted as CTLD family) classified into 17 groups in human genome (Varki et al. 2009). All the Syk-CLRs are type II transmembrane proteins belonging to either group II or group V of CTLD family. The type II C-type lectins contain four Syk-CLRs [CLEC4B (DCAR2), CLEC4C (BDCA-2), CLEC4E (Mincle), and CLEC6A (Dectin-2)]. In addition, the type V C-type lectins (also denoted as NK receptor-like lectin) contains four Syk-CLRs [CLEC2, CLEC5A (MDL-1), CLEC7A (Dectin-1), CLEC9A (CD370)] and 11 NK receptors [CLEC5B (NKR-P1A, CD161), CLEC5C (KLRF1), CLEC15B (KLRG2), MAFA (KLRG1), NKG2A (KLRC1, CD159a), NKG2B (KLRC1), NKG2C (KLRC2, CD159c), NKG2D (CD314), NKG2E (KLRC3), NKG2F (KLRC4), NKG2I (KLRE1)].

3.1.1 Syk-CLRs with ITAM Motif

CLEC7A (Dectin-1) and CLEC9A (CD370) possess a cytoplasmic immunoreceptor tyrosine-based activation-like YxxL motif (ITAM, YxxL-(X)10-12-YxxL) that can recruit Syk kinase. CLEC2 has a YxxL motif in its cytoplasmic tail, which is crucial for CLEC2-mediated signal transduction. This sequence is called hemi-ITAM, since its cytoplasmic domain only has one YxxL motif. CLEC1 contains a tyrosine residue in the sequence YSST, which may represent a novel signaling motif.

3.1.2 Syk-CLR Associated with Adaptor Fc Receptor Gamma-Chain (FcRγ)

All group II Syk-CLRs transduce signal via association with the ITAM-containing adaptor protein FcRγ: CLEC4B (DCAR2), CLEC4C (BDCA-2), CLEC4E (Mincle), and CLEC6A (Dectin-2).

3.1.3 Syk-CLR Associated with Adaptor DNAX-Associated Protein 12 (DAP12)

There are 12 DAP12-associated proteins in the group V CTLD family. While 11 members belong to NK receptors, the myeloid C-type lectin CLEC5A (MDL-1) is abundantly expressed in monocytes, macrophages, and polymorphonuclear cells (PMNs), but not in NK cells, dendritic cells, lymphocytes, or platelets. While human CLEC5A is only reported to associate with DAP12 (Chen et al. 2008), mouse CLEC5A is able to associate with DAP12 (Bakker et al. 1999) (in monocytes, macrophages, PMNs) and DAP10 (Inui et al. 2009) (in osteoclast).

3.2 CLEC5A (MDL-1) and Its Known Ligands

There are four myeloid Syk-CLRs [CLEC7A (Dectin-1), CLEC6A (Dectin-2), CLEC4E (Mincle), and CLEC5A (MDL-1)] involved in recognition of fungi (Taylor et al. 2007; Saijo et al. 2007; Wells et al. 2008), mycobacterial glycolipid (Ishikawa et al. 2009), virus (Chen et al. 2008, 2012), and bacteria (Rabes et al. 2015). Compared to the other three myeloid Syk-CLRs, the functions of CLEC5A are relatively un-explored yet. In this chapter, we will discuss the potential functions of CLEC5A in host-pathogen interaction, autoimmunity, and septic shock.

3.2.1 Discovery of CLEC5A

CLEC5A, initially named as "myeloid DAP12-associating lectin-1 (MDL-1)," was cloned in Dr. Lanier's group by indirect expression cloning in 1999 (Bakker et al. 1999). DAP12 is located in the cytosol without the presence of DAP12-associating receptor, while it was translocated to cell surface when cells where co-transfected with DAP12 and DAP12-associating proteins. By transfecting mouse cDNA library with flag-tagged DAP12 cDNA, they identified CLEC5A, a novel cDNA clone, associated with DAP12 (Bakker et al. 1999). Cross-linking of CLEC5A is able to induce calcium mobilization, suggesting CLEC5A is able to activate cells via DAP12-mediated signaling. Furthermore, CLEC5A is expressed in the myeloid lineage, including monocytes, macrophages, neutrophils, microglia, and osteoclasts, but not in monocyte-derived dendritic cells. The specific expression of CLEC5A in myeloid lineage is further confirmed by the observation that CLEC5A is a novel PU.1 transcriptional target during myeloid differentiation (Batliner et al. 2011). Furthermore, crystallization and X-ray diffraction analysis of human CLEC5A revealed that CLEC5A forms a homodimer on cell surface (Watson and O'Callaghan 2010). However, the ligand and function of CLEC5A were unknown until the discovery of CLEC5A as the dengue recognition receptor (Chen et al. 2008).

NCBI (AAI13100) 188 aa(CLEC5A):

```
  1 MNWHMIISGL IVVVLKVVGM TLFLLYFPQI FNKSNDGFTT TRSYGTVSQI FGSSSPSPNG
 61 FITTRSYGTV CPKDWEFYQA RCFFLSTSES SWNESRDFCK GKGSTLAIVN TPEKLKFLQD
121 ITDAEKYFIG LIYHREEKRW RWINNSVFNG NVTNQNQNFN CATIGLTKTF DAASCDISYR
181 RICEKNAK
```

NCBI (NP_037384) 165 aa(CLEC5A_S)

```
  1 MNWHMIISGL IVVVLKVVGM TLFLLYFPQI FNKSNDGFTT TRSYGTVCPK DWEFYQARCF
 61 FLSTSESSWN ESRDFCKGKG STLAIVNTPE KLKFLQDITD AEKYFIGLIY HREEKRWRWI
121 NNSVFNGNVT NQNQNFNCAT IGLTKTFDAA SCDISYRRIC EKNAK
```

Fig. 3.1 Alignment of human CLEC5A and its variant (CLEC5A_S). Human CLEC5A is a 188 a.a. polypeptide belonging to type II transmembrane protein. The transmembrane domain is located in a.a. 3–25, the neck region is located in a.a. 26–70, and the CTLD is located in a.a. 71–188. The "K_{16}" is critical for association with DAP12. The CLEC5A_S is short of 23 a.a. ($_{48}$SQIFGSSSPSPNGFITTRSYGTV$_{70}$) in the neck region

3.2.2 Length Polymorphism of CLEC5A

The full-length CLEC5A cDNA encodes a 188 a.a. peptide belonging to type II transmembrane protein with a putative N-linked glycosylation site in the extracellular domain. In addition, the alternative splicing variant CLEC5A_S1 (165 a.a.) was found in human peripheral blood cells. The CLEC5A_S1 lacks 23 a.a. ($_{48}$SQIFGSSSPSPNGFITTRSYGTV$_{70}$) in the neck region (Fig. 3.1).

It is interesting to note that length polymorphism in the neck region was also found in other CLRs. The 23 a.a. imperfect repeat segment (QNLTQLKAAVGELSEKSKLQEIY) is responsible for the length polymorphism of CLEC4L (DC-SIGN/CD209) and CLEC4M (DC-SIGNR/CD299). The predominant form of CLEC4L (DC-SIGN/CD209) and CLEC4M (DC-SIGNR/CD299) contains 7 repeats of largely conserved 23 amino acid segment. It has been shown that the neck region of CLR is responsible for homo-oligomerization and thus allows the CLR to bind multivalent ligands with high avidity, thus the length polymorphism in neck region would affect its binding affinity and specificity to their ligands. This speculation is supported by the observation that variable repeats in the neck region of CLEC4M (DC-SIGNR/CD299) affect its interaction with HIV-1 (Gramberg et al. 2006) and associates with susceptibility to HCV infection (Nattermann et al. 2006). The crystallography shows that CLEC5A forms homodimer, and twisting of neck region can elicit conformation change to activate CLEC5A (Watson et al. 2011). Whether the length polymorphism in CLEC5A neck region has significant impact on ligand binding needs to be further characterized.

3.2.3 CLEC5A Is the Pattern Recognition Receptor to Flaviviruses

Even though the other three myeloid lectins (CLEC6A, CLEC7A, and CLEC4E) have been shown to interact with various microbes, the ligand of CLEC5A was unknown until Chen et al. showed that CLEC5A interacts with dengue virus directly and plays a critical role in the pathogenesis of dengue hemorrhagic fever and dengue shock syndrome (Chen et al. 2008; Hsu et al. 2009).

Previous studies on host-virus interaction focused on the identification of "viral entry receptor." In contrast, whether virus can bind and activate innate immunity receptor via the "pathogen-associated molecular patterns" (PAMPs) has never been tested until we showed that CLEC5A can interact with members of flaviviruses. Similar to the outer most structures of microorganisms, most of the viral envelope proteins are also glycosylated. Thus, the distinct distribution of glycans on either the helical or icosahedral structures of each virus would be ideal PAMPs to bind and activate pattern recognition receptors. To address this question, we cloned more than 30 pattern recognition receptors, including C-type lectins, Siglecs, TLRs, TREMs (triggering receptor expressed on myeloid cells), and TREM-like receptors (TLTs) (Chen et al. 2008). The extracellular receptors of these receptors were fused with the Fc portion of human IgG1 and the recombinant receptor. Fc fusion proteins were expressed on mammalian 293 T cells. The conformation of these recombinant proteins was examined by sugar completion assay to confirm their glycan-binding specificity (Hsu et al. 2009).

By the ELISA-based binding assay, we found that the intact dengue virions bind to CLEC4L (DC-SIGN/CD209), CLEC4M (DC-SIGNR/CD299), and CLEC5A, specifically. Moreover, both live DV and UV-inactivated DV can induce DAP12 phosphorylation via CLEC5A in human macrophages, suggesting DV can trigger DAP12 phosphorylation via CLEC5A without replication. These results suggest that virions can act as ligands to bind and activate CLEC5A directly. We further clone CLEC5A_S (the short form of CLEC5A) and compare the binding specificity of CLEC5A.Fc and CLEC5A_S_Fc binding to DV, JEV, and WNV. Similar binding results were found between CLEC5A.Fc and CLEC5A_S_Fc, suggesting the neck domain length polymorphism did not have significant influence of binding specificity to members of flaviviruses. In addition, we also find that CLEC5A binds specifically to influenza virus H5N1, but not H1N1, specifically.

To further understand how CLEC5A interacts with dengue virus, we developed a novel detection method based on nanostructured hemisphere-based biosensor for lectin-virion interaction (Fig. 3.2) (Tung et al. 2014). Since glycans on icosahedral virion is not on a flat plan, the interaction between viral glycans and immobilized lectin.Fc is limited. The weak interaction between lectins and viral glycans makes it extremely difficult in determining the **stringency** of **washing** to discriminate specific vs. nonspecific interaction by conventional ELISA. To overcome this problem, we immobilized the lectin.Fc fusion proteins on the nanostructured hemisphere surface to form multiple-valence binding between lectins and virions.

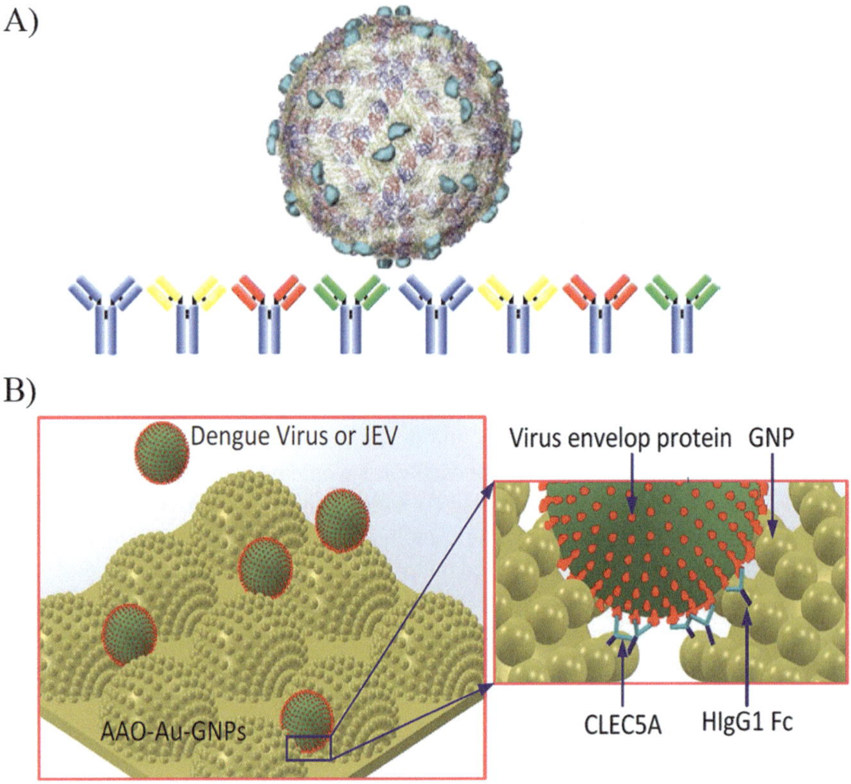

Fig. 3.2 Interaction of virions with immobilized lectin.Fc fusion protein. (**a**) The interaction between immobilized recombinant lectin.Fc on flat surface with glycans on virus surface is limited. (**b**) The nanostructured hemisphere creates a three-dimensional space to all the multiple-valence binding between immobilized lectin.Fc and glycans on viral surfaces (Figure b is provided by Dr. Gou-Jen Wang and Yen-Ting Wang (National Chung-Hsing University))

By comparing the binding affinity between the wild-type (WT) and mutant deficient of glycosylation site (Asn67 mutant and Asn153 mutant), CLEC5A was found to interact with Asn153, while CLEC4L (DC-SIGN/CD209) interacts with Asn153 of dengue virus (DV) envelope protein (un-published data). Since crystallography has shown that DC-SIGN interacts with the glycans on Asn153 (Pokidysheva et al. 2006), this observation suggests that DV can bind to both CLEC5A and DC-SIGN via glycans on Asn67 and Asn153, respectively, to trigger downstream signaling pathways.

3.3 CLEC5A in Flaviviral Infections

The *Flavivirus* genus includes the mosquito-borne dengue, Japanese encephalitis, and yellow fever viruses (Mukhopadhyay et al. 2005), and infections of flaviviruses can result in diverse clinical syndromes such as encephalitis, hemorrhagic fever, and shock syndrome (Mackenzie et al. 2004). There are four serotypes of DV, which can give rise to severe hemorrhagic syndrome (dengue hemorrhagic fever/DHF) and capillary leakage induced-hypovolemic shock (dengue shock syndrome/DSS) (Wilder-Smith and Schwartz 2005). Dengue is a major public health problem, with ~50 million people infected each year (of whom ~20,000 die) and ~2.5 billion people worldwide being at risk of infection. On the other hand, the Japanese encephalitis virus (JEV) serological group, which includes West Nile virus (WNV) and St. Louis encephalitis virus, is a major contributor to the occurrence of viral encephalitis worldwide (Weaver and Barrett 2004), with 50,000 new cases and 15,000 deaths per annum (Hollidge et al. 2010). Myeloid cells, including monocytes, macrophages, and dendritic cells, are the major targets of flaviviruses, and overactivation of macrophages by DV causes massive release for cytokines known as "cytokine storm," which is believed to be responsible for DHF and DSS. However, how DV activates macrophages to trigger release of inflammatory cytokines is unknown till we showed that CLEC5A is critical for DV-induced lethality in mouse model (Chen et al. 2008).

3.3.1 CLEC5A in the Pathogenesis of DV Infection

Dengue is a mosquito-borne infection caused by four serotypes of *dengue virus* (DV1-4) and is currently the most common arboviral disease worldwide (Wilder-Smith and Schwartz 2005). Primary infection with any of the four DV serotypes typically results in mild dengue fever (DF) and provides lifelong immunity to the infecting strain. However, secondary infection with different DV serotypes is associated with increased risk of developing DHF (characterized by thrombocytopenia and capillary leakage) and can progress to life-threatening hypovolemic DSS (Wilder-Smith and Schwartz 2005). The pathogenesis of DHF/DSS remains unclear, but massive cytokine secretion (cytokine storm) is believed to be one of the major contributory factors (Varki et al. 2009). Unfortunately, no DV-specific therapies or vaccines are available (Wilder-Smith and Schwartz 2005).

Since wild-type mice are resistant to DV infection, we utilized the STAT1-deficient mice as a model system to demonstrate the role of CLEC5A in flaviviral infection. We found that DV induces cytokine storms and causes systemic permeability leakage, subcutaneous hemorrhage, and systemic shock syndrome (Chen et al. 2008). Blockade of CLEC5A by antagonistic monoclonal antibodies (mAbs) suppressed DV-induced cytokine storm and rescued mice from DV-induced lethality (Chen et al. 2008). To further evaluate the critical genes involved in DHF,

Gomez et al. analyzed the gene expression in peripheral blood mononuclear cells of dengue patients using the "support vector machines" (SM) algorithm. They found that CLEC5A was associated with disease severity among the 28 dengue patients they analyzed (Gomes et al. 2010). Furthermore, genetic variations at CLEC5A were shown to increase the risk and regulate TNF secretion in dengue severity among Brazilians (Xavier-Carvalho et al. 2013). These observations demonstrated the critical role of CLEC5A in the pathogenesis of DHF and DSS.

3.3.2 CLEC5A in the Pathogenesis of JEV Infection

JEV is the most prevalent cause of encephalitis worldwide, even though both inactivated (Hoke et al. 1988) and live-attenuated JEV vaccines (Xin et al. 1988) have been used in Asia for decades. In fact, these vaccines are not completely effective against all the clinical isolates (Ku et al. 1994), and there are still ~35,000 reported cases of Japanese encephalitis (JE) and 10,000 deaths each year (Olsen et al. 2010). Unlike DHF and DSS, JE victims experience permanent neuropsychiatric sequelae, including persistent motor defects and severe cognitive and language impairments (Mackenzie et al. 2004). However, the molecular mechanism for the pathogenesis of JEV infection is still unclear.

We found that JEV also interacted with the CLEC5A and CLEC5A_S directly. Moreover, JEV can infect and activate microglia to induce the secretion of pro-inflammatory cytokines and neurotoxic substances. In contrast, blockade of CLEC5A by antagonistic mAb can suppress the release of pro-inflammatory cytokines and neurotoxic substances (Chen et al. 2012). Injection of antagonistic anti-CLEC5A mAbs maintains the intact of blood brain barrier and prevents JEV-induced permeability change and neurotoxicity. The protective effect is not from inhibiting JEV infection to neuronal cells, but via the inhibition of neuronal inflammation and cell infiltration from peripheral blood. Notably, mice receiving anti-CLEC5A mAbs are still able to develop protective cellular and humoral immunity against JEV infection. This observation demonstrated that blockade of CLEC5A during acute infection does not interfere the development of host immune response against JEV (Chen et al. 2012).

3.4 CLEC5A in Shock Syndrome

It is interesting to find that CLEC5A-positive cells play critical role in concanavalin A (ConA)-induced liver injury and lethal shock in mouse model system (Cheung et al. 2011). ConA-induced hepatic injury closely resembles the pathophysiology of T cell-mediated liver diseases; it therefore has been used extensively as an animal model for autoimmune and viral hepatitis (Tiegs et al. 1992). After ConA injection, massive CLEC5A-positive cells were found to infiltrate into mouse liver. The expression of CLEC5A from high to low was found in $CD11b^{+}Ly6G^{+}Ly6C^{lo}$ cells, followed by the $CD11b^{+}Ly6G^{-}Ly6C^{hi}$ and $CD11b^{+}Ly6G^{-}Ly6C^{lo}$ cells. Flow cytometry analysis further demonstrated that the CLEC5A-positive cells express CD33 and MHC class I, but not CD34 or MHC class II, suggesting the ConA-induced CLEC5A-positive cells in liver are the immature myeloid cells. Interestingly, activation of CLEC5A-positive cells by agonistic anti-CLEC5A mAb induced NO and TNF-α release in vitro, while injection of agonistic anti-CLEC5A mAb or DV into ConA-treated mice caused lethal shock in wild-type mice within 60 min and 24 h, respectively. This observation suggests that CLEC5A-mediated signaling may contribute to lethal shock syndrome during systemic inflammatory response syndrome (SIRS).

3.5 CLEC5A in Autoimmunity

It is interesting to find that CLEC5A is not only limited to host defense against flaviviral infection but also involved in the development of autoimmune diseases (Joyce-Shaikh et al. 2010; Chen et al. 2014). CLEC5A is highly expressed in mouse bone marrow and bone tissues, and highest levels of CLEC5A were found in the $CD11b^{+}Ly6G^{+}$ cells. In the collagen antibody-induced arthritis (CAIA) model, the $CD11b^{+}Ly6G^{+}$ cells increased up to 50 % and 40 % in bone marrow and peripheral blood after injection of anti-collagen mAb, and co-injection of arthrogenic and anti-CLEC5A mAb exacerbated disease severity. In addition, CLEC5A-deficient mice are resistant to collagen-induced arthritis (CIA), and this observation provides direct evidence that CLEC5A is critical for the pathogenesis of autoimmune arthritis. Since injection of recombinant CLEC5A.Fc is able to suppress bone erosion in CIA mice model, the CLEC5A-positive cells may recognize yet-identified endogenous ligands expressed in synovia to cause tissue damage.

To further understand the role of CLEC5A in autoimmune arthritis, we examine the expression of CLEC5A in peripheral blood isolated from arthritis patients (Chen et al. 2014). We found that the number of CLEC5A-positive monocyte is much higher in patients with active (53.6 %) and inactive (31 %) rheumatoid arthritis than osteoarthritis (27.9 %) and healthy control (21.2 %). Moreover, CLEC5A expression level in synovia was positively correlated with parameters of disease activity, articular damage, and levels of pro-inflammatory cytokines (Chen et al. 2014).

3.6 Cross Talk Between CLEC5A and TLRs in Host Defense Mechanism

Several intracellular receptors (TLR 3, 7, 8, 9) (Mikula et al. 2010) and sensors (MDA5, RIG-I, and AIM) (Kanneganti 2010) have been shown to be responsible for virus-induced inflammatory reactions via recognition of viral nucleic acids. In addition, surface receptors TLR2 (Kim et al. 2012) and TLR4 (Kurt-Jones et al. 2000) are also involved in host recognition to herpes simplex virus (HSV) and respiratory syncytial virus, respectively. However, whether intact virions physically interact with TLR2 and TLR4 is still unknown. It is interestingly observed that even HSV-1 can signal via the TLR2, this receptor does not mediate recognition of HSV glycoproteins (Reske et al. 2008). This observation suggests that TLR2 may associate with other receptor on cell surface to recognize the glycoproteins on HSV. We have shown that CLEC5A physically interacts with flaviviruses, and flaviviruses can activate CLEC5A directly to trigger downstream signaling. This observation suggests the possibility that virus may co-activate CLEC5A with members of TLRs to trigger downstream signaling.

3.6.1 Cross Talk Between CLEC5A and TLR4 in NALP3 Inflammasome Activation

Fever is one of the most typical symptoms for all DV patients, but how DV causes the severe high fever in patients is not clear, even though human macrophages have been regarded as the major source of inflammatory cytokines (Chen and Wang 2002) and the major target cells for DV replication (Jessie et al. 2004). Macrophages are heterogeneous, and their phenotypes depend on the presence of cytokines and other factors during its differentiation process. While M-CSF is essential for the development of "resting macrophages" (denoted as M-Mϕ), high level of GM-CSF is critical for monocyte differentiation into "inflammatory macrophages" (also denoted as GM-Mϕ) during inflammatory reaction (Wu et al. 2013a). Since IL-1β and TNF-α are the most potent endogenous pyrogens (Netea et al. 2000), we are interested to identify the source of these two cytokines from human macrophages. We found that DV can induce pro-inflammatory cytokines (TNF-α, IL-1β, and IL-18) from GM-Mϕ, but not M-Mϕ, even though the expression levels of CLEC5A and TLR4 are similar between these two populations (Wu et al. 2013b).

Unlike TNF-α, the production of IL-1α and IL-18 relies on the activation of inflammasome (Schroder and Tschopp 2010). Thus, we asked whether CLEC5A is also responsible for DV-induced inflammasome activation. While activation of CLEC5A induces low levels of IL-1β and IL-18, co-activation of CLEC5A and TLR4 induces high levels of IL-1β and IL-18 via NALP3 inflammasome. This observation implies that GM-Mϕ can produce high levels of endogenous pyrogens upon DV and gram-negative bacteria infection and can explain why high mortality and morbidity were observed in dengue patients with concurrent bacterial infections (Lee et al. 2005).

3.6.2 *Cross Talk Between CLEC5A and Other TLRs*

It has been shown that DV can activate macrophages via CLEC5A and TLR7, and co-activation of CLEC5A and TLR enhances the secretion of pro-inflammatory cytokines such as TNF-α, IL-6, IL-8, and IP-10, but not IFN-α (Chen et al. 2008). In contrast, knockdown of CLEC4L/DC-SIGN/CD209 did not affect DV-induced inflammatory cytokine release significantly. Interestingly, blockade of CLEC5A only suppressed the inflammatory cytokines, but not IFN-α. This observation suggests that co-activation of CLEC5A and TLR7 has synergistic effect on TNF-α production, and CLEC5A does not contribute to IFN-α secretion during DV infection. Thus, targeting CLEC5A is able to attenuate inflammatory reaction without attenuating antiviral immunity.

It has been shown that murine peritoneal neutrophils are the major source of IL-10 during sepsis by using the cecum ligation model (Kasten et al. 2010). Interestingly, co-activation of peritoneal neutrophils in $Gr\text{-}1^{hi}/Ly6G^{hi}CD^{155-/low}$ mice by immobilized anti-CLEC5A mAb, and TLR2 ligand Pam3 upregulated the secretion of IL-10 dramatically (Zhang et al. 2009). This observation demonstrates that the co-activation of CLEC5A and TLR2 is important for IL-10 secretion from murine peritoneal neutrophils. However, human neutrophils do not produce IL-10 (Davey et al. 2011); thus the effect of CLEC5A and TLR2 co-activation in human neutrophils needs to be further characterized in the future.

Above observations suggest that CLEC5A may form functional complex with specific TLRs in different cell lineages to generate combinatorial repertoires to recognize diverse PAMPs on pathogens. Genetic approach using CLEC5A and TLRs double knockout mice will be able to answer the significance of CLEC5A and TLRs in host defense mechanisms in the future.

3.7 Potential Functions of CLEC5A in Inflammatory Diseases

Previous studies have demonstrated that CLEC5A is involved in flaviviral infections, autoimmune arthritis, and aseptic shock syndrome. While flaviviruses have been shown to be CLEC5A exogenous ligands, the endogenous ligands of CLEC5A in autoimmunity are still unknown.

In addition to macrophages, CLEC5A is also upregulated in eosinophils and neutrophils. It has been shown that CLEC5A is upregulated in eosinophils isolated from asthma patients (Esnault et al. 2013), but whether CLEC5A is involved in allergen recognition is still unknown. In contrast, the abundant expression of CLEC5A in neutrophils (Chen et al. 2008; Zhang et al. 2009) suggests that CLEC5A may play a critical role in neutrophil-mediated defense mechanisms, such as phagocytosis, cytotoxicity, and the formation of neutrophil extracellular traps (NETs). Since neutrophils are involved in host defense against bacteria, fungi, and other pathogens, CLEC5A may be also involved in host defense against nonviral pathogens. This argument is supported by the observation that CLEC5A binds preferentially to mannans and fucose. Since mannans are the essential components of fungal and bacterial cell walls, CLEC5A may also play important roles in host defense against bacterial and fungi. In contrast, the fucose-containing glycans play important role in blood transfusion reactions, leukocyte-endothelial adhesion, host-microbe interactions, and Notch receptor signaling. Moreover, alterations in the expression of fucosylated oligosaccharides were observed in cancer and atherosclerosis. Therefore, CLEC5A may be involved in the recognition of fucosylated glycans upregulated during microbe invasion, inflammation, cancer progression, and atherosclerosis. Thus, CLEC5A is not only critical for flaviviral infections and may be also involved in the pathogenesis of nonviral infections, transfusion, and autoimmunity.

Acknowledgements This work was supported by the National Science Council (NSC 103-2321-B-001-044 and NSC 103-2320-B-001-010-MY3) and Summit and Thematic Research Project (AS-101-TP-B06-2) of Academia Sinica.

References

Akira S, Takeda K (2004) Toll-like receptor signalling. Nat Rev Immunol 4:499–511

Athman R, Philpott D (2004) Innate immunity via toll-like receptors and Nod proteins. Curr Opin Microbiol 7:25–32

Bakker AB, Baker E, Sutherland GR, Phillips JH, Lanier LL (1999) Myeloid DAP12-associating lectin (MDL)-1 is a cell surface receptor involved in the activation of myeloid cells. Proc Natl Acad Sci U S A 96:9792–9796

Batliner J et al (2011) CLEC5A (MDL-1) is a novel PU.1 transcriptional target during myeloid differentiation. Mol Immunol 48:714–719

Chen YC, Wang SY (2002) Activation of terminally differentiated human monocytes/macrophages by dengue virus: productive infection, hierarchical production of innate cytokines and chemokines, and the synergistic effect of lipopolysaccharide. J Virol 76:9877–9887

Chen ST et al (2008) CLEC5A is critical for dengue-virus-induced lethal disease. Nature 453:672–676

Chen ST et al (2012) CLEC5A regulates Japanese encephalitis virus-induced neuroinflammation and lethality. PLoS Pathog 8:e1002655

Chen DY et al (2014) A potential role of myeloid DAP12-associating lectin (MDL)-1 in the regulation of inflammation in rheumatoid arthritis patients. PLoS One 9:e86105

Cheung R et al (2011) Activation of MDL-1 (CLEC5A) on immature myeloid cells triggers lethal shock in mice. J Clin Invest 121:4446–4461

Davey MS et al (2011) Failure to detect production of IL-10 by activated human neutrophils. Nat Immunol 12:1017–1018; author reply 1018-1020

Esnault S et al (2013) Identification of genes expressed by human airway eosinophils after an in vivo allergen challenge. PLoS One 8:e67560
Gomes AL et al (2010) Classification of dengue fever patients based on gene expression data using support vector machines. PLoS One 5:e11267
Gramberg T et al (2006) Impact of polymorphisms in the DC-SIGNR neck domain on the interaction with pathogens. Virology 347:354–363
Hoke CH et al (1988) Protection against Japanese encephalitis by inactivated vaccines. N Engl J Med 319:608–614
Hollidge BS, Gonzalez-Scarano F, Soldan SS (2010) Arboviral encephalitides: transmission, emergence, and pathogenesis. J Neuroimmune Pharmacol 5:428–442
Hsu TL et al (2009) Profiling carbohydrate-receptor interaction with recombinant innate immunity receptor-Fc fusion proteins. J Biol Chem 284:34479–34489
Inui M et al (2009) Signal adaptor DAP10 associates with MDL-1 and triggers osteoclastogenesis in cooperation with DAP12. Proc Natl Acad Sci U S A 106:4816–4821
Ishikawa E et al (2009) Direct recognition of the mycobacterial glycolipid, trehalose dimycolate, by C-type lectin Mincle. J Exp Med 206:2879–2888
Jessie K, Fong MY, Devi S, Lam SK, Wong KT (2004) Localization of dengue virus in naturally infected human tissues, by immunohistochemistry and in situ hybridization. J Infect Dis 189:1411–1418
Joyce-Shaikh B et al (2010) Myeloid DAP12-associating lectin (MDL)-1 regulates synovial inflammation and bone erosion associated with autoimmune arthritis. J Exp Med 207:579–589
Kanneganti TD (2010) Central roles of NLRs and inflammasomes in viral infection. Nat Rev Immunol 10:688–698
Kasten KR, Muenzer JT, Caldwell CC (2010) Neutrophils are significant producers of IL-10 during sepsis. Biochem Biophys Res Commun 393:28–31
Kim M et al (2012) Herpes simplex virus antigens directly activate NK cells via TLR2, thus facilitating their presentation to CD4 T lymphocytes. J Immunol 188:4158–4170
Ku CC et al (1994) Homologous and heterologous neutralization antibody responses after immunization with Japanese encephalitis vaccine among Taiwan children. J Med Virol 44:122–131
Kurt-Jones EA et al (2000) Pattern recognition receptors TLR4 and CD14 mediate response to respiratory syncytial virus. Nat Immunol 1:398–401
Lee IK, Liu JW, Yang KD (2005) Clinical characteristics and risk factors for concurrent bacteremia in adults with dengue hemorrhagic fever. Am J Trop Med Hyg 72:221–226
Mackenzie JS, Gubler DJ, Petersen LR (2004) Emerging flaviviruses: the spread and resurgence of Japanese encephalitis, West Nile and dengue viruses. Nat Med 10:S98–S109
Mikula I Jr, Pastorekova S, Mikula I Sr (2010) Toll-like receptors in immune response to the viral infections. Acta Virol 54:231–245
Mukhopadhyay S, Kuhn RJ, Rossmann MG (2005) A structural perspective of the flavivirus life cycle. Nat Rev Microbiol 3:13–22
Nattermann J et al (2006) The tandem-repeat polymorphism of the DC-SIGNR gene in HCV infection. J Viral Hepat 13:42–46
Netea MG, Kullberg BJ, Van der Meer JW (2000) Circulating cytokines as mediators of fever. Clin Infect Dis 31(Suppl 5):S178–S184
Olsen SJ et al (2010) Japanese encephalitis virus remains an important cause of encephalitis in Thailand. Int J Infect Dis 14:e888–e892
Osorio F, Reis E, Sousa C (2011) Myeloid C-type lectin receptors in pathogen recognition and host defense. Immunity 34:651–664
Pokidysheva E et al (2006) Cryo-EM reconstruction of dengue virus in complex with the carbohydrate recognition domain of DC-SIGN. Cell 124:485–493
Rabes A et al (2015) The C-type lectin receptor Mincle binds to Streptococcus pneumoniae but plays a limited role in the anti-pneumococcal innate immune response. PLoS One 10:e0117022

Reske A, Pollara G, Krummenacher C, Katz DR, Chain BM (2008) Glycoprotein-dependent and TLR2-independent innate immune recognition of herpes simplex virus-1 by dendritic cells. J Immunol 180:7525–7536

Saijo S et al (2007) Dectin-1 is required for host defense against Pneumocystis carinii but not against Candida albicans. Nat Immunol 8:39–46

Sancho D, Reis E, Sousa C (2012) Signaling by myeloid C-type lectin receptors in immunity and homeostasis. Annu Rev Immunol 30:491–529

Schroder K, Tschopp J (2010) The inflammasomes. Cell 140:821–832

Taylor PR et al (2007) Dectin-1 is required for beta-glucan recognition and control of fungal infection. Nat Immunol 8:31–38

Tiegs G, Hentschel J, Wendel A (1992) A T cell-dependent experimental liver injury in mice inducible by concanavalin A. J Clin Invest 90:196–203

Tung YT, Wu MF, Wang GJ, Hsieh SL (2014) Nanostructured electrochemical biosensor for the detection of the weak binding between the dengue virus and the CLEC5A receptor. Nanomed: Nanotechnol Biol Med 10:1335–1341

Varki A, Cummings RD, Esko JD et al (2009) Chapter 31: C-type lectins. In: Essentials of glycobiology, Cold Spring Harbor Press, New York, USA

Watson AA, O'Callaghan CA (2010) Crystallization and X-ray diffraction analysis of human CLEC5A (MDL-1), a dengue virus receptor. Acta Crystallogr Sect F: Struct Biol Cryst Commun 66:29–31

Watson AA et al (2011) Structural flexibility of the macrophage dengue virus receptor CLEC5A: implications for ligand binding and signaling. J Biol Chem 286:24208–24218

Weaver SC, Barrett AD (2004) Transmission cycles, host range, evolution and emergence of arboviral disease. Nat Rev Microbiol 2:789–801

Wells CA et al (2008) The macrophage-inducible C-type lectin, mincle, is an essential component of the innate immune response to Candida albicans. J Immunol 180:7404–7413

Wilder-Smith A, Schwartz E (2005) Dengue in travelers. N Engl J Med 353:924–932

Wu MF, Chen ST, Hsieh SL (2013a) Distinct regulation of dengue virus-induced inflammasome activation in human macrophage subsets. J Biomed Sci 20:36

Wu MF et al (2013b) CLEC5A is critical for dengue virus-induced inflammasome activation in human macrophages. Blood 121:95–106

Xavier-Carvalho C et al (2013) Single nucleotide polymorphisms in candidate genes and dengue severity in children: a case-control, functional and meta-analysis study. Infect Genet Evol: J Mol Epidemiol Evol Genet Infect Dis 20:197–205

Xin YY, Ming ZG, Peng GY, Jian A, Min LH (1988) Safety of a live-attenuated Japanese encephalitis virus vaccine (SA14-14-2) for children. Am J Trop Med Hyg 39:214–217

Zhang X, Majlessi L, Deriaud E, Leclerc C, Lo-Man R (2009) Coactivation of Syk kinase and MyD88 adaptor protein pathways by bacteria promotes regulatory properties of neutrophils. Immunity 31:761–771

Part II
HemITAM-Bearing Receptors

Chapter 4
Dectin-1 (CLEC7A, BGR, CLECSF12)

Patawee Asamaphan, Janet A. Willment, and Gordon D. Brown

Abstract Dectin-1 is the archetypical example of the C-type lectin receptor (CLR) family of pattern recognition receptors (PRRs). Expressed primarily by cells of the innate immune system, this receptor is best known for its role in antifungal immunity through its ability to recognise cell wall β-glucans. Upon recognition of these carbohydrates, Dectin-1 transduces intracellular signalling through several pathways activating or regulating numerous cellular responses such as phagocytosis, the respiratory burst, neutrophil extracellular trap formation, inflammasome activation and cytokine and chemokine production. Moreover, like the Toll-like receptors (TLRs), Dectin-1 is able to instruct the development of adaptive immunity, promoting Th1- and Th17-type responses. Dectin-1 collaborates with other PRRs to synergise and regulate innate and adaptive immune responses. More recently, Dectin-1 has been found to recognise a broader range of microbial pathogens, including bacteria, as well as endogenous ligands, influencing autoimmune and other diseases, including rheumatoid arthritis, ulcerative colitis and cancer. In this chapter, we will discuss the structure, expression and ligands of Dectin-1, as well as the intracellular signalling pathways and cellular responses that this receptor can induce. We will describe the role of Dectin-1 in antifungal immunity and in immunity to other pathogens. We will briefly discuss the interaction of Dectin-1 with other PRRs and its broader role in immunity, through recognition of endogenous ligands, for example.

Keywords Dectin-1 • C-type lectin • ITAM-like • Antifungal immunity • Beta-glucan

P. Asamaphan • J.A. Willment • G.D. Brown (✉)
Aberdeen Fungal Group, Institute of Medical Sciences, University of Aberdeen, Aberdeen AB25 2ZD, Scotland, UK
e-mail: gordon.brown@abdn.ac.uk

S. Yamasaki (ed.), *C-Type Lectin Receptors in Immunity*,
DOI 10.1007/978-4-431-56015-9_4

4.1 Dectin-1 Structure, Expression and Ligands

Dectin-1 is a member of the group V C-type lectin family and encoded within the Dectin-1 gene cluster located in the natural gene complex (NKC) on chromosome 6 in mouse and chromosome 12 in human (Zelensky and Gready 2005). This type II transmembrane receptor has a single carbohydrate recognition domain (CRD), a stalk region, a transmembrane region and a cytoplasmic tail (Fig. 4.1a) (Drummond and Brown 2011). The N-terminal cytoplasmic tail has an immunoreceptor tyrosine-based activation motif (ITAM)-like, YXXL, which can activate downstream signalling pathways (Drummond and Brown 2011). The extracellular domains of Dectin-1 possess several putative N- and O-glycosylation sites, and N-glycosylation has been shown to be involved in cell surface expression and function of this receptor (Kato et al. 2006). Dectin-1 is alternatively spliced into two major isoforms (isoforms A and B, the latter of which lacks the stalk region) and several minor isoforms (isoform C–H) (Willment et al. 2001; Heinsbroek et al. 2006). Only the two major isoforms (A and B; Fig. 4.1a) are able to interact with extracellular ligands (Willment et al. 2001; Heinsbroek et al. 2006). Dectin-1 minor isoform E, which contains the CRD and ITAM-like domain, has been shown to be located in the cytoplasm and interact with a Ran binding protein, RanBPM, although the functional significance of this interaction is unclear (Xie et al. 2006).

Dectin-1 is primarily expressed by innate immune cells including neutrophils, monocytes, macrophages and dendritic cells (DCs) (Drummond and Brown 2011). This receptor is also expressed on microglia, eosinophils, mast cells and certain lymphocytes, including B cells and γδ T-cells (Drummond and Brown 2011). On myeloid cells, the expression of this receptor can be regulated by cytokines and growth factors. For example, expression of Dectin-1 is upregulated by granulocyte-macrophage colony-stimulating factor (GM-CSF), IL-4 and IL-13, but downregulated by IL-10, LPS and dexamethasone (Willment et al. 2003). There is evidence that expression of Dectin-1 can be induced on epithelial cells (Lee et al. 2009; Cohen-Kedar et al. 2014).

Originally identified through recognition of an endogenous ligand on T-cells, Dectin-1 is best known for its ability to recognise β-1,3-glucans (Ariizumi et al. 2000; Drummond and Brown 2011). These carbohydrates are commonly found in the cell walls of fungi, and consequently, Dectin-1 has been found to play a key role in antifungal immunity. Indeed, Dectin-1 has been shown to recognise numerous fungal species including pathogens such as *Candida albicans*, *Aspergillus fumigatus*, *Coccidioides immitis* and *Pneumocystis carinii* (Hardison and Brown 2012; Viriyakosol et al. 2013). Although Dectin-1 is structurally similar to classical carbohydrate-binding CLRs, the receptor lacks the residues typically involved in sugar recognition, and the mechanisms by which Dectin-1 actually recognises β-1,3-glucans is unknown. It has been shown that the minimum unit ligand for Dectin-1 is between 11 and 13 glucose monomers and that the affinity of its interaction with these carbohydrates is influenced by side chain branching (Palma et al. 2006; Adams et al. 2008). Structural and mutation analyses have

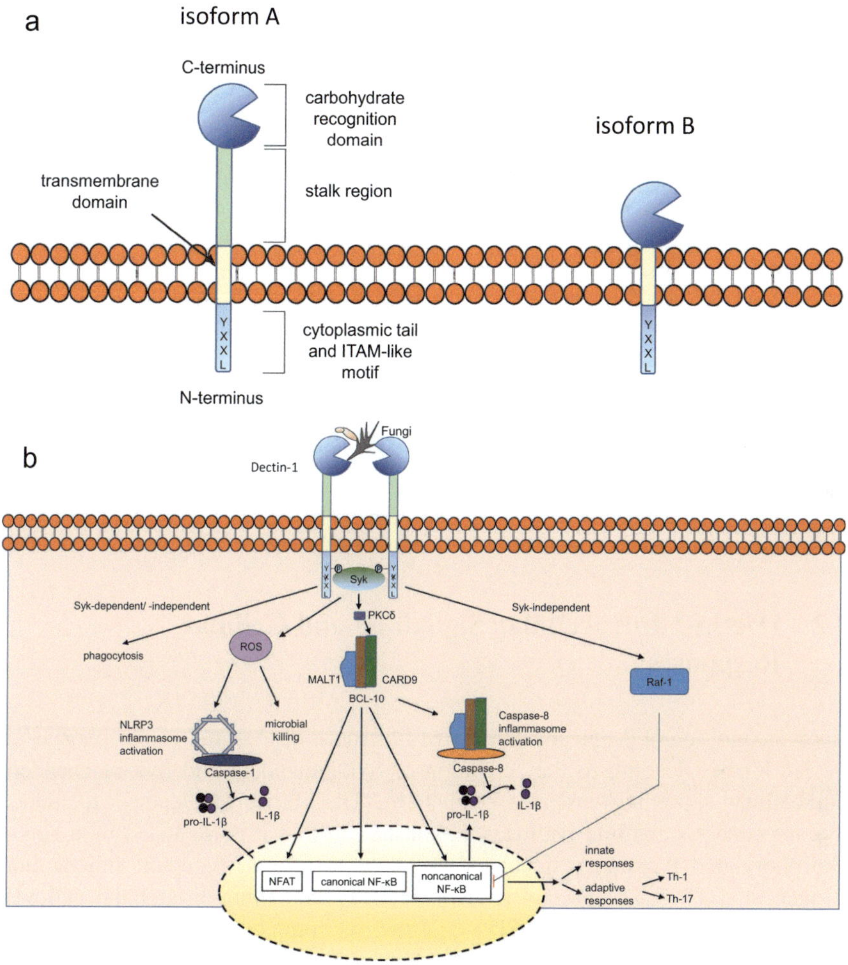

Fig. 4.1 Dectin-1 recognises β-glucan on fungi and mediates signalling through Syk-dependent and Syk-independent pathways. (**a**) Dectin-1 consists of a single carbohydrate recognition domain, a stalk region, a transmembrane and a cytoplasmic tail that contains an ITAM-like motif. Dectin-1 has two major isoforms in mice and humans, isoform A and isoform B, and six minor isoforms (not shown). These two isoforms differ in the presence of the stalk region. (**b**) Upon recognition of β-glucan on fungi, Dectin-1 mediates Syk-dependent NFAT and NF-κB pathways and the Syk-independent Raf-1 pathway to promote the production of cytokine and chemokine productions which in turn promotes antifungal defence through Th1 and Th17 responses. In addition, Dectin-1 signalling also promotes early release of arachidonic acid and eicosanoid production. Dectin-1 also mediates antifungal responses through phagocytosis, reactive oxygen species (ROS) production and inflammasome activation, which are essential in the cleavage and activation of inactive pro-IL-1β to active IL-1β

revealed a shallow groove on the surface of Dectin-1, which may be the ligand-binding site, and that this groove is flanked by two resides, Trp221 and His223, that are indispensable for ligand binding (Adachi et al. 2004; Brown et al. 2007). Dectin-1 can interact with other pathogens, but the structures involved are unknown (discussed below).

In addition to microbial pathogens, there is evidence that Dectin-1 interacts with endogenous ligands. Dectin-1 has been reported to bind the intermediate filament protein, vimentin, through which the receptor may be involved in driving lipid oxidation in atherosclerosis (Thiagarajan et al. 2013). Dectin-1 was also shown to be required for reverse transcytosis of secretory IgA-antigen complexes by intestinal M cells and the induction of subsequent mucosal and systemic antibody responses (Rochereau et al. 2013). In addition, Dectin-1 was shown to recognise galactosylated IgG1, in part through association with FcgammaRIIB, which resulted in inhibition of complement-mediated inflammation (Karsten et al. 2012). Recently, Dectin-1 was found to recognise N-glycans present on the surface of tumour cells and play a role in antitumour immunity (Chiba et al. 2014).

4.2 Dectin-1 Intracellular Signalling and Cellular Responses

Upon recognition of β-glucans, the ITAM-like motif of Dectin-1 is phosphorylated by Src kinases leading to the activation of Syk-dependent and Syk-independent intracellular signalling cascades (Fig. 4.1b). The ability of Dectin-1 to induce Syk-dependent signalling pathways is unusual, in that it is mediated by a single phosphorylated tyrosine residue and is likely to require receptor dimerization (Rogers et al. 2005; Drummond and Brown 2011). Signalling through this pathway involves PKC delta and the CARD9-Bcl10-Malt1 complex and leads to the induction of canonical and non-canonical NF-κB subunits (p65/c-REL and RelB, respectively) and interferon regulatory factor (IRF) 1, resulting in gene transcription (Drummond and Brown 2013; Wevers et al. 2014). Recently, however, CARD9 was found to be dispensable for NF-κB activation, but regulated extracellular signal-regulated kinase (ERK) activation by linking Ras-GRF1 to H-Ras (Jia et al. 2014). Syk activation by Dectin-1 also induces IRF5 and nuclear factor of activated T-cells (NFAT), through phospholipase C gamma and Calcineurin; a pathway inhibited by immunosuppressive drugs, such as cyclosporine (Goodridge et al. 2007; del Fresno et al. 2013). The Syk-independent pathway from Dectin-1 involves activation of Raf-1, which integrates with the Syk-dependent pathway at the point of NF-κB activation (Gringhuis et al. 2009). The ability of Dectin-1 to induce productive intracellular signalling (i.e. leading to cellular responses) requires receptor clustering into a ‘phagocytic synapse’ and exclusion of regulatory tyrosine phosphatases (Goodridge et al. 2011). Moreover, the ability of Dectin-1 to induce productive responses to purified agonists can be cell-type specific, an effect linked to differential utilisation of CARD9 (Rosas et al. 2008; Goodridge et al. 2009).

Activation of Dectin-1 signalling pathways can induce or regulate multiple cellular responses including actin-mediated phagocytosis, neutrophil extracellular trap (NET) formation, activation of the respiratory burst and DC maturation and antigen presentation, in part through the use of autophagy machinery such as light chain 3 protein (LC3) (Drummond and Brown 2011; Hardison and Brown 2012; Ma et al. 2012; Branzk et al. 2014). This latter process was recently shown to involve the FYVE and coiled-coil domain containing 1 (FYCO1) protein, which facilitated the maturation of Dectin-1 induced phagosomes (Ma et al. 2014). Dectin-1 also induces the production of eicosanoids, several cytokines and chemokines (such as TNF, IL-10, IL-6, IL-2, IL-23, IFN-β, CCL2, CCL3) and can modulate cytokine production induced by other PRRs (see later) (Drummond and Brown 2011; del Fresno et al. 2013). Dectin-1 is able to activate inflammasomes, facilitating the production of IL-1β. Indeed, this receptor activates the NLRP3/caspase-1 inflammasome upon recognition of β-glucans, in a Syk-dependent manner (Drummond and Brown 2011; Ganesan et al. 2014). Dectin-1 can also directly induce a non-canonical caspase-8 inflammasome through Syk, CARD9, MALT1 and the non-receptor tyrosine kinase Tec (Gringhuis et al. 2012; Zwolanek et al. 2014). Although most cellular responses described here involve signalling through Syk, this is not always the case in all cells. The induction of phagocytosis by Dectin-1 in macrophages, for example, does not require Syk but rather Bruton's tyrosine kinase (Btk) and Vav-1 (Herre et al. 2004; Strijbis et al. 2013).

4.3 Dectin-1 in Antifungal Immunity

Dectin-1 plays an essential role in antifungal immunity in both mouse and human. Polymorphisms of human Dectin-1, such as the Y238X polymorphism which results in a truncated protein that is not expressed at the cell surface, have been linked to increased susceptibility to infections with *A. fumigatus*, *Trichophyton rubrum* and *C. albicans* (Ferwerda et al. 2009; Cunha et al. 2010). However, the high prevalence of this polymorphism in European and African populations does not correlate with disease prevalence, suggesting that there are other factors contributing to susceptibility in affected individuals (Ferwerda et al. 2009). This likely explains the lack of an effect of the Y238X polymorphism on fungal susceptibility that has been reported in other studies. Deficiency of CARD9 on the other hand, renders both humans and mice extremely susceptible to fungal infection highlighting the importance of this pathway in antifungal immunity (Hardison and Brown 2012). In mice, Dectin-1 is required for protective immunity to several pathogens including *C. albicans*, *A. fumigatus*, *Pneumocystis carinii*, *Coccoidiodes posdasii* and most recently *Paracoccidioides brasiliensis* (Loures et al. 2014; Dambuza and Brown 2015). Notably, the requirement for Dectin-1 in protective immune responses is dependent on fungal strain, at least in mouse models of candidiasis (Marakalala et al. 2013).

The increased susceptibility to fungal infections caused by Dectin-1 deficiency results from defective innate and adaptive antifungal immune responses. Indeed, loss of Dectin-1 correlates with aberrant or defective cellular responses such as fungal phagocytosis and killing, inflammasome activation and induction of inflammatory mediators. Recently, for example, NETs were shown to be selectively released in response to large non-ingestible fungal hyphae (Branzk et al. 2014). Dectin-1 acted as the sensor of microbial size, reducing NET formation following phagocytosis of smaller-sized yeasts, and loss of this receptor led to aberrant NET release and pathology (Branzk et al. 2014). Another example is the autophagy machinery where Dectin-1 induces the phagosomal recruitment of LC3, an autophagic factor required for the fungicidal activity of leukocytes. Loss of Dectin-1 led to failure of LC3 recruitment and reduced fungal killing (Kyrmizi et al. 2013; Tam et al. 2014).

Like the TLRs, Dectin-1 is capable of instructing the development of adaptive immune responses, particularly Th1 and Th17 immunity (LeibundGut-Landmann et al. 2007; Drummond and Brown 2011). While Th1 responses are important for the control of systemic fungal infections, Th17 responses are critical for controlling fungal infections at the mucosa (Kashem et al. 2015). Indeed, a number of human diseases associated with chronic mucocutaneous candidiasis, including CARD9 deficiency, have been linked to alterations in components of the Th17 response (Hernandez-Santos and Gaffen 2012). Several studies have shown that Dectin-1 is required for Th17 polarisation during infection, such as during mucocutaneous infections with *C. albicans* (Drummond and Brown 2011; Kashem et al. 2015). How Dectin-1 promotes Th17 responses is incompletely understood, but involves MALT1-dependent activation of the NF-κB subunit c-REL, which is required for the induction of polarising cytokines, such as IL-1β (through activation of the inflammasome) and IL-23p19 (Drummond and Brown 2011; Gringhuis et al. 2011). The ability of Dectin-1 to drive these responses is dependent on the specific morphology of *C. albicans* (i.e. yeast versus hyphae) (Drummond and Brown 2011; Kashem et al. 2015). Expression of Dectin-1 on lymphocytes, such as γδ T-cells, provides an important innate source for the rapid production of IL-17 and other key cytokines during infection (Martin et al. 2009; Drummond and Brown 2011). Dectin-1 can induce humoral responses, stimulate cytotoxic T-cell responses and induce Th17 cells in response to some fungi, such as *Paracoccidioides brasiliensis* (Drummond and Brown 2011; Loures et al. 2014). Triggering of Dectin-1 can induce innate memory (or trained immunity), through the epigenetic reprogramming of monocytes and the induction of neutrophilic myeloid-derived suppressor cells (Quintin et al. 2012; Rieber et al. 2015).

Fungal pathogens have evolved several mechanisms to avoid immune recognition by Dectin-1. For example, the surface conidial hydrophobin layer and cell wall galactosaminogalactan mask Dectin-1 recognition of *A. fumigatus* (Carrion Sde et al. 2013; Gravelat et al. 2013). Changes in cell wall structure during fungal morphological switching can reduce immune recognition by Dectin-1, such as occurs with yeast versus hyphae in *C. albicans* (Gantner et al. 2005). In some pathogens, there is active masking of β-glucans upon infection, such as the switch to α-glucan production following infection with *Histoplasma capsulatum* (Rappleye et al. 2007). The differential exposure of fungal β-glucans can have substantial impact on host immunity, such as the allergic Th2 responses induced by intact *Cladosporium cladosporioides* (*C. cladosporioides*), which expose little of this carbohydrate at the surface (Mintz-Cole et al. 2012). In contrast, heat-killed *C. cladosporioides* and live *Aspergillus versicolor* both have exposed β-glucans and induce Dectin-1-dependent pulmonary Th17 responses (Mintz-Cole et al. 2012).

4.4 Role of Dectin-1 in Immunity to Other Pathogens

While Dectin-1 is best known for its role in antifungal immunity, it can recognise several other pathogens including *Haemophilus influenzae*, *Salmonella typhimurium*, *Mycobacterium tuberculosis* and *Leishmania infantum* (Drummond and Brown 2013; Lefevre et al. 2013; Heyl et al. 2014). For example, Dectin-1 was found to be required for the induction of macrophage microbicidal and inflammatory responses to *L. infantum* in vitro, and for the control pathogen growth in vivo (Lefevre et al. 2013). However, the role of Dectin-1 in immunity to most of these organisms is not well understood. For example, Dectin-1 was shown to be required for IL-12 responses to mycobacteria in vitro, but loss of this receptor had no effect on anti-mycobacterial immunity in vivo (Drummond and Brown 2013). How Dectin-1 even recognises any of these organisms is still unclear.

4.5 Interaction of Dectin-1 with Other PRRs

The recognition of pathogens by leukocytes involves multiple PRRs, which interact to promote pathogen-specific responses. Dectin-1 was one of the first receptors shown to be capable of such 'crosstalk'. For example, optimal responses to fungi requires signalling from both Dectin-1 and TLRs to synergistically induce the production of cytokines, such as TNF and IL-23, while repressing the production of others, such as IL-12 (Drummond and Brown 2013). Such interactions can be cell-type specific. For example, in DCs, Dectin-1 alone is sufficient for the production of TNF-α, whereas in macrophages, the production of this cytokine requires co-stimulation with TLRs (Drummond and Brown 2013). Dectin-1 interacts with other surface proteins, such as the tetraspanins CD63 and CD37, which regulate the surface expression and functional responses of Dectin-1 (Meyer-Wentrup et al. 2007). Other examples include the indirect activation of complement receptor 3 by Dectin-1, through Vav signalling, that is required for effective neutrophil anti-candida effector functions (Li et al. 2011).

Dectin-1 can interact with other CLRs. For example, an optimal Th17 response to *C. albicans* was found to require signalling from both Dectin-1 and Dectin-2 (Robinson et al. 2009). Dectin-1, in combination with TLR2, was shown to be able to amplify mannose receptor-induced IL-17 production (van de Veerdonk et al. 2009). However, these interactions can also have negative consequences. For example, engagement of the CLR mincle during chromoblastomycoses was recently shown to promote non-protective Th2 immunity by suppressing Dectin-1-mediated Th1 responses (Wevers et al. 2014).

4.6 A Broader Role for Dectin-1 in Immunity

There is emerging evidence that Dectin-1 plays a broader role in immunity. As discussed above, Dectin-1 recognises endogenous molecules, such as galactosylated IgG1 complexes, and regulates the subsequent immune responses induced by these ligands. Dectin-1 has been implicated in the regulation of autoimmune diseases, such as arthritis and colitis. For example, stimulation of Dectin-1 can trigger the development of severe chronic arthritis in genetically susceptible mice (Yoshitomi et al. 2005). In contrast, signalling from Dectin-1 was found to be required for the control of excessive inflammation induced by pathogenic fungi during ulcerative colitis (Iliev et al. 2012). These functions of Dectin-1 can have therapeutic potential, such as promoting innate memory described above. Other examples include administration of the Dectin-1 ligand, β-glucan, to protect against type 1 diabetes, enhance antitumor immune responses, promote wound healing, and CNS axon regeneration, for example (Tian et al. 2013; Karumuthil-Melethil et al. 2014; van den Berg et al. 2014; Baldwin et al. 2015). Even the CRD of Dectin-1 has been used to develop novel therapeutic strategies, when engineered into soluble fusion proteins or incorporated into chimeric T-cell receptors, for example (Ricks et al. 2013; Kumaresan et al. 2014).

4.7 Conclusion

Dectin-1 is one of the best studied receptors, yet we are still discovering new roles and functions for this archetypical PRR. Studies in mice and in humans have clearly demonstrated that the innate and adaptive immune functions mediated by this receptor play a key role in antifungal immunity. Less well understood is the role of Dectin-1 in autoimmunity and its broader immune functions in other infectious and noninfectious diseases. The ability of Dectin-1 to collaborate with other PRRs adds significant complexity yet will be key to fully understand the physiological roles of this CLR. Excitingly, our growing knowledge is already suggesting how Dectin-1 can be targeted or used to provide novel therapeutic strategies for the future.

Acknowledgements We thank the Wellcome Trust and University of Aberdeen for funding.

References

Adachi Y, Ishii T, Ikeda Y, Hoshino A, Tamura H, Aketagawa J, Tanaka S, Ohno N (2004) Characterization of beta-glucan recognition site on C-type lectin, Dectin 1. Infect Immun 72 (7):4159–4171

Adams EL, Rice PJ, Graves B, Ensley HE, Yu H, Brown GD, Gordon S, Monteiro MA, Papp-Szabo E, Lowman DW, Power TD, Wempe MF, Williams DL (2008) Differential high affinity interaction of Dectin-1 with natural or synthetic glucans is dependent upon primary structure and is influenced by polymer chain length and side chain branching. J Pharmacol Exp Ther 325 (1):115–123

Ariizumi K, Shen GL, Shikano S, Xu S, Ritter R, Kumamoto T, Edelbaum D, Morita A, Bergstresser PR, Takashima A (2000) Identification of a novel, dendritic cell-associated molecule, Dectin-1, by subtractive cDNA cloning. J Biol Chem 275(26):20157–20167

Baldwin KT, Carbajal KS, Segal BM, Giger RJ (2015) Neuroinflammation triggered by beta-glucan/Dectin-1 signaling enables CNS axon regeneration. Proc Natl Acad Sci U S A 112 (8):2581–2586

Branzk N, Lubojemska A, Hardison SE, Wang Q, Gutierrez MG, Brown GD, Papayannopoulos V (2014) Neutrophils sense microbe size and selectively release neutrophil extracellular traps in response to large pathogens. Nat Immunol 15(11):1017–1025

Brown J, O'Callaghan CA, Marshall ASJ, Gilbert RJC, Siebold C, Gordon S, Brown GD, Jones EY (2007) Structure of the fungal beta-glucan-binding immune receptor Dectin-1: implications for function. Protein Sci 16(6):1042–1052

Carrion Sde J, Leal SM Jr, Ghannoum MA, Aimanianda V, Latge JP, Pearlman E (2013) The RodA hydrophobin on Aspergillus fumigatus spores masks Dectin-1- and Dectin-2-dependent responses and enhances fungal survival in vivo. J Immunol 191(5):2581–2588

Chiba S, Ikushima H, Ueki H, Yanai H, Kimura Y, Hangai S, Nishio J, Negishi H, Tamura T, Saijo S, Iwakura Y, Taniguchi T (2014) Recognition of tumor cells by Dectin-1 orchestrates innate immune cells for anti-tumor responses. Elife 3, e04177

Cohen-Kedar S, Baram L, Elad H, Brazowski E, Guzner-Gur H, Dotan I (2014) Human intestinal epithelial cells respond to beta-glucans via Dectin-1 and Syk. Eur J Immunol 44 (12):3729–3740

Cunha C, Di Ianni M, Bozza S, Giovannini G, Zagarella S, Zelante T, D'Angelo C, Pierini A, Pitzurra L, Falzetti F, Carotti A, Perruccio K, Latge J-P, Rodrigues F, Velardi A, Aversa F, Romani L, Carvalho A (2010) Dectin-1 Y238X polymorphism associates with susceptibility to invasive aspergillosis in hematopoietic transplantation through impairment of both recipient- and donor-dependent mechanisms of antifungal immunity. Blood 116(24):5394–5402

Dambuza IM, Brown GD (2015) C-type lectins in immunity: recent developments. Curr Opin Immunol 32C:21–27

del Fresno C, Soulat D, Roth S, Blazek K, Udalova I, Sancho D, Ruland J, Ardavin C (2013) Interferon-beta production via Dectin-1-Syk-IRF5 signaling in dendritic cells is crucial for immunity to C. albicans. Immunity 38(6):1176–1186

Drummond RA, Brown GD (2011) The role of Dectin-1 in the host defence against fungal infections. Curr Opin Microbiol 14(4):392–399

Drummond RA, Brown GD (2013) Signalling C-type lectins in antimicrobial immunity. PloS Pathog 9(7), e1003417

Ferwerda B, Ferwerda G, Plantinga TS, Willment JA, van Spriel AB, Venselaar H, Elbers CC, Johnson MD, Cambi A, Huysamen C, Jacobs L, Jansen T, Verheijen K, Masthoff L, Morre SA, Vriend G, Williams DL, Perfect JR, Joosten LAB, Wijmenga C, van der Meer JWM, Adema GJ, Kullberg BJ, Brown GD, Netea MG (2009) Human Dectin-1 deficiency and mucocutaneous fungal infections. N Engl J Med 361(18):1760–1767

Ganesan S, Rathinam VAK, Bossaller L, Army K, Kaiser WJ, Mocarski ES, Dillon CP, Green DR, Mayadas TN, Levitz SM, Hise AG, Silverman N, Fitzgerald KA (2014) Caspase-8 modulates Dectin-1 and complement receptor 3-driven IL-1 beta production in response to beta-glucans and the fungal pathogen, Candida albicans. J Immunol 193(5):2519–2530

Gantner BN, Simmons RM, Underhill DM (2005) Dectin-1 mediates macrophage recognition of Candida albicans yeast but not filaments. EMBO J 24(6):1277–1286

Goodridge HS, Simmons RM, Underhill DM (2007) Dectin-1 stimulation by Candida albicans yeast or zymosan triggers NFAT activation in macrophages and dendritic cells. J Immunol 178 (5):3107–3115

Goodridge HS, Shimada T, Wolf AJ, Hsu YM, Becker CA, Lin X, Underhill DM (2009) Differential use of CARD9 by Dectin-1 in macrophages and dendritic cells. J Immunol 182 (2):1146–1154

Goodridge HS, Reyes CN, Becker CA, Katsumoto TR, Ma J, Wolf AJ, Bose N, Chan AS, Magee AS, Danielson ME, Weiss A, Vasilakos JP, Underhill DM (2011) Activation of the innate immune receptor Dectin-1 upon formation of a 'phagocytic synapse'. Nature 472 (7344):471–475

Gravelat FN, Beauvais A, Liu H, Lee MJ, Snarr BD, Chen D, Xu W, Kravtsov I, Hoareau CMQ, Vanier G, Urb M, Campoli P, Al Abdallah Q, Lehoux M, Chabot JC, Ouimet M-C, Baptista SD, Fritz JH, Nierman WC, Latge JP, Mitchell AP, Filler SG, Fontaine T, Sheppard DC (2013) Aspergillus galactosaminogalactan mediates adherence to host constituents and conceals hyphal beta-glucan from the immune system. PloS Pathog 9(8), e1003575

Gringhuis SI, den Dunnen J, Litjens M, van der Vlist M, Wevers B, Bruijns SCM, Geijtenbeek TBH (2009) Dectin-1 directs T helper cell differentiation by controlling noncanonical NF-kappa B activation through Raf-1 and Syk. Nat Immunol 10(2):203–213

Gringhuis SI, Wevers BA, Kaptein TM, van Capel TMM, Theelen B, Boekhout T, de Jong EC, Geijtenbeek TBH (2011) Selective C-Rel activation via Malt1 controls anti-fungal T-H-17 immunity by Dectin-1 and Dectin-2. PloS Pathog 7(1), e1001259

Gringhuis SI, Kaptein TM, Wevers BA, Theelen B, van der Vlist M, Boekhout T, Geijtenbeek TBH (2012) Dectin-1 is an extracellular pathogen sensor for the induction and processing of IL-1 beta via a noncanonical caspase-8 inflammasome. Nat Immunol 13(3):246–U63

Hardison SE, Brown GD (2012) C-type lectin receptors orchestrate antifungal immunity. Nat Immunol 13(9):817–822

Heinsbroek SEM, Taylor PR, Rosas M, Willment JA, Williams DL, Gordon S, Brown GD (2006) Expression of functionally different Dectin-1 isoforms by murine macrophages. J Immunol 176(9):5513–5518

Hernandez-Santos N, Gaffen SL (2012) Th17 cells in immunity to Candida albicans. Cell Host Microbe 11(5):425–435

Herre J, Marshall AJ, Caron E, Edwards AD, Williams DL, Schweighoffer E, Tybulewicz VL, Reis e Sousa C, Gordon S, Brown GD (2004) Dectin-1 utilizes novel mechanisms for yeast phagocytosis in macrophages. Blood 104(13):4038–4045

Heyl KA, Klassert TE, Heinrich A, Muller MM, Klaile E, Dienemann H, Grunewald C, Bals R, Singer BB, Slevogt H (2014) Dectin-1 is expressed in human lung and mediates the proinflammatory immune response to nontypeable Haemophilus influenzae. Am Soc Microbiol 5(5):e01492–14

Iliev ID, Funari VA, Taylor KD, Quoclinh N, Reyes CN, Strom SP, Brown J, Becker CA, Fleshner PR, Dubinsky M, Rotter JI, Wang HL, McGovern DPB, Brown GD, Underhill DM (2012) Interactions between commensal fungi and the C-type lectin receptor Dectin-1 influence colitis. Science 336(6086):1314–1317

Jia X-M, Tang B, Zhu L-L, Liu Y-H, Zhao X-Q, Gorjestani S, Hsu Y-MS, Yang L, Guan J-H, Xu G-T, Lin X (2014) CARD9 mediates Dectin-1-induced ERK activation by linking Ras-GRF1 to H-Ras for antifungal immunity. J Exp Med 211(11):2307–2321

Karsten CM, Pandey MK, Figge J, Kilchenstein R, Taylor PR, Rosas M, McDonald JU, Orr SJ, Berger M, Petzold D, Blanchard V, Winkler A, Hess C, Reid DM, Majoul IV, Strait RT, Harris

NL, Kohl G, Wex E, Ludwig R, Zillikens D, Nimmerjahn F, Finkelman FD, Brown GD, Ehlers M, Kohl J (2012) Anti-inflammatory activity of IgG1 mediated by Fc galactosylation and association of FcgammaRIIB and Dectin-1. Nat Med 18(9):1401–1406

Karumuthil-Melethil S, Gudi R, Johnson BM, Perez N, Vasu C (2014) Fungal beta-glucan, a Dectin-1 ligand, promotes protection from type 1 diabetes by inducing regulatory innate immune response. J Immunol 193(7):3308–3321

Kashem SW, Igyarto BZ, Gerami-Nejad M, Kumamoto Y, Mohammed J, Jarrett E, Drummond RA, Zurawski SM, Zurawski G, Berman J, Iwasaki A, Brown GD, Kaplan DH (2015) Candida albicans morphology and dendritic cell subsets determine T helper cell differentiation. Immunity 42(2):356–366

Kato Y, Adachi Y, Ohno N (2006) Contribution of N-linked oligosaccharides to the expression and functions of beta-glucan receptor, Dectin-1. Biol Pharm Bull 29(8):1580–1586

Kumaresan PR, Manuri PR, Albert ND, Maiti S, Singh H, Mi T, Roszik J, Rabinovich B, Olivares S, Krishnamurthy J, Zhang L, Najjar AM, Huls MH, Lee DA, Champlin RE, Kontoyiannis DP, Cooper LJ (2014) Bioengineering T cells to target carbohydrate to treat opportunistic fungal infection. Proc Natl Acad Sci U S A 111(29):10660–10665

Kyrmizi I, Gresnigt MS, Akoumianaki T, Samonis G, Sidiropoulos P, Boumpas D, Netea MG, van de Veerdonk FL, Kontoyiannis DP, Chamilos G (2013) Corticosteroids block autophagy protein recruitment in Aspergillus fumigatus phagosomes via targeting Dectin-1/Syk kinase signaling. J Immunol 191(3):1287–1299

Lee HM, Yuk JM, Shin DM, Jo EK (2009) Dectin-1 is inducible and plays an essential role for mycobacteria-induced innate immune responses in airway epithelial cells. J Clin Immunol 29 (6):795–805

Lefevre L, Lugo-Villarino G, Meunier E, Valentin A, Olagnier D, Authier H, Duval C, Dardenne C, Bernad J, Lemesre JL, Auwerx J, Neyrolles O, Pipy B, Coste A (2013) The C-type lectin receptors Dectin-1, MR, and SIGNR3 contribute both positively and negatively to the macrophage response to Leishmania infantum. Immunity 38(5):1038–1049

LeibundGut-Landmann S, Gross O, Robinson MJ, Osorio F, Slack EC, Tsoni SV, Schweighoffer E, Tybulewicz V, Brown GD, Ruland J, Sousa CRE (2007) Syk- and CARD9-dependent coupling of innate immunity to the induction of T helper cells that produce interleukin 17. Nat Immunol 8(6):630–638

Li X, Utomo A, Cullere X, Choi MM, Milner DA Jr, Venkatesh D, Yun SH, Mayadas TN (2011) The beta-glucan receptor Dectin-1 activates the integrin Mac-1 in neutrophils via Vav protein signaling to promote Candida albicans clearance. Cell Host Microbe 10(6):603–615

Loures FV, Araujo EF, Feriotti C, Bazan SB, Costa TA, Brown GD, Calich VLG (2014) Dectin-1 induces M1 macrophages and prominent expansion of CD8(+)IL-17(+) cells in pulmonary paracoccidioidomycosis. J Infect Dis 210(5):762–773

Ma J, Becker C, Lowell CA, Underhill DM (2012) Dectin-1-triggered recruitment of light chain 3 protein to phagosomes facilitates major histocompatibility complex class II presentation of fungal-derived antigens. J Biol Chem 287(41):34149–34156

Ma J, Becker C, Reyes C, Underhill DM (2014) Cutting edge: FYCO1 recruitment to Dectin-1 phagosomes is accelerated by light chain 3 protein and regulates phagosome maturation and reactive oxygen production. J Immunol 192(4):1356–1360

Marakalala MJ, Vautier S, Potrykus J, Walker LA, Shepardson KM, Hopke A, Mora-Montes HM, Kerrigan A, Netea MG, Murray GI, MacCallum DM, Wheeler R, Munro CA, Gow NA, Cramer RA, Brown AJ, Brown GD (2013) Differential adaptation of Candida albicans in vivo modulates immune recognition by Dectin-1. PloS Pathog 9(4), e1003315

Martin B, Hirota K, Cua DJ, Stockinger B, Veldhoen M (2009) Interleukin-17-producing gamma delta T cells selectively expand in response to pathogen products and environmental signals. Immunity 31(2):321–330

Meyer-Wentrup F, Figdor CG, Ansems M, Brossart P, Wright MD, Adema GJ, van Spriel AB (2007) Dectin-1 interaction with tetraspanin CD37 inhibits IL-6 production. J Immunol 178 (1):154–162

Mintz-Cole RA, Gibson AM, Bass SA, Budelsky AL, Reponen T, Hershey GK (2012) Dectin-1 and IL-17A suppress murine asthma induced by Aspergillus versicolor but not Cladosporium cladosporioides due to differences in beta-glucan surface exposure. J Immunol 189 (7):3609–3617

Palma AS, Feizi T, Zhang Y, Stoll MS, Lawson AM, Diaz-Rodriguez E, Campanero-Rhodes MA, Costa J, Gordon S, Brown GD, Chai W (2006) Ligands for the beta-glucan receptor, Dectin-1, assigned using "designer" microarrays of oligosaccharide probes (neoglycolipids) generated from glucan polysaccharides. J Biol Chem 281(9):5771–5779

Quintin J, Saeed S, Martens JH, Giamarellos-Bourboulis EJ, Ifrim DC, Logie C, Jacobs L, Jansen T, Kullberg BJ, Wijmenga C, Joosten LA, Xavier RJ, van der Meer JW, Stunnenberg HG, Netea MG (2012) Candida albicans infection affords protection against reinfection via functional reprogramming of monocytes. Cell Host Microbe 12(2):223–232

Rappleye CA, Eissenberg LG, Goldman WE (2007) Histoplasma capsulatum alpha-(1,3)-glucan blocks innate immune recognition by the beta-glucan receptor. Proc Natl Acad Sci U S A 104 (4):1366–1370

Ricks DM, Chen K, Zheng M, Steele C, Kolls JK (2013) Dectin immunoadhesins and pneumocystis pneumonia. Infect Immun 81(9):3451–3462

Rieber N, Singh A, Oz H, Carevic M, Bouzani M, Amich J, Ost M, Ye Z, Ballbach M, Schafer I, Mezger M, Klimosch SN, Weber AN, Handgretinger R, Krappmann S, Liese J, Engeholm M, Schule R, Salih HR, Marodi L, Speckmann C, Grimbacher B, Ruland J, Brown GD, Beilhack A, Loeffler J, Hartl D (2015) Pathogenic fungi regulate immunity by inducing neutrophilic myeloid-derived suppressor cells. Cell Host Microbe 17:507–514

Robinson MJ, Osorio F, Rosas M, Freitas RP, Schweighoffer E, Gross O, SjefVerbeek J, Ruland J, Tybulewicz V, Brown GD, Moita LF, Taylor PR, Reis e Sousa C (2009) Dectin-2 is a Syk-coupled pattern recognition receptor crucial for Th17 responses to fungal infection. J Exp Med 206(9):2037–2051

Rochereau N, Drocourt D, Perouzel E, Pavot V, Redelinghuys P, Brown GD, Tiraby G, Roblin X, Verrier B, Genin C, Corthesy B, Paul S (2013) Dectin-1 is essential for reverse transcytosis of glycosylated SIgA-antigen complexes by intestinal M cells. PloS Biol 11(9), e1001658

Rogers NC, Slack EC, Edwards AD, Nolte MA, Schulz O, Schweighoffer E, Williams DL, Gordon S, Tybulewicz VL, Brown GD, Sousa CRE (2005) Syk-dependent cytokine induction by Dectin-1 reveals a novel pattern recognition pathway for C type lectins. Immunity 22 (4):507–517

Rosas M, Liddiard K, Kimberg M, Faro-Trindade I, McDonald JU, Williams DL, Brown GD, Taylor PR (2008) The induction of inflammation by Dectin-1 in vivo is dependent on myeloid cell programming and the progression of phagocytosis. J Immunol 181(5):3549–3557

Strijbis K, Tafesse FG, Fairn GD, Witte MD, Dougan SK, Watson N, Spooner E, Esteban A, Vyas VK, Fink GR, Grinstein S, Ploegh HL (2013) Bruton's tyrosine kinase (BTK) and Vav1 contribute to Dectin1-dependent phagocytosis of Candida albicans in macrophages. PloS Pathog 9(6), e1003446

Tam JM, Mansour MK, Khan NS, Seward M, Puranam S, Tanne A, Sokolovska A, Becker CE, Acharya M, Baird MA, Choi AMK, Davidson MW, Segal BH, Lacy-Hulbert A, Stuart LM, Xavier RJ, Vyas JM (2014) Dectin-1-dependent LC3 recruitment to phagosomes enhances fungicidal activity in macrophages. J Infect Dis 210(11):1844–1854

Thiagarajan PS, Yakubenko VP, Elsori DH, Yadav SP, Willard B, Tan CD, Rodriguez ER, Febbraio M, Cathcart MK (2013) Vimentin is an endogenous ligand for the pattern recognition receptor Dectin-1. Cardiovasc Res 99(3):494–504

Tian J, Ma J, Ma K, Guo H, Baidoo SE, Zhang Y, Yan J, Lu L, Xu H, Wang S (2013) Beta-glucan enhances antitumor immune responses by regulating differentiation and function of monocytic myeloid-derived suppressor cells. Eur J Immunol 43(5):1220–1230

van de Veerdonk FL, Marijnissen RJ, Kullberg BJ, Koenen HJ, Cheng SC, Joosten I, van den Berg WB, Williams DL, van der Meer JW, Joosten LA, Netea MG (2009) The macrophage mannose receptor induces IL-17 in response to Candida albicans. Cell Host Microbe 5(4):329–340

van den Berg LM, Zijlstra-Willems EM, Richters CD, Ulrich MM, Geijtenbeek TB (2014) Dectin-1 activation induces proliferation and migration of human keratinocytes enhancing wound re-epithelialization. Cell Immunol 289(1–2):49–54

Viriyakosol S, del Pilar Jimenez M, Gurney MA, Ashbaugh ME, Fierer J (2013) Dectin-1 is required for resistance to coccidioidomycosis in mice. Am Soc Microbiol 4(1), e00597

Wevers BA, Kaptein TM, Zijlstra-Willems EM, Theelen B, Boekhout T, Geijtenbeek TBH, Gringhuis SI (2014) Fungal engagement of the C-type lectin mincle suppresses Dectin-1-induced antifungal immunity. Cell Host Microbe 15(4):494–505

Willment JA, Gordon S, Brown GD (2001) Characterization of the human beta-glucan receptor and its alternatively spliced isoforms. J Biol Chem 276(47):43818–43823

Willment JA, Lin HH, Reid DM, Taylor PR, Williams DL, Wong SYC, Gordon S, Brown GD (2003) Dectin-1 expression and function are enhanced on alternatively activated and GM-CSF-treated macrophages and are negatively regulated by IL-10, dexamethasone, and lipopolysaccharide. J Immunol 171(9):4569–4573

Xie J, Sun M, Guo L, Liu W, Jiang J, Chen X, Zhou L, Gu J (2006) Human Dectin-1 isoform E is a cytoplasmic protein and interacts with RanBPM. Biochem Biophys Res Commun 347 (4):1067–1073

Yoshitomi H, Sakaguchi N, Kobayashi K, Brown GD, Tagami T, Sakihama T, Hirota K, Tanaka S, Nomura T, Miki I, Gordon S, Akira S, Nakamura T, Sakaguchi S (2005) A role for fungal beta-glucans and their receptor Dectin-1 in the induction of autoimmune arthritis in genetically susceptible mice. J Exp Med 201(6):949–960

Zelensky AN, Gready JE (2005) The C-type lectin-like domain superfamily. FEBS J 272 (24):6179–6217

Zwolanek F, Riedelberger M, Stolz V, Jenull S, Istel F, Koepruelue AD, Ellmeier W, Kuchler K (2014) The non-receptor tyrosine kinase Tec controls assembly and activity of the noncanonical caspase-8 inflammasome. PloS Pathog 10(12), e1004525

Chapter 5
DNGR-1, an F-Actin-Binding C-Type Lectin Receptor Involved in Cross-Presentation of Dead Cell-Associated Antigens by Dendritic Cells

Pavel Hanč, Salvador Iborra, Santiago Zelenay, Janneke van Blijswijk, David Sancho, and Caetano Reis e Sousa

Keywords Dendritic cells • Necrosis • Damage-associated molecular patterns • DNGR-1 • CLEC9A • F-actin • Cross-presentation • Cytotoxic T cells • Innate immunity

5.1 Introduction

DNGR-1 (also known as CLEC9A) is a C-type lectin homodimeric innate immune receptor that detects the presence of dead cells (Sancho et al. 2009). In both mouse and Man, DNGR-1 is specifically expressed at high levels by dendritic cells (DCs), in particular the $XCR1^+$ '$CD8\alpha^+$-like' subtype that excels in the ability to cross-present cell-associated antigens to $CD8^+$ T cells (Caminschi et al. 2008; Huysamen et al. 2008; Poulin et al. 2010, 2012; Sancho et al. 2008). DNGR-1 plays a key role in the latter process. The receptor is found both at the DC surface and within endosomes, with its ligand-binding domains facing the exterior milieu or the endosomal lumen. This topology allows for effective surveillance of cell corpses with which DCs come into contact, as well as internalised dead cell debris. DNGR-1 signalling, in part via Syk, in response to engagement by ligand, acts to delay

P. Hanč • J. van Blijswijk • C. Reis e Sousa (✉)
Immunobiology Laboratory, Lincoln's Inn Fields Laboratory, The Francis Crick Institute, 44 Lincoln's Inn Fields, London WC2A 3LY, UK
e-mail: caetano@crick.ac.uk

S. Iborra • D. Sancho
Centro Nacional de Investigaciones Cardiovasculares Carlos III (CNIC), Madrid, Spain

S. Zelenay
Immunobiology Laboratory, Lincoln's Inn Fields Laboratory, The Francis Crick Institute, 44 Lincoln's Inn Fields, London WC2A 3LY, UK

Cancer Inflammation and Immunity Laboratory, Cancer Research UK Manchester Institute, The University of Manchester, Wilmslow Road, Manchester M20 4BX, UK

S. Yamasaki (ed.), *C-Type Lectin Receptors in Immunity*,
DOI 10.1007/978-4-431-56015-9_5

endosomal maturation and slow cargo degradation, ensuring the preservation of dead cell-associated antigens (Iborra et al. 2012; Sancho et al. 2009; Zelenay et al. 2012). Whilst the actual mechanism underlying the regulation of cross-presentation of antigens from cellular corpses by DNGR-1 remains a mystery, progress has been made in identifying the ligand. This turns out to be F-actin, an abundant intracellular component that is inaccessible in intact cells and whose exposure therefore serves as a damage-associated molecular pattern (DAMP) (Ahrens et al. 2012; Zhang et al. 2012). The recently solved structure of the DNGR-1: F-actin complex shows that each ligand-binding domain of the receptor nests in the groove between two F-actin protofilaments and contacts three actin subunits. The structure clearly explains specificity of the receptor for F-actin and sheds light into the mode of binding and its biophysical characteristics (Hanč et al. 2015). Here, we review the properties and biology of DNGR-1 and the importance of cytoskeletal recognition in innate immune detection of cell damage.

5.2 Cellular Distribution of DNGR-1

DCs comprise a heterogeneous group of cells and, therefore, are commonly subdivided into subsets based on select functional attributes, differences in levels of expression of certain cell-surface markers, localisation and ontogenetic criteria (Hashimoto et al. 2011; Heath and Carbone 2009; Schraml and Reis e Sousa 2015; Steinman and Idoyaga 2010). Plasmacytoid DCs (pDCs) comprise a subgroup of DCs dependent on E2-2 and characterised, in part, by their rapid production of interferon-α upon toll-like receptor 7, 8 and 9 engagement (Reizis et al. 2011). Conventional DCs (cDCs) are another broad class of DCs best known for their efficiency in initiating and directing T cell responses (Hashimoto et al. 2011; Heath and Carbone 2009; Steinman and Idoyaga 2010). In the mouse, cDCs can be stratified into $CD11b^-$ and $CD11b^+$ subsets. $CD11b^-$ '$CD8\alpha^+$-like' cDCs comprise the $CD8\alpha^+$ DCs in lymphoid organs and their $CD103^+$ $CD11b^-$ DC counterparts in tissues, which can migrate to lymph nodes both in the steady state and during inflammation. All '$CD8\alpha^+$-like' mouse cDCs share common gene expression patterns, depend for their development on IRF8 as well as members of the Batf transcription factor family and possess a well-developed capacity to cross-present exogenous antigens to $CD8^+$ T cells (Edelson et al. 2010; Hashimoto et al. 2011; Heath and Carbone 2009; Miller et al. 2012; Steinman and Idoyaga 2010; Hildner et al. 2008; Tussiwand et al. 2012). $CD11b^+$ cDCs are more heterogeneous and comprise multiple subtypes that share a common dependence on IRF4 for their development (Persson et al. 2013; Schlitzer et al. 2013; Suzuki et al. 2004).

In mice, DNGR-1 is highly expressed by the $CD8\alpha^+$-like subset of cDCs and at lower levels by pDCs but not by other leukocytes (Fig. 5.1) (Caminschi et al. 2008; Sancho et al. 2008). In human, DNGR-1 identifies a $BDCA\text{-}3^+$ $CD11b^-$ DC subset in the blood and spleen that shows high resemblance to the murine $CD8\alpha^+$-like cDCs and can also develop in vitro and in humanised mice, but it is not detected in

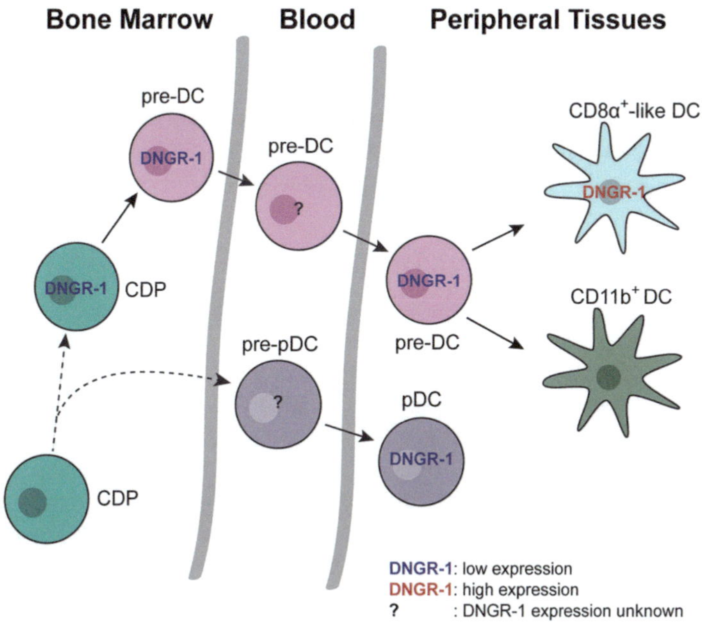

Fig. 5.1 Cellular distribution of DNGR-1 in the mouse. The cellular distribution of DNGR-1 in the bone marrow, blood and peripheral tissues of mouse is depicted. DNGR-1 written in *red* identifies cells expressing high levels of DNGR-1 and *blue* identifies cells with low DNGR-1 expression. A *question mark* denotes that DNGR-1 expression is unknown for that cell type. *CDP* common DC precursor, *pDC* plasmacytoid DC, *DC* dendritic cell

human pDC (Huysamen et al. 2008; Poulin et al. 2010, 2012; Sancho et al. 2008). In this respect, high expression of DNGR-1 identifies the 'CD8α$^+$-like' cDC family across species and largely mimics the expression pattern of XCR1 (Bachem et al. 2010, 2012; Crozat et al. 2011; Vu Manh et al. 2015a). However, unlike DNGR-1, XCR1 is not expressed by mouse pDCs. Furthermore, XCR1 restriction to CD8α$^+$-like cDCs is conserved across mouse, Man, sheep and pig, whereas that of DNGR-1 may fail in the latter species. For this reason, the CD8α$^+$-like cDCs are now often termed XCR1$^+$ DCs rather than DNGR-1$^+$ DCs (Vu Manh et al. 2015b).

Recently, DC precursors were found to also express DNGR-1, albeit at lower levels than 'CD8α$^+$-like' cDCs (Schraml et al. 2013). DCs originate from haematopoietic stem cells in the bone marrow that gradually lose the ability to generate other leukocytes until a stage is reached where the precursor can only give rise to DCs. This cell is called the 'common DC precursor' (CDP) and its progeny, the pre-DC, exits the bone marrow and seeds all tissues (Onai et al. 2007; Naik et al. 2006, 2007; Liu et al. 2009). Interestingly, DNGR-1 marks CDPs and pre-DCs in the bone marrow that have lost the ability to generate pDCs and only give rise to cDCs (Schraml et al. 2013) (Fig. 5.1). This finding was exploited to construct a fate-mapping system allowing for the indelible genetic marking of mouse DNGR-1$^+$ cells and their progeny (Schraml et al. 2013). In such mice, the progeny of DNGR-1

$^+$ CDPs/pre-DCs expresses yellow fluorescent protein (YFP) and thereby serves as a means for identifying cDCs based on ontogenetic criteria (Schraml and Reis e Sousa 2015). Analysis of such mice formally demonstrates that DNGR-1 expression is restricted to the DC lineage as no other cells, including ones of non-haematopoietic origin, are labelled with YFP.

5.3 Antigen Targeting to DNGR-1

DNGR-1 at the plasma membrane is internalised upon binding ligand or antibodies (Hanč et al. 2015; Sancho et al. 2008). The latter finding, together with its DC-restricted expression pattern, suggested that DNGR-1 could constitute a useful target for antibody-mediated antigen delivery to DCs to facilitate immunisation or tolerisation, much like DEC-205, another C-type lectin receptor (Bonifaz et al. 2002; Hawiger et al. 2001). Indeed, antigenic peptides covalently coupled to anti-mouse DNGR-1 are selectively cross-presented on MHC class I by DNGR-1 $^+$ CD8α^+-like DCs. In the presence of adjuvants, this induces potent CTL responses in vivo, which can prevent development, as well as mediate eradication, of transplantable melanoma (Lahoud et al. 2011; Picco et al. 2014; Sancho et al. 2008). In addition to cross-presentation, antibody targeting to DNGR-1 can also lead to activation of CD4$^+$ T cells through antigen presentation on MHC class II, resulting in cellular as well as humoral responses in mice and nonhuman primates (Caminschi et al. 2008, 2012; Joffre et al. 2010; Lahoud et al. 2011; Li et al. 2015; Park et al. 2013). Interestingly, the requirement for adjuvant, indispensable for the induction of cytotoxic T cells (Lahoud et al. 2011; Picco et al. 2014; Sancho et al. 2008), is not universally seen when using DNGR-1 targeting to generate CD4$^+$ T cells (Caminschi et al. 2008, 2012; Joffre et al. 2010; Lahoud et al. 2011; Li et al. 2015; Park et al. 2013). This may be a function of the antibody used for targeting (Li et al. 2015) and allows for great versatility in modulating the CD4$^+$ T cell responses (Joffre et al. 2010). Thus, using certain anti-DNGR-1 antibodies with or without different types of adjuvants, it is possible, at least in the mouse, to induce antigen-specific regulatory T cells, Th1 cells or Th17 responses (Joffre et al. 2010). In sum, targeting antigens to DNGR-1 constitutes a promising approach for therapeutic or prophylactic vaccination (Caminschi and Shortman 2012; Radford and Caminschi 2013) and might also be useful for tolerisation.

5.4 Structure of DNGR-1

In human, DNGR-1 is encoded by the *CLEC9A* gene consisting of six exons with a single splice variant (Huysamen et al. 2008; Sancho et al. 2008). The mouse *Clec9a* gene contains seven exons that were numbered 1–7 by Sancho et al. (2008) whilst

Huysamen et al. (2008) refer to the fourth and fifth exons as 4′ and 4, respectively. In all, five splice variants of mouse DNGR-1 have been reported. Sancho et al. (2008) described three isoforms, termed *long* (containing all seven exons), *short* (lacking exon 4) and *very short* (containing only exons 1–3 and 7). Huysamen et al. (2008) described two additional isoforms – one lacking exon 2, and the other lacking exons 4 and 6 and containing a premature stop codon in exon 7 (for clarity, we use the exon and splice variant notation as per Sancho et al. (2008)).

The genomic structure of DNGR-1 is conserved between mouse and human and, in both species, exon 1 encodes a short cytoplasmic domain containing the hemITAM signalling motif (see below), exon 2 encodes the transmembrane domain and exons 3–6 (human) or 3–7 (mouse) are the extracellular domain composed of a single C-type lectin-like domain (CTLD) and a membrane-proximal 'neck' region. The neck region contains a conserved cysteine residue at position 94 (mouse) or 96 (human), which is thought to be involved in receptor dimerisation via formation of a disulphide bond (Huysamen et al. 2008; Sancho et al. 2008). The CTLD contains six canonical cysteine residues involved in forming three disulphide bonds that stabilise the domain (Zelensky and Gready 2005) but lacks the conserved residues involved in calcium coordination and carbohydrate binding that are present in classical C-type lectins (Huysamen et al. 2008; Sancho et al. 2008). As such, it belongs to group V of the C-type lectin superfamily, which is primarily known for including NK cell receptors (Zelensky and Gready 2005).

The murine isoform that most closely resembles the human receptor is the *short* isoform. However, whilst human DNGR-1 is expressed as a glycosylated disulphide-bonded homodimer, the *short* mouse isoform was initially reported to be expressed as a non-glycosylated monomer (Huysamen et al. 2008). This would be surprising given that the *short* isoform still contains the putative dimerisation cysteine, and there is a high degree of homology between the mouse and human variants. Our data suggest, in fact, that the mouse *short* receptor also expresses as a glycosylated dimer (unpublished observations), much like its *long* counterpart (Sancho et al. 2008).

The crystal structure of a glycosylation-deficient S225D mutant CTLD of human DNGR-1 has been published (PDB ID 3VPP) (Zhang et al. 2012). It covers residues S111-L236 with a missing stretch of five amino acids between E202 and N208 that was, presumably, disordered and did not generate electron density (Fig. 5.2a). Using electron cryomicroscopy, we have recently found that this missing region can be observed in ligand-bound DNGR-1, forming a loop that extends out of the CTLD (Hanč et al. 2015; Fig. 5.2b). This is in line with the predicted flexibility of the region and suggests that the interaction of DNGR-1 with ligand helps stabilise the loop, even though its removal results in no significant decrease in DNGR-1 binding (Hanč et al. 2015). The crystal structure of the CTLD of human DNGR-1 further reveals a calcium ion involved in formation of a salt bridge between α2 helix and β1/β5 sheet (Zhang et al. 2012) in the region corresponding to the calcium coordination site 4 of other C-type lectins (nomenclature as per Zelensky and Gready (2005); Fig. 5.2a). This is consistent with the notion that the ion helps stabilise the CTLD fold but is not involved in ligand binding (Zhang et al. 2012).

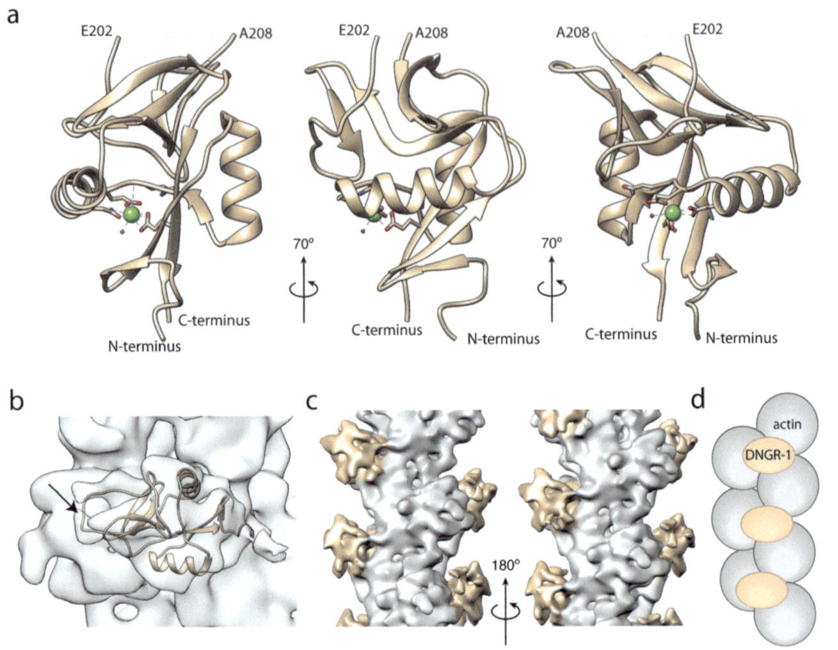

Fig. 5.2 Structure of DNGR-1 CTLD and DNGR-1 in complex with F-actin. (**a**) Structure of hDNGR-1 CTLD (PDB ID 3VPP, (Zhang et al. 2012)) with the C- and N-termini as well as the missing region labelled. Ca^{2+} ion (*green ball*) and Ca^{2+}-coordinating residues (*sticks*) are highlighted. (**b**) DNGR-1 CTLD structure fitted into the electron density of DNGR-1: F-actin complex with the flexible loop missing from crystal structure indicated by the arrow (PDB ID 3J82, (Hanč et al. 2015)) (**c**) DNGR-1-decorated F-actin filament, actin in *grey*, DNGR-1 in *gold* (Electron Microscopy Database ID EMD-6102, (Hanč et al. 2015)). (**d**) Schematic depiction of the mode of DNGR-1 binding to F-actin. Golden ovals represent DNGR-1 and *grey circles* represent monomers of G-actin

In the published DNGR-1 crystal, the asymmetric unit contained a dimer of CTLDs (Zhang et al. 2012). The dimeric interface was, however, highly fenestrated and buried within it was a loop, which is normally accessible to antibodies, implying that the interface was likely an artefact of crystal packing (Zhang et al. 2012). Structural homology of DNGR-1 to lectin-like oxidised low-density lipoprotein (LDL) receptor-1 (LOX-1), also known as CLEC8A (PDB ID 1YPU) (Park et al. 2005), led Zhang et al. to build a model of a 'LOX-1-like' dimer of DNGR-1 CTLDs (Zhang et al. 2012). A subsequent study, however, showed that a tryptophan residue in LOX-1 (W150) is crucial for maintaining the dimeric interface as its loss results in abrogation of all inter-domain interactions between the two LOX-1 CTLDs, even in the continued presence of the neck region dimerisation cysteine (Nakano et al. 2012). Notably, this tryptophan is not conserved between LOX-1 and DNGR-1, casting doubt on the validity of the 'LOX-1-like' model of DNGR-1 CTLD dimerisation.

5.5 DNGR-1 Ligand: F-Actin

DNGR-1 was originally found to bind a ubiquitous intracellular molecule exposed in cells that have lost membrane integrity (Sancho et al. 2009). Further analysis showed that the ligand is predominantly cytoplasmic, resistant to nucleases and glycosidases and susceptible to protease and acid treatment and heat denaturation (Sancho et al. 2009). Subsequently, the ligand was determined to be the filamentous form of actin (F-actin) with no binding detected to G-actin monomers (Ahrens et al. 2012; Zhang et al. 2012). Actin is evolutionarily highly conserved, and human, turkey, trout, *Drosophila* and yeast F-actin can all be recognised by mouse DNGR-1 (Ahrens et al. 2012).

It was suggested that DNGR-1 binds only to F-actin complexed with calponin homology domain-containing actin-binding proteins (ABPs) such as β-spectrin and α-actinin (Zhang et al. 2012). However, this notion was disproved by the recent structure of the extracellular domain of DNGR-1 bound to F-actin, which clearly showed that individual actin filaments act as ligands for receptor binding irrespective of decoration by ABPs (Hanč et al. 2015). However, it is important to note that F-actin depolymerises rapidly under a critical concentration of actin monomers, a process that can be delayed or reversed by presence of certain ABPs. Hence, filaments decorated with such proteins are more stable and, possibly, also more prone to form bundles, providing a plausible explanation for the apparent increase in binding of DNGR-1 to filaments containing ABPs (Ahrens et al. 2012).

The structure of the extracellular domain of DNGR-1 bound to F-actin reveals that the binding site for each CTLD is within the groove between two F-actin protofilaments and that the receptor contacts two actin subunits across the filament, and two subunits along one protofilament (Fig. 5.2c and d) (Hanč et al. 2015). The composite nature of the binding site clearly explains why DNGR-1 binds to F-actin but not G-actin (Ahrens et al. 2012; Zhang et al. 2012). Notably, the above is a very uncommon mode of binding amongst ABPs, and, consequently, decoration of filaments is unlikely to directly interfere with DNGR-1 binding (Hanč et al. 2015). However, it is possible that proteins, which introduce conformational changes into the filament, can still affect DNGR-1 binding.

Interestingly, only a handful of DNGR-1 residues is involved in the interaction with ligand (Hanč et al. 2015). Accordingly, DNGR-1 shows only moderate affinity for F-actin with fast association and dissociation and a K_d in the low micromolar range (Ahrens et al. 2012; Hanč et al. 2015). Avidity of the multivalent interaction between membrane-bound DNGR-1 and the polymeric ligand, however, decreases the off-rate, effectively increasing the overall strength of binding by as much as three orders of magnitude, leading to a K_d in the nanomolar range and accounting for the ability of DNGR-1 to efficiently recognise F-actin under physiological conditions (Hanč et al. 2015).

It remains to be established whether one DNGR-1 dimer can bind with both ligand-binding domains to a single actin filament (*cis*) and/or to cross-link two adjacent filaments (*trans*). Electron cryomicroscopy experiments where DNGR-1 did not appear to induce filament cross-linking would suggest that DNGR-1 possesses the ability to bind in *cis* (Hanč et al. 2015). Such mode of binding would have interesting implications for the structure of the protein, as binding sites on the actin filament create a pattern that repeats every 5.5 nm (Hanč et al. 2015). In order for one dimer of DNGR-1 to be able to span this distance, the neck region would have to be exceptionally flexible, and the ligand-binding domains would need to be able to move and bind independently, confirming that the 'LOX-1-like' dimer model (Zhang et al. 2012) is likely incorrect and suggesting a possible lack of CTLD dimerisation altogether.

5.6 DNGR-1 in Cross-Presentation of Dead Cell-Associated Antigens

DNGR-1 bears an intracellular tail with a modified immunoreceptor tyrosine-based activation motif (termed hemITAM) that couples to spleen tyrosine kinase (Syk) (Huysamen et al. 2008; Sancho et al. 2009). Another structurally related C-type lectin receptor, Dectin-1 (CLEC7A), possesses an analogous hemITAM (Rogers et al. 2005) and, when engaged by fungal-derived β-glucans, signals via Syk to activate NF-κB, MAPKs and NFAT (LeibundGut-Landmann et al. 2007). This results in DC activation and allows for coupling of innate sensing of fungal particles to the induction of an antifungal adaptive immune response (LeibundGut-Landmann et al. 2007, 2008), thereby fulfilling the criteria for a pattern recognition receptor (Janeway 1989). The finding that DNGR-1 binds to dead cells (Sancho et al. 2009), together with a suggestion that the intracellular signalling domain of DNGR-1 could induce pro-inflammatory cytokine production in macrophages (Huysamen et al. 2008), led to the hypothesis that DNGR-1 might behave as a danger receptor, i.e. one that mediates DC activation in response to contact with necrotic cells (Matzinger 1994). This hypothesis led to the christening of the receptor, later adapted to constitute an acronym for '***DC***, ***NK lectin group receptor-1***' (Sancho et al. 2008).

Supporting the hypothesis, DNGR-1 deficiency was found to negatively impact priming of $CD8^+$ T cells against model antigens carried by dead cells in a model of immunisation with ultraviolet radiation-treated cells (Sancho et al. 2009). However, DNGR-1 was then found to be dispensable for inducing DC activation by dying cells, which is likely triggered by unknown innate immune receptors and is marginal compared to that induced by microbes (Zelenay et al. 2012). Furthermore, triggering of DNGR-1 with antibodies or dead cells did not activate DCs and neither did a chimeric receptor containing a Dectin-1 ectodomain fused to the intracellular domain of DNGR-1 upon treatment with Dectin-1 agonist (Zelenay et al. 2012). A DEDG sequence immediately upstream of the hemITAM tyrosine contributes to signalling by CLEC-2, another Syk-coupled C-type lectin receptor (Fuller et al. 2007). This sequence is shared with Dectin-1 but, notably, in DNGR-1 from all species, the glycine is replaced by an isoleucine. Mutation of the isoleucine to glycine potentiated the ability of Dectin-1/DNGR-1 chimaeras to stimulate pro-inflammatory cytokine production by myeloid cells (Zelenay et al. 2012). These results indicate that the isoleucine and, likely, additional residues directly preceding the hemITAM tyrosine of DNGR-1, dampen downstream signalling and do not allow, for example, stimulation of NF-κB, which is a driver of DC activation. The effect of these residues may be to decrease hemITAM tyrosine phosphorylation by Src family kinases and/or the interaction with Syk as suggested for other C-type lectin receptors such as CLEC-2 or NKp80 (Fuller et al. 2007; Ruckrich and Steinle 2013). The net result is that DNGR-1 does not act as a DC activatory receptor, unlike Dectin-1.

Given that DNGR-1 does not activate DCs yet is necessary for efficient cross-priming of $CD8^+$ T cells against dead cell-associated antigens, it began to be suspected that it might impact directly on the mechanism of dead cell handling and, specifically, cross-presentation, i.e. the means by which exogenous antigens are processed and presented by MHC class I molecules. Consistent with that notion, and further supporting the concept that DNGR-1 does not function by stimulating DCs, DNGR-1 deficiency did not impair the ability of DCs to prime $CD4^+$ T cells specific for dead cell-associated antigens in vitro, nor impacted on the $CD4^+$ T cell-dependent herpes simplex virus type 1-specific antibody in vivo (Zelenay et al. 2012). Therefore, the hypothesis was pursued that DNGR-1 signals to preserve antigenic information present within cell corpses and makes it available for cross-presentation. Interestingly, DNGR-1-bound antibodies are directed to non-lysosomal compartments (Sancho et al. 2009; Zelenay et al. 2012), similar to ones accessed by the mannose receptor to promote cross-presentation (Burgdorf et al. 2007). Using dead cells as ligands, it was found that DNGR-1 diverts endocytic cargo to a poorly degradative early endosomal compartment marked by expression of EEA1, Rab5a and Rab27a and, at later time points, to recycling endosomes that stain with anti-Rab11 (Zelenay et al. 2012; Fig. 5.3). These compartments have limited acidification potential and low proteolytic activity, generating partially degraded antigens that are suitable substrates for cross-presentation by MHC class I molecules (Burgdorf et al. 2007; Joffre et al. 2012; Schuette and Burgdorf 2014). Notably, impaired cross-presentation of dead cell-associated antigens by DNGR-1-deficient '$CD8\alpha^+$-like' cDCs could be reversed by blockade of lysosomal acidification or inhibition of lysosomal proteases (Iborra et al. 2012), confirming that DNGR-1 functions primarily by retaining necrotic cell cargo in endosomal compartments protected from lysosomal degradation. Thus, DNGR-1 sensing of cell corpses impacts the endocytic fate of the cargo and thereby, indirectly, the process of cross-presentation (Fig. 5.3). Other dead cell-sensing pathways involving toll-like receptors, opsonins or the complement system have been likewise implicated in the routing of dead cell cargo to non-degradative endosomes and/or regulation of antigen processing (Apetoh et al. 2007; Baudino et al. 2014; Peng and Elkon 2011).

In principle, the ability of DNGR-1 to regulate cross-presentation means that it could impact cross-tolerance as much as cross-priming. Yet, DNGR-1 deficiency

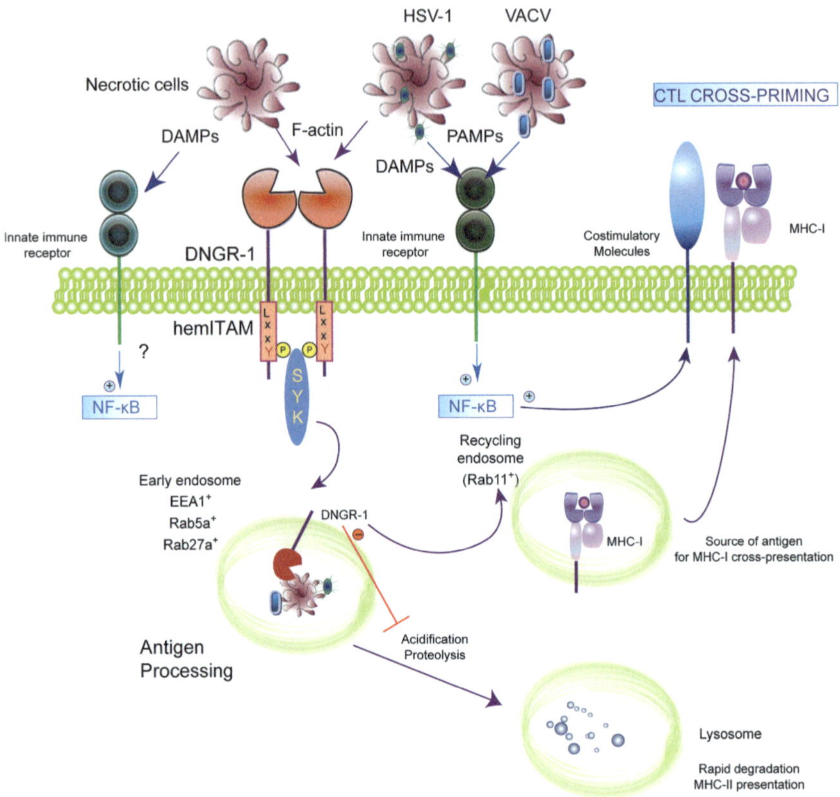

Fig. 5.3 DNGR-1 regulates cross-presentation of dead cell-associated antigens. DNGR-1 binds to exposed filamentous actin (F-actin) on cells undergoing primary or secondary necrosis or cells dying upon infection with cytophatic viruses (HSV-1 or VACV). DCs are also able to recognise DAMPs released by necrotic cells and/or PAMPs from the virus through different innate immune receptors that promote activation of the DCs. DNGR-1 bears an intracellular tail with a modified immunoreceptor tyrosine-based activation motif (termed hemITAM) that couples and signals through the spleen tyrosine kinase (Syk); however, there is no evidence that DNGR-1 can trigger or modulate activation pathways in DCs. DNGR-1 sequesters cargo in a poorly degradative early endocytic compartment. Inhibition of acidification-mediated proteolysis reverts defective cross-presentation in DNGR-1-deficient cells. This suggests that DNGR-1 promotes CTL cross-priming by preventing the rapid/complete degradation of antigenic cargo, allowing its further processing, transport and binding to MHC class I molecules

did not affect cross-tolerance of OVA-specific $CD8^+$ T cells to pancreatic ovalbumin in RIP-mOVA transgenic mice, believed to be driven by antigens contained within apoptotic β cells (Zelenay et al. 2012). This may be because apoptotic cells do not expose F-actin (Ahrens et al. 2012) and, therefore, do not engage DNGR-1. Indeed, in all models tested thus far, DNGR-1 appears to regulate immunity rather than tolerance to dead cell-associated antigens. DNGR-1 deficiency in mice leads to impairment of $CD8^+$ T cell priming to viral antigens in models of cytopathic

infection with vaccinia virus or herpes simplex virus–1 (HSV-1) (Iborra et al. 2012; Zelenay et al. 2012). This, in turn, indicates that the ability of DNGR-1 to facilitate cross-presentation of dead cell-associated antigens is not redundant with DC activation induced by viral signals. These results suggest a novel control point in immunity to infection where, in addition to receptors that detect pathogen-associated molecular patterns (PAMPs) and damage-associated molecular patterns (DAMPs) to promote DC activation, there are others such as DNGR-1 that sense DAMPs to identify 'antigenically interesting' substrates for antigen processing and presentation. This is relevant for vaccination as, in addition to novel adjuvants, tissue damage agonists might enhance antigen cross-presentation and help elicit CTL responses.

5.7 Open Questions

The molecular mechanism by which DNGR-1 regulates cross-presentation remains unclear. The evidence that DNGR-1 favours localisation of cargo to a poorly degradative compartment, and that impaired cross-presentation by DNGR-1-deficient DCs can be rescued by inhibition of lysosomal proteases or acidification, implies that DNGR-1 could function by diverting dead cell-derived cargo towards a compartment devoted to cross-presentation. Alternatively, it could facilitate maintenance of the dead cell material-containing compartment in a poorly degradative state, as well as controlling the cross-presentation process itself. The latter could involve promoting fusion of dead cell-containing compartments with the ER-Golgi intermediate compartment (Cebrian et al. 2011) or with the MHC class I rich endosomal recycling compartment (Nair-Gupta et al. 2014). In addition, DNGR-1 could also assist translocation of the ligand to the cytosol as shown for another myeloid CLR, the mannose receptor (Zehner et al. 2011). In this regard, the downstream signals used by DNGR-1 to regulate cross-presentation are an active area of research. Cross-presentation ability is impaired in Syk-deficient DCs (Iborra et al. 2012; Sancho et al. 2009), suggesting that signalling through Syk is involved in the process. This conclusion is additionally supported by the inability of a tyrosine hemITAM mutant (Y7F) of DNGR-1 to promote cross-presentation of dead cell-associated antigens, even though this mutant is still localised to the recycling endosomal compartment ((Sancho et al. 2009) and unpublished observations).

The regulation of dead cell-associated antigen cross-presentation by DNGR-1 provides a model for testing the role of cross-priming in vivo. Cross-priming is often considered dispensable for generation of $CD8^+$ effector T cell responses to viruses, as many of these infect DCs directly (Zinkernagel 2002; Joffre et al. 2012). However, whether cross-priming might play a more subtle role, for example, influencing $CD8^+$ T cell effector versus memory cell fate decisions or shaping different memory T cell compartments, has not been explored. Our recent data point towards such a function for DNGR-1 in immune responses to viruses (Iborra and Sancho, unpublished observations).

Although regulation of cross-presentation is the main currently characterised DNGR-1 function, it is possible that additional DC functions could also be affected by DNGR-1 signalling. Further studies are required, for instance, to determine if DNGR-1 expression by CD8α$^+$ DCs, a prevalent DC subset in the thymus, affects T cell selection. Furthermore, the function of DNGR-1, if any, in DC precursors is presently unknown. Notably, DC differentiation from bone marrow precursors deficient for DNGR-1 is normal in vitro. Similarly, DC accumulation in lymphoid and non-lymphoid tissues appears unperturbed in DNGR-1-deficient mice, at least under steady state conditions. Whether DNGR-1 engagement in DC precursors can take place and modulate DC development in particular inflammatory settings is an area of active current interest.

Regarding the ligand, F-actin even from fungi is detected by human or mouse DNGR-1. This opens the possibility that F-actin from some eukaryote pathogens could be sensed in a similar fashion to dead cells, allowing such pathogens to modulate immunity through DNGR-1. In addition, some CLRs that bind protein ligands exhibit a certain level of plasticity in their ligand binding, and it is conceivable that, similar to Mincle (Yamasaki et al. 2008, 2009), DNGR-1 might bind additional ligands, even of a different nature. However, a glycan array using a DNGR-1 ectodomain did not reveal any specific binding to glycans (unpublished observations).

This points to the reason of why DNGR-1 may have been selected in evolution. Current data support the notion that DNGR-1 confers an advantage in immunity against viral infection (Iborra et al. 2012; Zelenay et al. 2012). It is thus conceivable that DNGR1 has evolved to detect cells infected with intracellular pathogens (e.g. vaccinia, *Salmonella*, *Shigella*), which often move inside or even between cells by subverting host actin cytoskeleton (Goldberg 2001; Handa et al. 2013; Lee et al. 2014; Patel and Galan 2005). In addition, DNGR-1 might play an important role in the interdependence between DAMP and PAMP recognition. For instance, a hierarchy in the nucleic acid-mediated activation of immune responses has been reported, where promiscuous sensing of nucleic acids by HMGBs promotes selective activation of nucleic acid-sensing receptors (Yanai et al. 2009). HMGB-1 also potentiates the pro-inflammatory activity of LPS (Qin et al. 2009). The study of hierarchical, synergistic and/or antagonistic connections of PAMP and DAMP sensing on innate and adaptive immune responses is, therefore, an important area for further investigation.

5.8 Concluding Remarks

The study of DNGR-1 during the past years has uncovered several features that make it unique amongst C-type lectin receptors involved in immunity. First, DNGR-1 expression is restricted to the DC lineage with high levels found almost exclusively amongst the CD8α$^+$-like DC family. Second, its expression levels do not change under inflammatory conditions, in contrast to those of other myeloid cell-specific members of this group. Third, it has a dedicated role in controlling

dead cell cargo fate, favouring presentation of dead cell-associated antigens to CD8 $^+$ T cells. This feature, in turn, revealed the existence of specific innate immune receptors able to decode antigenicity independently of adjuvanticity (Zelenay and Reis e Sousa 2013) and underscored a nonredundant role for DAMP recognition in both sterile and non-sterile settings. Lastly, the abundant nature of the DNGR-1 ligand, F-actin, the exposure of which, in damaged cells, acts as a 'present me' signal, adds to the concept of 'find me' and 'eat me' signals as critical components of innate sensing mechanisms that instruct adaptive immunity. In fact, the unusual way of DNGR-1:F-actin interaction compared to other ABPs and the notion that the exposure of essential and highly conserved cytoskeletal proteins can be detected by the immune system open new areas of investigation beyond the instruction of adaptive immune responses.

References

Ahrens S, Zelenay S, Sancho D, Hanč P, Kjaer S, Feest C, Fletcher G, Durkin C, Postigo A, Skehel M, Batista F, Thompson B, Way M, Reis e Sousa C, Schulz O (2012) F-actin is an evolutionarily conserved damage-associated molecular pattern recognized by DNGR-1, a receptor for dead cells. Immunity 36(4):635–645. doi:10.1016/j.immuni.2012.03.008

Apetoh L, Ghiringhelli F, Tesniere A, Criollo A, Ortiz C, Lidereau R, Mariette C, Chaput N, Mira JP, Delaloge S, Andre F, Tursz T, Kroemer G, Zitvogel L (2007) The interaction between HMGB1 and TLR4 dictates the outcome of anticancer chemotherapy and radiotherapy. Immunol Rev 220:47–59. doi:10.1111/j.1600-065X.2007.00573.x

Bachem A, Guttler S, Hartung E, Ebstein F, Schaefer M, Tannert A, Salama A, Movassaghi K, Opitz C, Mages HW, Henn V, Kloetzel PM, Gurka S, Kroczek RA (2010) Superior antigen cross-presentation and XCR1 expression define human CD11c+CD141+ cells as homologues of mouse CD8+ dendritic cells. J Exp Med 207(6):1273–1281. doi:10.1084/jem.20100348

Bachem A, Hartung E, Guttler S, Mora A, Zhou X, Hegemann A, Plantinga M, Mazzini E, Stoitzner P, Gurka S, Henn V, Mages HW, Kroczek RA (2012) Expression of XCR1 characterizes the Batf3-dependent lineage of dendritic cells capable of antigen cross-presentation. Front Immunol 3:214. doi:10.3389/fimmu.2012.00214

Baudino L, Sardini A, Ruseva MM, Fossati-Jimack L, Cook HT, Scott D, Simpson E, Botto M (2014) C3 opsonization regulates endocytic handling of apoptotic cells resulting in enhanced T-cell responses to cargo-derived antigens. Proc Natl Acad Sci U S A 111(4):1503–1508. doi:10.1073/pnas.1316877111

Bonifaz L, Bonnyay D, Mahnke K, Rivera M, Nussenzweig MC, Steinman RM (2002) Efficient targeting of protein antigen to the dendritic cell receptor DEC-205 in the steady state leads to antigen presentation on major histocompatibility complex class I products and peripheral CD8+ T cell tolerance. J Exp Med 196(12):1627–1638

Burgdorf S, Kautz A, Bohnert V, Knolle PA, Kurts C (2007) Distinct pathways of antigen uptake and intracellular routing in CD4 and CD8 T cell activation. Science 316(5824):612–616. doi:10.1126/science.1137971

Caminschi I, Shortman K (2012) Boosting antibody responses by targeting antigens to dendritic cells. Trends Immunol 33(2):71–77. doi:10.1016/j.it.2011.10.007

Caminschi I, Proietto AI, Ahmet F, Kitsoulis S, Shin Teh J, Lo JC, Rizzitelli A, Wu L, Vremec D, van Dommelen SL, Campbell IK, Maraskovsky E, Braley H, Davey GM, Mottram P, van de Velde N, Jensen K, Lew AM, Wright MD, Heath WR, Shortman K, Lahoud MH (2008) The dendritic cell subtype-restricted C-type lectin Clec9A is a target for vaccine enhancement. Blood 112(8):3264–3273. doi:10.1182/blood-2008-05-155176

Caminschi I, Vremec D, Ahmet F, Lahoud MH, Villadangos JA, Murphy KM, Heath WR, Shortman K (2012) Antibody responses initiated by Clec9A-bearing dendritic cells in normal and Batf3 (−/−) mice. Mol Immunol 50(1–2):9–17. doi:10.1016/j.molimm.2011.11.008

Cebrian I, Visentin G, Blanchard N, Jouve M, Bobard A, Moita C, Enninga J, Moita LF, Amigorena S, Savina A (2011) Sec22b regulates phagosomal maturation and antigen crosspresentation by dendritic cells. Cell 147(6):1355–1368. doi:10.1016/j.cell.2011.11.021

Crozat K, Tamoutounour S, Vu Manh TP, Fossum E, Luche H, Ardouin L, Guilliams M, Azukizawa H, Bogen B, Malissen B, Henri S, Dalod M (2011) Cutting edge: expression of XCR1 defines mouse lymphoid-tissue resident and migratory dendritic cells of the CD8alpha+ type. J Immunol 187(9):4411–4415. doi:10.4049/jimmunol.1101717

Edelson BT, Kc W, Juang R, Kohyama M, Benoit LA, Klekotka PA, Moon C, Albring JC, Ise W, Michael DG, Bhattacharya D, Stappenbeck TS, Holtzman MJ, Sung SS, Murphy TL, Hildner K, Murphy KM (2010) Peripheral CD103+ dendritic cells form a unified subset developmentally related to CD8alpha+ conventional dendritic cells. J Exp Med 207 (4):823–836. doi:10.1084/jem.20091627

Fuller GL, Williams JA, Tomlinson MG, Eble JA, Hanna SL, Pohlmann S, Suzuki-Inoue K, Ozaki Y, Watson SP, Pearce AC (2007) The C-type lectin receptors CLEC-2 and Dectin-1, but not DC-SIGN, signal via a novel YXXL-dependent signaling cascade. J Biol Chem 282 (17):12397–12409. doi:10.1074/jbc.M609558200

Goldberg MB (2001) Actin-based motility of intracellular microbial pathogens. Microbiol Mol Biol Rev MMBR 65(4):595–626. doi:10.1128/MMBR.65.4.595-626.2001

Hanč P, Fujii T, Iborra S, Yamada Y, Huotari J, Schulz O, Ahrens S, Kjaer S, Way M, Sancho D, Namba K, Reis ESC (2015) Structure of the complex of F-actin and DNGR-1, a C-type lectin receptor involved in dendritic cell cross-presentation of dead cell-associated antigens. Immunity 42(5):839–849. doi:10.1016/j.immuni.2015.04.009

Handa Y, Durkin CH, Dodding MP, Way M (2013) Vaccinia virus F11 promotes viral spread by acting as a PDZ-containing scaffolding protein to bind myosin-9A and inhibit RhoA signaling. Cell Host Microbe 14(1):51–62. doi:10.1016/j.chom.2013.06.006

Hashimoto D, Miller J, Merad M (2011) Dendritic cell and macrophage heterogeneity in vivo. Immunity 35(3):323–335. doi:10.1016/j.immuni.2011.09.007

Hawiger D, Inaba K, Dorsett Y, Guo M, Mahnke K, Rivera M, Ravetch JV, Steinman RM, Nussenzweig MC (2001) Dendritic cells induce peripheral T cell unresponsiveness under steady state conditions in vivo. J Exp Med 194(6):769–779

Heath WR, Carbone FR (2009) Dendritic cell subsets in primary and secondary T cell responses at body surfaces. Nat Immunol 10(12):1237–1244. doi:10.1038/ni.1822

Hildner K, Edelson BT, Purtha WE, Diamond M, Matsushita H, Kohyama M, Calderon B, Schraml BU, Unanue ER, Diamond MS, Schreiber RD, Murphy TL, Murphy KM (2008) Batf3 deficiency reveals a critical role for CD8alpha+ dendritic cells in cytotoxic T cell immunity. Science 322(5904):1097–1100. doi:10.1126/science.1164206

Huysamen C, Willment JA, Dennehy KM, Brown GD (2008) CLEC9A is a novel activation C-type lectin-like receptor expressed on BDCA3+ dendritic cells and a subset of monocytes. J Biol Chem 283(24):16693–16701. doi:10.1074/jbc.M709923200

Iborra S, Izquierdo HM, Martinez-Lopez M, Blanco-Menendez N, Reis e Sousa C, Sancho D (2012) The DC receptor DNGR-1 mediates cross-priming of CTLs during vaccinia virus infection in mice. J Clin Invest 122(5):1628–1643. doi:10.1172/JCI60660

Janeway CA Jr (1989) Approaching the asymptote? Evolution and revolution in immunology. Cold Spring Harb Symp Quant Biol 54(Pt 1):1–13

Joffre OP, Sancho D, Zelenay S, Keller AM, Reis e Sousa C (2010) Efficient and versatile manipulation of the peripheral CD4+ T-cell compartment by antigen targeting to DNGR-1/CLEC9A. Eur J Immunol 40(5):1255–1265. doi:10.1002/eji.201040419

Joffre OP, Segura E, Savina A, Amigorena S (2012) Cross-presentation by dendritic cells. Nat Rev Immunol 12(8):557–569. doi:10.1038/nri3254

Lahoud MH, Ahmet F, Kitsoulis S, Wan SS, Vremec D, Lee CN, Phipson B, Shi W, Smyth GK, Lew AM, Kato Y, Mueller SN, Davey GM, Heath WR, Shortman K, Caminschi I (2011) Targeting antigen to mouse dendritic cells via Clec9A induces potent CD4 T cell responses

biased toward a follicular helper phenotype. J Immunol 187(2):842–850. doi:10.4049/jimmunol.1101176

Lee JH, Park H, Park YH (2014) Molecular mechanisms of host cytoskeletal rearrangements by Shigella invasins. Int J Mol Sci 15(10):18253–18266. doi:10.3390/ijms151018253

LeibundGut-Landmann S, Gross O, Robinson MJ, Osorio F, Slack EC, Tsoni SV, Schweighoffer E, Tybulewicz V, Brown GD, Ruland J, Reis e Sousa C (2007) Syk- and CARD9-dependent coupling of innate immunity to the induction of T helper cells that produce interleukin 17. Nat Immunol 8(6):630–638. doi:10.1038/ni1460

Leibundgut-Landmann S, Osorio F, Brown GD, Reis e Sousa C (2008) Stimulation of dendritic cells via the dectin-1/Syk pathway allows priming of cytotoxic T-cell responses. Blood 112 (13):4971–4980. doi:10.1182/blood-2008-05-158469

Li J, Ahmet F, Sullivan LC, Brooks AG, Kent SJ, De Rose R, Salazar AM, Reis ESC, Shortman K, Lahoud MH, Heath WR, Caminschi I (2015) Antibodies targeting Clec9A promote strong humoral immunity without adjuvant in mice and non-human primates. Eur J Immunol 45 (3):854–864. doi:10.1002/eji.201445127

Liu K, Victora GD, Schwickert TA, Guermonprez P, Meredith MM, Yao K, Chu FF, Randolph GJ, Rudensky AY, Nussenzweig M (2009) In vivo analysis of dendritic cell development and homeostasis. Science 324(5925):392–397. doi:10.1126/science.1170540

Matzinger P (1994) Tolerance, danger, and the extended family. Annu Rev Immunol 12:991–1045. doi:10.1146/annurev.iy.12.040194.005015

Miller JC, Brown BD, Shay T, Gautier EL, Jojic V, Cohain A, Pandey G, Leboeuf M, Elpek KG, Helft J, Hashimoto D, Chow A, Price J, Greter M, Bogunovic M, Bellemare-Pelletier A, Frenette PS, Randolph GJ, Turley SJ, Merad M, Immunological Genome Consortium (2012) Deciphering the transcriptional network of the dendritic cell lineage. Nat Immunol 13 (9):888–899. doi:10.1038/ni.2370

Naik SH, Metcalf D, van Nieuwenhuijze A, Wicks I, Wu L, O'Keeffe M, Shortman K (2006) Intrasplenic steady-state dendritic cell precursors that are distinct from monocytes. Nat Immunol 7(6):663–671. doi:10.1038/ni1340

Naik SH, Sathe P, Park HY, Metcalf D, Proietto AI, Dakic A, Carotta S, O'Keeffe M, Bahlo M, Papenfuss A, Kwak JY, Wu L, Shortman K (2007) Development of plasmacytoid and conventional dendritic cell subtypes from single precursor cells derived in vitro and in vivo. Nat Immunol 8(11):1217–1226. doi:10.1038/ni1522

Nair-Gupta P, Baccarini A, Tung N, Seyffer F, Florey O, Huang Y, Banerjee M, Overholtzer M, Roche PA, Tampe R, Brown BD, Amsen D, Whiteheart SW, Blander JM (2014) TLR signals induce phagosomal MHC-I delivery from the endosomal recycling compartment to allow cross-presentation. Cell 158(3):506–521. doi:10.1016/j.cell.2014.04.054

Nakano S, Sugihara M, Yamada R, Katayanagi K, Tate S (2012) Structural implication for the impaired binding of W150A mutant LOX-1 to oxidized low density lipoprotein, OxLDL. Biochim Biophys Acta 1824(5):739–749. doi:10.1016/j.bbapap.2012.02.003

Onai N, Obata-Onai A, Schmid MA, Ohteki T, Jarrossay D, Manz MG (2007) Identification of clonogenic common Flt3+M-CSFR+ plasmacytoid and conventional dendritic cell progenitors in mouse bone marrow. Nat Immunol 8(11):1207–1216. doi:10.1038/ni1518

Park H, Adsit FG, Boyington JC (2005) The 1.4 angstrom crystal structure of the human oxidized low density lipoprotein receptor lox-1. J Biol Chem 280(14):13593–13599. doi:10.1074/jbc.M500768200

Park HY, Light A, Lahoud MH, Caminschi I, Tarlinton DM, Shortman K (2013) Evolution of B cell responses to Clec9A-targeted antigen. J Immunol 191(10):4919–4925. doi:10.4049/jimmunol.1301947

Patel JC, Galan JE (2005) Manipulation of the host actin cytoskeleton by Salmonella – all in the name of entry. Curr Opin Microbiol 8(1):10–15. doi:10.1016/j.mib.2004.09.001

Peng Y, Elkon KB (2011) Autoimmunity in MFG-E8-deficient mice is associated with altered trafficking and enhanced cross-presentation of apoptotic cell antigens. J Clin Invest 121 (6):2221–2241. doi:10.1172/JCI43254

Persson EK, Uronen-Hansson H, Semmrich M, Rivollier A, Hagerbrand K, Marsal J, Gudjonsson S, Hakansson U, Reizis B, Kotarsky K, Agace WW (2013) IRF4 transcription-factor-dependent CD103(+)CD11b(+) dendritic cells drive mucosal T helper 17 cell differentiation. Immunity 38(5):958–969. doi:10.1016/j.immuni.2013.03.009

Picco G, Beatson R, Taylor-Papadimitriou J, Burchell JM (2014) Targeting DNGR-1 (CLEC9A) with antibody/MUC1 peptide conjugates as a vaccine for carcinomas. Eur J Immunol 44 (7):1947–1955. doi:10.1002/eji.201344076

Poulin LF, Salio M, Griessinger E, Anjos-Afonso F, Craciun L, Chen JL, Keller AM, Joffre O, Zelenay S, Nye E, Le Moine A, Faure F, Donckier V, Sancho D, Cerundolo V, Bonnet D, Reis e Sousa C (2010) Characterization of human DNGR-1+ BDCA3+ leukocytes as putative equivalents of mouse CD8alpha+ dendritic cells. J Exp Med 207(6):1261–1271. doi:10.1084/jem.20092618

Poulin LF, Reyal Y, Uronen-Hansson H, Schraml BU, Sancho D, Murphy KM, Hakansson UK, Moita LF, Agace WW, Bonnet D, Reis e Sousa C (2012) DNGR-1 is a specific and universal marker of mouse and human Batf3-dependent dendritic cells in lymphoid and nonlymphoid tissues. Blood 119(25):6052–6062. doi:10.1182/blood-2012-01-406967

Qin YH, Dai SM, Tang GS, Zhang J, Ren D, Wang ZW, Shen Q (2009) HMGB1 enhances the proinflammatory activity of lipopolysaccharide by promoting the phosphorylation of MAPK p38 through receptor for advanced glycation end products. J Immunol 183(10):6244–6250. doi:10.4049/jimmunol.0900390

Radford KJ, Caminschi I (2013) New generation of dendritic cell vaccines. Hum Vaccines Immunother 9(2):259–264

Reizis B, Bunin A, Ghosh HS, Lewis KL, Sisirak V (2011) Plasmacytoid dendritic cells: recent progress and open questions. Annu Rev Immunol 29:163–183. doi:10.1146/annurev-immunol-031210-101345

Rogers NC, Slack EC, Edwards AD, Nolte MA, Schulz O, Schweighoffer E, Williams DL, Gordon S, Tybulewicz VL, Brown GD, Reis e Sousa C (2005) Syk-dependent cytokine induction by Dectin-1 reveals a novel pattern recognition pathway for C type lectins. Immunity 22(4):507–517. doi:10.1016/j.immuni.2005.03.004

Ruckrich T, Steinle A (2013) Attenuated natural killer (NK) cell activation through C-type lectin-like receptor NKp80 is due to an anomalous hemi-immunoreceptor tyrosine-based activation motif (HemITAM) with impaired Syk kinase recruitment capacity. J Biol Chem 288 (24):17725–17733. doi:10.1074/jbc.M113.453548

Sancho D, Mourao-Sa D, Joffre OP, Schulz O, Rogers NC, Pennington DJ, Carlyle JR, Reis e Sousa C (2008) Tumor therapy in mice via antigen targeting to a novel, DC-restricted C-type lectin. J Clin Invest 118(6):2098–2110. doi:10.1172/JCI34584

Sancho D, Joffre OP, Keller AM, Rogers NC, Martinez D, Hernanz-Falcon P, Rosewell I, Reis e Sousa C (2009) Identification of a dendritic cell receptor that couples sensing of necrosis to immunity. Nature 458(7240):899–903. doi:10.1038/nature07750

Schlitzer A, McGovern N, Teo P, Zelante T, Atarashi K, Low D, Ho AW, See P, Shin A, Wasan PS, Hoeffel G, Malleret B, Heiseke A, Chew S, Jardine L, Purvis HA, Hilkens CM, Tam J, Poidinger M, Stanley ER, Krug AB, Renia L, Sivasankar B, Ng LG, Collin M, Ricciardi-Castagnoli P, Honda K, Haniffa M, Ginhoux F (2013) IRF4 transcription factor-dependent CD11b+ dendritic cells in human and mouse control mucosal IL-17 cytokine responses. Immunity 38(5):970–983. doi:10.1016/j.immuni.2013.04.011

Schraml BU, Reis e Sousa C (2015) Defining dendritic cells. Curr Opin Immunol 32:13–20. doi:10.1016/j.coi.2014.11.001

Schraml BU, van Blijswijk J, Zelenay S, Whitney PG, Filby A, Acton SE, Rogers NC, Moncaut N, Carvajal JJ, Reis e Sousa C (2013) Genetic tracing via DNGR-1 expression history defines dendritic cells as a hematopoietic lineage. Cell 154(4):843–858. doi:10.1016/j.cell.2013.07.014

Schuette V, Burgdorf S (2014) The ins-and-outs of endosomal antigens for cross-presentation. Curr Opin Immunol 26:63–68. doi:10.1016/j.coi.2013.11.001

Steinman RM, Idoyaga J (2010) Features of the dendritic cell lineage. Immunol Rev 234(1):5–17. doi:10.1111/j.0105-2896.2009.00888.x

Suzuki S, Honma K, Matsuyama T, Suzuki K, Toriyama K, Akitoyo I, Yamamoto K, Suematsu T, Nakamura M, Yui K, Kumatori A (2004) Critical roles of interferon regulatory factor 4 in CD11bhighCD8alpha- dendritic cell development. Proc Natl Acad Sci U S A 101 (24):8981–8986. doi:10.1073/pnas.0402139101

Tussiwand R, Lee WL, Murphy TL, Mashayekhi M, Kc W, Albring JC, Satpathy AT, Rotondo JA, Edelson BT, Kretzer NM, Wu X, Weiss LA, Glasmacher E, Li P, Liao W, Behnke M, Lam SS, Aurthur CT, Leonard WJ, Singh H, Stallings CL, Sibley LD, Schreiber RD, Murphy KM (2012) Compensatory dendritic cell development mediated by BATF-IRF interactions. Nature 490(7421):502–507. doi:10.1038/nature11531

Vu Manh T, Elhmouzi-Younes J, Urien C, Ruscanu S, Jouneau L, Bourge M, Moroldo M, Foucras G, Salmon H, Marty H, Quéré P, Bertho N, Boudinot P, Dalod M, Schwartz-Cornil I (2015a) Defining mononuclear phagocyte subset homology across several distant warm-blooded vertebrates through comparative transcriptomics. Front Immunol 6:299. doi:10.3389/fimmu.2015.00299

Vu Manh T-P, Bertho N, Hosmalin A, Schwartz-Cornil I, Dalod M (2015b) Investigating evolutionary conservation of dendritic cell subset identity and functions. Front Immunol 6:260. doi:10.3389/fimmu.2015.00260

Yamasaki S, Ishikawa E, Sakuma M, Hara H, Ogata K, Saito T (2008) Mincle is an ITAM-coupled activating receptor that senses damaged cells. Nat Immunol 9(10):1179–1188. doi:10.1038/ni.1651

Yamasaki S, Matsumoto M, Takeuchi O, Matsuzawa T, Ishikawa E, Sakuma M, Tateno H, Uno J, Hirabayashi J, Mikami Y, Takeda K, Akira S, Saito T (2009) C-type lectin Mincle is an activating receptor for pathogenic fungus, Malassezia. Proc Natl Acad Sci U S A 106 (6):1897–1902. doi:10.1073/pnas.0805177106

Yanai H, Ban T, Wang Z, Choi MK, Kawamura T, Negishi H, Nakasato M, Lu Y, Hangai S, Koshiba R, Savitsky D, Ronfani L, Akira S, Bianchi ME, Honda K, Tamura T, Kodama T, Taniguchi T (2009) HMGB proteins function as universal sentinels for nucleic-acid-mediated innate immune responses. Nature 462(7269):99–103. doi:10.1038/nature08512

Zehner M, Chasan AI, Schuette V, Embgenbroich M, Quast T, Kolanus W, Burgdorf S (2011) Mannose receptor polyubiquitination regulates endosomal recruitment of p97 and cytosolic antigen translocation for cross-presentation. Proc Natl Acad Sci U S A 108(24):9933–9938. doi:10.1073/pnas.1102397108

Zelenay S, Reis e Sousa C (2013) Adaptive immunity after cell death. Trends Immunol 34 (7):329–335. doi:10.1016/j.it.2013.03.005

Zelenay S, Keller AM, Whitney PG, Schraml BU, Deddouche S, Rogers NC, Schulz O, Sancho D, Reis e Sousa C (2012) The dendritic cell receptor DNGR-1 controls endocytic handling of necrotic cell antigens to favor cross-priming of CTLs in virus-infected mice. J Clin Invest 122 (5):1615–1627. doi:10.1172/JCI60644

Zelensky AN, Gready JE (2005) The C-type lectin-like domain superfamily. FEBS J 272 (24):6179–6217. doi:10.1111/j.1742-4658.2005.05031.x

Zhang JG, Czabotar PE, Policheni AN, Caminschi I, Wan SS, Kitsoulis S, Tullett KM, Robin AY, Brammananth R, van Delft MF, Lu J, O'Reilly LA, Josefsson EC, Kile BT, Chin WJ, Mintern JD, Olshina MA, Wong W, Baum J, Wright MD, Huang DC, Mohandas N, Coppel RL, Colman PM, Nicola NA, Shortman K, Lahoud MH (2012) The dendritic cell receptor Clec9A binds damaged cells via exposed actin filaments. Immunity 36(4):646–657. doi:10.1016/j.immuni.2012.03.009

Zinkernagel RM (2002) On cross-priming of MHC class I-specific CTL: rule or exception? Eur J Immunol 32(9):2385–2392. doi:10.1002/1521-4141(200209)32:9<2385::AID-IMMU2385>3.0.CO;2-V

Chapter 6
C-Type Lectin-Like Receptor 2 (CLEC-2)

Katsue Suzuki-Inoue

Abstract C-type lectin-like receptor 2 (CLEC-2) has been identified as a receptor for a platelet-activating snake venom, rhodocytin. CLEC-2 elicits powerful platelet activation signals in a manner dependent on Src, Syk kinases, and phospholipase Cγ2, similar to the collagen receptor glycoprotein (GP) VI/FcRγ-chain complex. In contrast to GPVI/FcRγ, which initiates platelet activation through tandem YxxL motifs called the immunoreceptor tyrosine-based activation motif (ITAM), CLEC-2 signals via a single YxxL motif called hemi-ITAM. An endogenous ligand of CLEC-2 has been identified as podoplanin, which is expressed on the surface of tumor cells and facilitates tumor metastasis by inducing platelet activation. CLEC-2 in platelets facilitates blood/lymphatic vessel separation by binding to podoplanin in lymphatic endothelial cells during development. In adults, platelet CLEC-2 prevents the backflow of blood into lymphatic vessels at the lymphovenous junction by forming thrombi. Moreover, platelet CLEC-2 maintains the integrity of high endothelial venules in lymph nodes by binding to podoplanin on stromal cells and in hyper-permeabilized capillaries during inflammation. In concert with GPVI, platelet CLEC-2 plays a role in thrombosis and hemostasis, although the precise mechanism remains unknown. Although CLEC-2 expression is almost specific to platelets/megakaryocytes in humans, CLEC-2 is also expressed in dendritic cells in mice, where it plays an important role in adaptive immunity. CLEC-2 may be a good target for a novel antiplatelet drug or antimetastatic drug, which could prevent arterial thrombosis and cancer metastasis, the main causes of death in developed countries. In this article, we review the signal transduction, structure, expression, and function of CLEC-2.

Keywords CLEC-2 • Platelets • Rhodocytin • Podoplanin • Lymphatic endothelial cells • Dendritic cells • Thrombus formation • Tumor metastasis • Lymphangiogenesis • Vascular integrity

---

K. Suzuki-Inoue (✉)
Department of Clinical and Laboratory Medicine, Faculty of Medicine, University of Yamanashi, 1110 Shimokato, Chuo, Yamanashi 409-3898, Japan
e-mail: katsuei@yamanashi.ac.jp

S. Yamasaki (ed.), *C-Type Lectin Receptors in Immunity*,
DOI 10.1007/978-4-431-56015-9_6

6.1 History

Snake venom contains a vast number of toxins that target proteins necessary for thrombosis and hemostasis. A platelet-activating snake venom, rhodocytin (also called aggretin), was purified from *Calloselasma rhodostoma* venom independently by the Huang (Huang et al. 1995) and Morita (Shin and Morita 1998) groups in the 1990s. Rhodocytin-induced and collagen-induced platelet aggregations are quite similar in that both agonists induce platelet aggregation with a long lag phase and are dependent on the Src family kinase, aspirin, and cytochalasin D (Suzuki-Inoue et al. 2001; Navdaev et al. 2001; Inoue et al. 1999). However, rhodocytin induces platelet aggregation independently of the collagen receptor GPVI/FcRγ-chain, a member of the immunoglobulin super family, since it can induce platelet aggregation in mice deficient in the receptor (Suzuki-Inoue et al. 2001; Navdaev et al. 2001). This indicates that there is another platelet activation receptor. Initial studies indicated that rhodocytin induces platelet aggregation by binding to integrin α2β1 and GPIb/IX/V (Suzuki-Inoue et al. 2001; Navdaev et al. 2001). However, later studies reported that rhodocytin stimulated the aggregation of platelets deficient in α2β1 or the extracellular domain of GPIb/IX/V (Bergmeier et al. 2001), indicating that there is another activation receptor for rhodocytin, although the toxin may also bind to α2β1 and/or GPIb/IX/V. Finally, by rhodocytin affinity chromatography and MS/MS-based analyses, C-type lectin-like receptor 2 (CLEC-2) was identified as a receptor for rhodocytin (Suzuki-Inoue et al. 2006). CLEC-2 was first cloned during a bioinformatic screen for C-type lectin-like receptors (Colonna et al. 2000). However, neither its function nor its ligand(s) had been elucidated for several years.

In this article, we review the signal transduction, structure, expression, and function of CLEC-2, updating our previous review (Suzuki-Inoue et al. 2011).

6.2 Mechanism of Signal Transduction

CLEC-2 has a YxxL motif in its cytoplasmic tail, which resembles the immunoreceptor tyrosine-based activation motif (ITAM; YxxL-(X)10-12-YxxL) that has two YxxL motifs. The single YxxL motif that is found in CLEC-2 and other receptors is called hemi-ITAM. ITAM is a signaling motif found in immune receptors, such as the T-cell receptor and the platelet collagen receptor GPVI/FcRγ-chain. Cross-linking of GPVI leads to tyrosine phosphorylation of ITAM in the cytoplasmic domain of the FcRγ-chain, which is constitutively associated with GPVI, by the Src family kinases, Fyn and Lyn. This leads to binding of the tandem SH2 domain of the tyrosine kinase, Syk, to the phosphorylated ITAM. Subsequent activation of Syk initiates downstream signaling events that culminate in the tyrosine phosphorylation of LAT, SLP-76, and Vav1/3 and activation of effecter enzymes including Btk, PI3-kinase, Rac/Cdc42, and phospholipaseCγ2 (PLCγ) (reviewed in Watson et al. (2005) (Fig. 6.1a)).

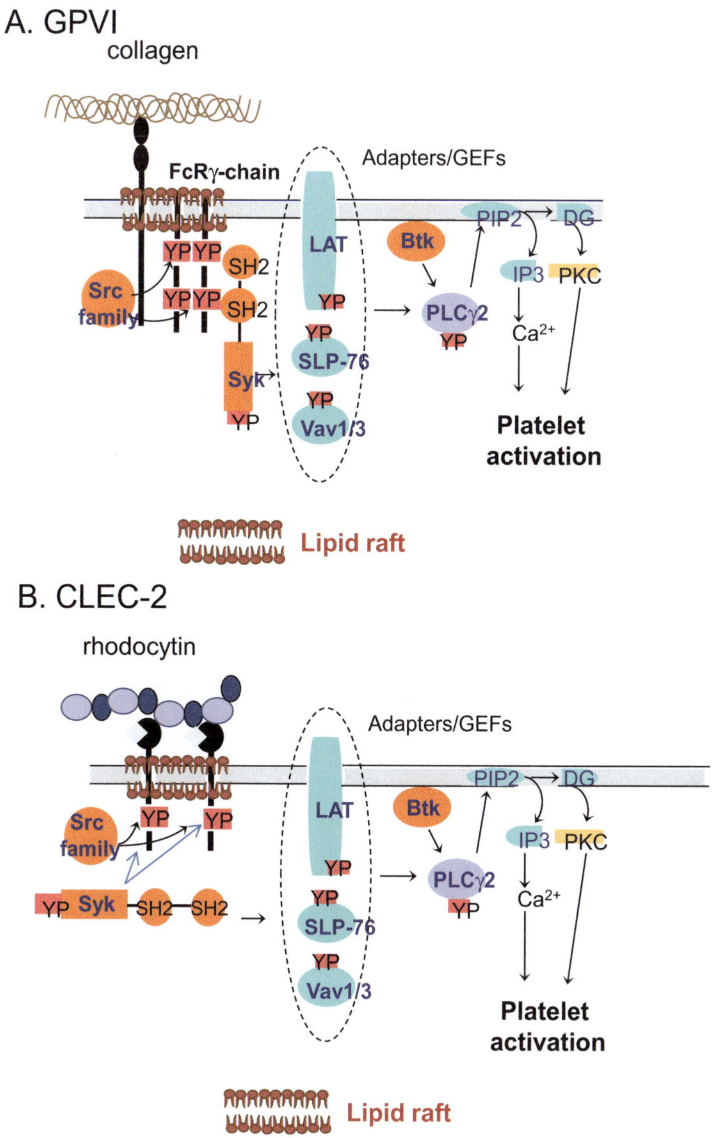

Fig. 6.1 The signal transduction pathway mediated through the GPVI/FcRγ-chain and CLEC-2. (**a**) CLEC-2 is cross-linked by its endogenous ligand, podoplanin, or by an exogenous ligand (rhodocytin) and then undergoes tyrosine phosphorylation of a single YITL motif called hemi-ITAM, which leads to activation of downstream signaling. (**b**) GPVI is cross-linked by its endogenous ligand, collagen, and undergoes tyrosine phosphorylation on tandem YxxL motifs called ITAM, which leads to activation of downstream signaling. Lipid rafts are necessary for CLEC-2 and GPVI-mediated signal transduction. YP indicates phosphorylated tyrosine

Rhodocytin stimulates the phosphorylation of the single YxxL motif in CLEC-2, and then the tandem SH2 domains of Syk bind to the phosphorylated YxxL, and the hemi-ITAM was shown to be necessary for CLEC-2 signal transduction (Suzuki-Inoue et al. 2006; Fuller et al. 2007). CLEC-2 has only a single YxxL; however, CLEC-2 is present as a dimer in resting platelets, and the tandem SH2 domains of Syk bind to the phosphorylated YxxLs of two CLEC-2 molecules with a stoichiometry of 2:1 (Watson et al. 2009; Hughes et al. 2010a). The ITAM of the GPVI/FcRγ-chain is tyrosine phosphorylated by the Src family kinases, Fyn and Lyn, which are constitutively associated with the cytoplasmic tail of GPVI. This is followed by the binding and subsequent activation of Syk. In the case of CLEC-2, hemi-ITAM is mainly phosphorylated by Syk itself (Spalton et al. 2009). A specific Src family kinase inhibitor, PP2, also inhibits CLEC-2 tyrosine phosphorylation in human platelets, but not in murine platelets, suggesting that CLEC-2 is phosphorylated by Syk and Src family kinases in human platelets, but only by Syk in murine platelets (Suzuki-Inoue et al. 2006; Severin et al. 2011).

The signaling pathway of CLEC-2 downstream of Syk is almost the same as that of GPVI and includes the tyrosine phosphorylation of LAT, SLP-76, and Vav1/3 and activation of effecter enzymes including Btk and PLCγ2 (Suzuki-Inoue et al. 2006; Fuller et al. 2007). Murine platelets deficient in Syk or PLCγ2 failed to respond even to the maximal concentration of rhodocytin, suggesting that Syk and PLCγ2 are crucial for CLEC-2-mediated signal transduction. On the other hand, those deficient in the adaptor proteins, LAT or SLP-76, and the guanine nucleotide exchange factor, Vav1/3, did not respond to the low concentration, but did respond to the high concentration of rhodocytin, suggesting that these molecules are necessary, but can be compensated for during CLEC-2 signaling (Suzuki-Inoue et al. 2006). Btk, a tyrosine kinase that is necessary for PLCγ2 activation, is tyrosine phosphorylated upon rhodocytin stimulation (Suzuki-Inoue et al. 2006), but the dependence of the CLEC-2 signaling pathway on this kinase is unknown (Fig. 6.1b).

6.3 CLEC-2 Expression

CLEC-2 was first identified from a bioinformatic screen for C-type lectin-like receptors. At that time, reverse transcriptase-PCR and Northern blot analyses indicated that CLEC-2 mRNA is expressed in the liver and several hematopoietic cell types, including monocytes, dendritic cells, NK cells, and granulocytes (Colonna et al. 2000), although platelets and megakaryocytes were not checked for expression. Later, the CLEC-2 protein expression was systematically analyzed, and the CLEC-2 protein was found to be expressed in platelets, megakaryocytic cell lines, liver sinusoidal endothelial cells (Chaipan et al. 2006), and liver Kupffer cells (Tang et al. 2010) in humans. In mice, however, it has been reported that CLEC-2 is also expressed in peripheral neutrophils (Kerrigan et al. 2009) and macrophages (Chang et al. 2010), where it mediates phagocytosis and increases the expression of proinflammatory cytokines, including tumor necrosis factor α (TNFα), as well as Kupffer cells (Tang et al. 2010). Taken together, these findings indicate that CLEC-2 is highly and relatively specifically expressed in platelets/megakaryocytes, but is also present in other types of cells at low levels, especially in mice.

6.4 Structure and the Mode of Ligand Binding to CLEC-2

C-type lectins can be classified as "classical" and "nonclassical" C-type lectins based on their ability to recognize carbohydrate and noncarbohydrate ligands, respectively. CLEC-2 belongs to the nonclassical C-type lectins, which contain a C-type lectin-like domain (CLTD) homologous to a carbohydrate recognition domain, but lack the consensus sequence for binding sugars and calcium (Colonna et al. 2000). In fact, Watson et al. reported that no glycosylation was observed on either of the subunits of the CLEC-2 ligand snake venom rhodocytin in the crystal structure, consistent with their bioinformatic, SDS-PAGE, and mass spectroscopy results (Watson et al. 2007). Podoplanin is a type I transmembrane sialomucin-like glycoprotein, which we found is an endogenous ligand for CLEC-2 (Suzuki-Inoue et al. 2007) (see the "Function" section for details) (Sect. 6.5). Although the association between CLEC-2 and podoplanin is dependent on the sialic acid on the O-glycans of podoplanin (Suzuki-Inoue et al. 2007), not only sialic acid but also the stereostructure of the podoplanin protein was found to be critical for the CLEC-2-binding activity of podoplanin (Kato et al. 2008). This finding is consistent with the characteristics of "nonclassical" C-type lectins.

CLEC-2 is also glycosylated and detected as 32- and 40-kDa forms in platelets, probably due to differential glycosylation. Consistent with this finding, there are two potential sites of N-glycosylation (at positions 120 and 134) (Watson et al. 2007). In addition, the double band collapses to a single band of 27 kDa, the molecular weight that is deduced from the amino acid sequence of CLEC-2 upon N-glycosidase treatment (Suzuki-Inoue et al. 2006). However, neither of the two potential *N*-linked glycosylation sites are supposed to play a role in at least rhodocytin binding (Watson et al. 2007). With regard to the interaction between CLEC-2 and podoplanin, the stereostructure of the CLTD in CLEC-2 is necessary for binding to podoplanin, since the extracellular domain of CLEC-2-Fc lacking only a small part of the CLTD lost the ability to bind to podoplanin (Kato et al. 2008).

6.5 Function

6.5.1 *Tumor Metastasis*

It has long been recognized that several kinds of tumor cells cause the aggregation of platelets, which facilitates tumor growth and metastasis (Katagiri et al. 1991; Kitagawa et al. 1989; Kato et al. 2003). Platelet aggregates surrounding tumor cells protect them from shear stress and/or NK cells (Nieswandt et al. 1999) in the blood stream and serve as a place for tumor cell nesting, and the growth factors released from activated platelets can stimulate angiogenesis and tumor growth. Podoplanin is a type I transmembrane sialomucin-like glycoprotein expressed on several kinds of tumor cells, including squamous cell carcinomas (Schacht et al. 2005; Kato et al. 2005), seminomas (Kato et al. 2004), and brain tumors (Mishima et al. 2006a, b; Kato et al. 2006), and it has been shown to induce platelet aggregation (reviewed in Tsuruo and Fujita 2008). Podoplanin expression is reportedly associated with tumor metastasis or malignant progression (Mishima et al. 2006b; Yuan et al. 2006).

Suzuki-Inoue et al. noticed that the profile of podoplanin-induced platelet aggregation is similar to that of rhodocytin-induced platelet aggregation, and they identified CLEC-2 as a receptor for podoplanin (Suzuki-Inoue et al. 2007). In an experimental mouse model of metastasis, an anti-podoplanin blocking antibody significantly inhibited the number of metastatic lung nodules, consisting of tumor cells expressing podoplanin (Kato et al. 2008). Conversely, the lung metastasis of podoplanin-expressing tumors greatly inhibited in CLEC-2-deficient bone marrow chimeric mice (Shirai et al. 2013), implying that CLEC-2/podoplanin may be a promising target for antimetastatic drugs. However, tumor growth and lymphatic metastasis were not inhibited in CLEC-2-deficient chimeric mice (Shirai et al. 2013), suggesting that the CLEC-2/podoplanin interaction only plays a role in hematogenous tumor metastasis, where tumors have access to platelets in the blood flow.

6.5.2 *Lymphatic/Blood Vessel Separation During Development*

Podoplanin is expressed not only in tumor cells but also in various kinds of normal tissues, including lymphatic endothelial cells, type I alveolar cells, and kidney podocytes, after which podoplanin was named (reviewed in Tsuruo and Fujita 2008). Podoplanin is expressed in lymphatic endothelial cells, but not in vascular endothelial cells, and hence is used as a marker for lymphatic endothelial cells. Under physiological conditions, the CLEC-2 in platelets cannot interact with the podoplanin in lymphatic endothelial cells. During organ development, however, the cluster of endothelial cells in the cardinal vein is committed to the lymphatic phenotype, and these sprout to form the primary lymphatic sacs from which part of the peripheral lymphatic vasculature is generated by further centrifugal growth (reviewed in Tammela and Alitalo 2010). At this stage, the CLEC-2 in platelets can interact with the podoplanin in lymphatic endothelial cells.

Studies in CLEC-2-deficient mice revealed that CLEC-2 facilitates blood/lymphatic vessel separation during development. CLEC-2-deficient mice died at the embryonic/neonatal stages, exhibiting disorganized and blood-filled lymphatic vessels and severe edema due to abnormal blood/lymphatic vessel separation (Bertozzi et al. 2010; Suzuki-Inoue et al. 2010) (Fig. 6.2). Platelet/megakaryocyte-specific CLEC-2-deficient mice also showed blood-filled lymphatics (Finney et al. 2011; Suzuki-Inoue et al. 2010), suggesting that the CLEC-2 in platelets is required for blood/lymphatic vessel separation. Podoplanin deficiency and endothelial cell *O*-glycan deficiency also caused the blood/lymphatic misconnections (Fu et al. 2008), and it had been previously shown that the sialic acid present on the *O*-glycans of podoplanin is essential for binding to CLEC-2 (Suzuki-Inoue et al. 2007). These findings suggest that the interaction between CLEC-2 on platelets and podoplanin on lymphatic endothelial cells is important for normal lymphatic vessel development.

Mice deficient in the signaling molecules downstream of CLEC-2, including Syk, SLP-76, and PLCγ2, showed blood/lymphatic vessel misconnection (Ichise et al. 2009; Abtahian et al. 2003). Inhibition of platelet activation by treatment of pregnant wild-type mice with acetyl salicylic acid resulted in half of the embryos exhibiting blood/lymphatic misconnection (Uhrin et al. 2010). These findings suggest that platelet activation is required for blood/lymphatic vessel separation. Platelet activation results in granule release and platelet aggregation. Platelet granules contain a vast number of angiogenic factors, growth factors, and extracellular matrix, implying that these factors could contribute to blood/lymphatic

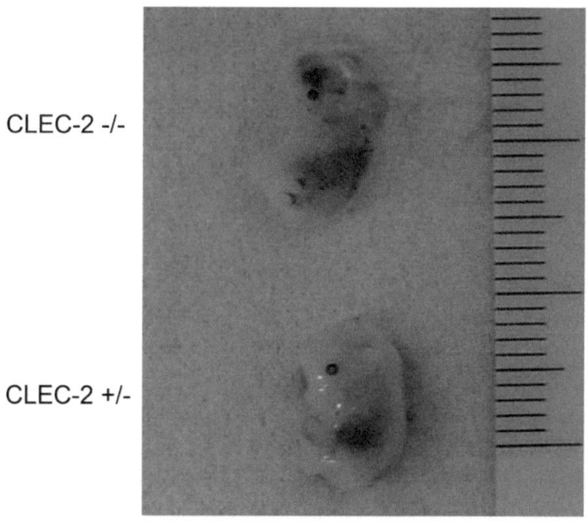

Fig. 6.2 CLEC-2−/− and CLEC-2+/− mouse embryos. CLEC-2-deficient mouse embryos show blood-filled lymphatic vessels in the skin

separation. Alternatively, platelet aggregates built up at the separation zone of lymph sacs and cardinal veins, which may physically help in the separation. In fact, Uhrin et al. and Bertozzi et al. reported that platelet aggregates build up at the separation zone of podoplanin-positive lymph sacs and cardinal veins in wild-type embryos, but not in podoplanin-deficient or SLP-76-deficient embryos (Uhrin et al. 2010; Bertozzi et al. 2010). However, mice deficient in integrin $\alpha IIb\beta 3$, which is necessary for platelet aggregation, but not for granule release, do not show the non-separation phenotype (Bertozzi et al. 2010), suggesting a role for granule release. In line with this notion, supernatants from activated platelets and their main content, TGFβ, inhibited the migration and proliferation of lymphatic endothelial cells (Osada et al. 2012). However, patients who lack sufficient platelet density or a-granules also do not exhibit defective blood/lymphatic vessel separation (Michelson 2013). Further studies are therefore necessary to address these mechanisms.

6.5.3 Maintenance of the Vascular Integrity in Adults

6.5.3.1 Prevention of the Backflow of Blood into Lymphatic Vessels at the Lymphovenous (LV) Junction

The vascular networks connect at the LV junction, where lymph drains into blood. An LV valve (LVV) prevents the backflow of blood into lymphatic vessels. Hess et al. reported that the loss of CLEC-2 resulted in backfilling of the lymphatic network with blood from the thoracic duct in both neonatal and mature mice, even when the LVVs were intact (Hess et al. 2014). Fibrin-containing platelet thrombi were observed at the LVV and in the terminal thoracic duct in wild-type mice, but not in CLEC-2-deficient mice. An analysis of mice lacking LVVs or lymphatic valves revealed that platelet-mediated thrombus formation limits LV backflow even under conditions of impaired valve function. This indicates that hemostasis via CLEC-2 functions with the LVV to safeguard the lymphatic vascular network throughout life.

6.5.3.2 Integrity of High Endothelial Venules in Lymph Nodes

Circulating lymphocytes continuously enter lymph nodes for immune surveillance through specialized blood vessels named high endothelial venules (HEV). This process increases markedly during immune responses. Herzog et al. revealed the mechanism by which HEVs permit lymphocyte transmigration while maintaining the vascular integrity (Herzog et al. 2013). The podoplanin expressed on fibroblastic reticular cells, which surround HEVs, stimulates platelets by binding to its receptor, CLEC-2. Sphingosine-1-phosphate released from activated platelets promotes the expression of VE-cadherin on HEVs, which is essential for overall vascular integrity. Mice deficient in CLEC-2, podoplanin, or sphingosine-1-phosphate exhibited spontaneous bleeding in mucosal lymph nodes and bleeding in the draining peripheral lymph nodes after immunization. A role for platelet activation via CLEC-2 in the maintenance of HEV integrity during immune responses was added to the list of the roles platelets play beyond clotting.

6.5.3.3 Integrity of Vessels During Inflammation

It is known that platelets play a critical role in maintaining the vascular integrity, especially during inflammation, when inflammatory cytokines enhance vascular permeability. For this reason, petechial hemorrhage was observed upon the induction of severe thrombocytopenia even without traumatic injury. However, how platelets maintain the vascular integrity during inflammation is unclear. Boulaftali et al. proved that ITAM signaling, but not G-protein-coupled receptor (GPCR) signaling, is critical for the prevention of inflammation-induced hemorrhage (Boulaftali et al. 2013). Inflammation-induced hemorrhage in thrombocytopenic mice was rescued by the transfusion of wild-type platelets or thrombin receptor-deficient platelets, but not by the transfusion of platelets deficient in CLEC-2, GPVI, or SLP-76. These results indicate that controlling the vascular integrity is a major function of immune-type receptors in platelets. On the other hand, signaling from GPCR is not necessary for controlling the vascular integrity, although it is important for hemostasis. Upon platelet leakage from vessels, GPVI signaling is likely activated at sites of inflammation by collagen and/or laminin, the two physiological ligands for GPVI found in the vessel wall (Ozaki et al. 2009) (Fig. 6.3b). How platelets are activated through CLEC-2 at the site of inflammation is difficult to explain, but Boulaftali et al. speculated that podoplanin on the surface of infiltrating macrophages, or an as yet unidentified ligand expressed in tissues around vessels or in the vessel wall, stimulates platelets via CLEC-2.

6.5.4 *Thrombosis and Hemostasis*

Studies to investigate the role of CLEC-2 in thrombosis and hemostasis demonstrated that CLEC-2-null mice exhibit mortality at the embryonic/neonatal stages, while irradiated chimeric animals were rescued by transplantation of a CLEC-2−/− fetal liver (Suzuki-Inoue et al. 2010). Antibody-induced CLEC-2-deficient mice were also utilized for this purpose. The injection of an anti-CLEC-2 antibody into mice leads to a specific loss of CLEC-2 in circulating platelets for several days, although the precise mechanism underlying this observation remains unknown. CLEC-2-deficient platelets resulting from both methods displayed normal adhesion under flow conditions, but the subsequent thrombus formation was severely impaired in vitro, although the in vitro platelet aggregation induced by agonists other than rhodocytin was normal (May et al. 2009; Suzuki-Inoue et al. 2010).

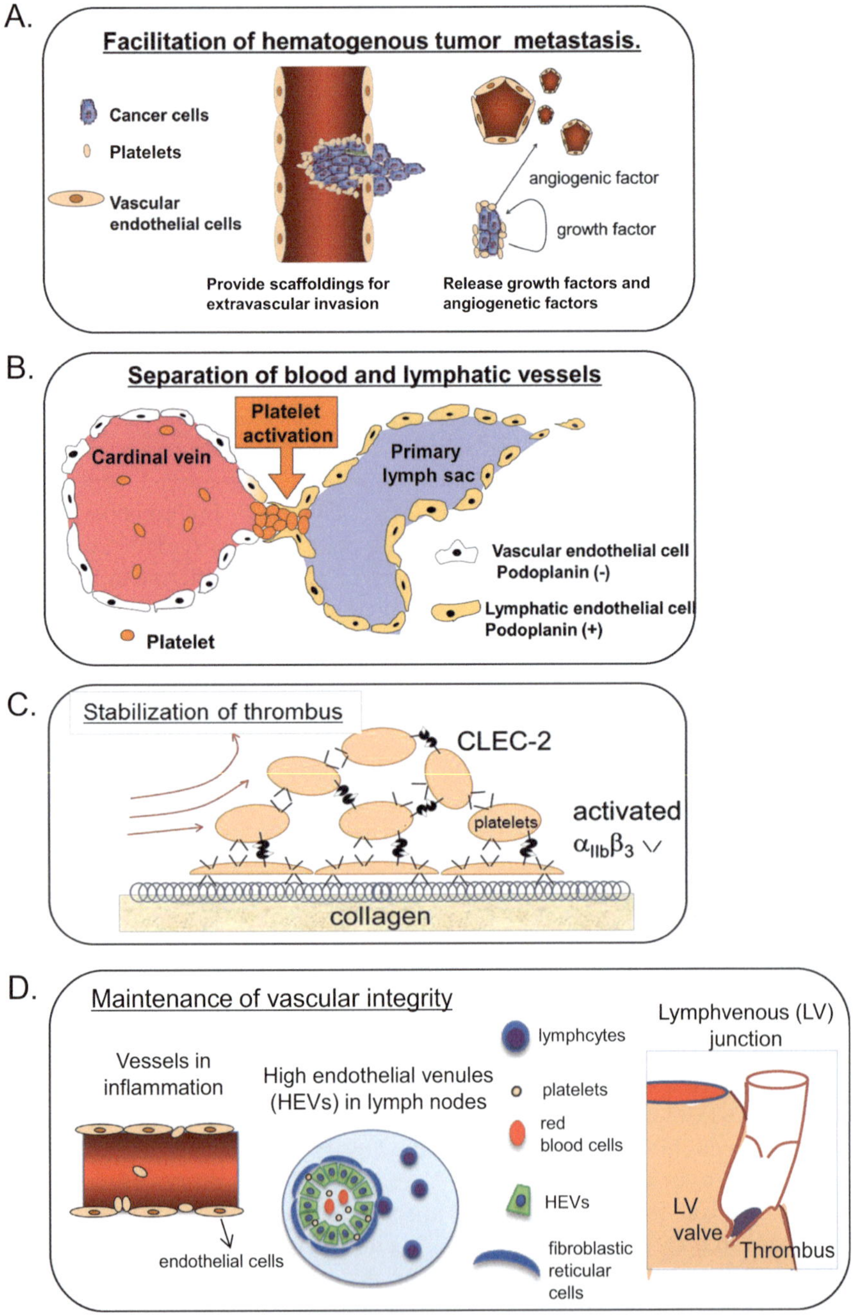

Fig. 6.3 The suggested physiological and pathological roles of platelet CLEC-2. (**a**) Podoplanin expressed in tumor cells facilitates tumor metastasis by inducing platelet activation through CLEC-2. Platelets adhering to tumor cells protect them from shear stress and/or NK cells, provide tumor cells with a scaffold for extravasation, and release growth factors or angiogenic factors to facilitate tumor growth. (**b**) Podoplanin expressed in lymphatic endothelial cells facilitates blood/

$FeCl_3$-induced thrombus formation in the artery was also inhibited in CLEC-2-deficient chimeras (May et al. 2009). These mice also showed mildly increased or mild but not significantly increased tail bleeding (Suzuki-Inoue et al. 2010; May et al. 2009; Tang et al. 2010). These findings suggest that CLEC-2 is involved in thrombus stabilization in vitro and in vivo.

In contrast to these reports, Hughes et al. reported that CLEC-2 is neither required for platelet aggregate stability in the presence of arteriolar shear nor for normal hemostasis (Hughes et al. 2010b). The discrepancies between studies may be because the role of CLEC-2 in thrombosis and hemostasis is relatively minor and depends on the experimental conditions. Simultaneous deletion of GPVI and CLEC-2 from mice, however, caused a severe bleeding tendency and profound impairment of arterial thrombus formation (Bender et al. 2013). These findings suggest that CLEC-2 plays a definite role in thrombosis and hemostasis, although the deletion of CLEC-2 alone results in a relatively minor phenotype. The mechanism by which CLEC-2 and GPVI play a complementary role in thrombosis and hemostasis is a topic of interest for future investigation.

6.5.5 A Role for CLEC-2 on the Surface of Dendritic Cells in Immunity

CLEC-2 is also expressed on the surface of dendritic cells in mice, but not in humans. Recently, accumulating evidence has suggested that the CLEC-2 in dendritic cells plays an important role in adaptive immunity.

To initiate adaptive immunity, dendritic cells move from parenchymal tissues to lymphoid organs by migrating along stromal scaffolds consisting of podoplanin-positive fibroblastic reticular cells (FRCs). Acton et al. reported that CLEC-2 deficiency in DCs impaired their entry into the lymphatics and their trafficking to and within lymph nodes, thereby reducing T-cell priming (Acton et al. 2012). CLEC-2 engagement by podoplanin induces the spreading and migration of dendritic cells along stromal surfaces. CLEC-2 activation triggered cell spreading via downregulation of RhoA activity and Rac1 activation. In turn, CLEC-2 in dendritic cells controls the fibroblastic reticular network tension and lymph node expansion by generating signals downstream of podoplanin in FRCs (Acton et al. 2014;

Fig. 6.3 (continued) lymphatic separation in the developmental stage. (**c**) CLEC-2 stabilizes thrombus formation under a normal blood flow, at least partly through homophilic interactions. (**d**) CLEC-2 maintains the vascular integrity. The vascular integrity under inflammatory conditions is maintained by signaling mediated through CLEC-2 and GPVI. High endothelial venule integrity is maintained by the interaction between the CLEC-2 on platelets and the podoplanin on fibroblastic reticular cells. The integrity of the lymphovenous junction is maintained by thrombus formation induced by the interaction between the CLEC-2 on platelets and the podoplanin on lymphatic endothelial cells

Astarita et al. 2015). Immunogenic challenge induces the infiltration and division of lymphocytes, which markedly increases the lymph node cellularity, leading to lymph node expansion. In the resting state, podoplanin signaling in stromal FRCs maintains the physical elasticity of the lymph nodes by inducing actomyosin contractility in FRCs via the activation of RhoA/C. Upon immunogenic challenge, the engagement of CLEC-2 in mature dendritic cells causes podoplanin clustering and rapidly uncouples podoplanin from RhoA/C activation, relaxing the actomyosin cytoskeleton and permitting FRC stretching. This process allows increased lymph node cellularity and provides places for antigen presentation by dendritic cells to lymphocytes. Thus, the CLEC-2/podoplanin axis induces bidirectional signaling between dendritic cells and FRCs, leading to increased motility of dendritic cells and FRC extension, both of which are necessary for efficient T-cell priming.

Benezech et al. showed that a constitutive lack of CLEC-2 expression leads to defective lymphatic cell proliferation, resulting in impaired development of the lymphatic vascular structures and involution of the LN anlagen in the embryo (Benezech et al. 2014). In contrast, the deletion of CLEC-2 on the megakaryocyte/platelet lineage still allows the formation of lymph nodes, although they are blood-filled lymph nodes, as previously reported (Herzog et al. 2013). Thus, the interaction between the CLEC-2 on dendritic cells and podoplanin on stromal cells may regulate lymph node development.

6.6 Concluding Remarks

The discovery of the novel platelet activation receptor, CLEC-2, revealed that platelets have various roles within and beyond clotting. Platelets regulate tumor metastasis, blood/lymphatic vessel separation, and the integrity of vessels, including HEVs in lymph nodes and the hyper-permeabilized capillaries that occur during inflammation through the interaction between CLEC-2 and its endogenous ligand, podoplanin. In concert with GPVI, the CLEC-2 in platelets plays a role in thrombosis and hemostasis, although the precise mechanism remains unknown. The roles of CLEC-2 in platelets are summarized in Fig. 6.3. CLEC-2 in dendritic cells plays an important role in adaptive immunity in mice. There is growing evidence that CLEC-2 has other endogenous and exogenous ligands, the discovery of which would also provide additional advances in the field of platelet biology. With regard to the clinical aspects of the research on CLEC-2, it could be a good target for an anti-hematogenous metastasis drug or as an antiplatelet drug. There is still a long way to go before there could be a practical use of these drugs, but research on CLEC-2 may lead to a better understanding of cancer and arterial thrombosis, which are the main causes of death in developed countries.

References

Abtahian F, Guerriero A, Sebzda E, Lu MM, Zhou R, Mocsai A, Myers EE, Huang B, Jackson DG, Ferrari VA, Tybulewicz V, Lowell CA, Lepore JJ, Koretzky GA, Kahn ML (2003) Regulation of blood and lymphatic vascular separation by signaling proteins SLP-76 and Syk. Science 299 (5604):247–251

Acton SE, Astarita JL, Malhotra D, Lukacs-Kornek V, Franz B, Hess PR, Jakus Z, Kuligowski M, Fletcher AL, Elpek KG, Bellemare-Pelletier A, Sceats L, Reynoso ED, Gonzalez SF, Graham DB, Chang J, Peters A, Woodruff M, Kim YA, Swat W, Morita T, Kuchroo V, Carroll MC, Kahn ML, Wucherpfennig KW, Turley SJ (2012) Podoplanin-rich stromal networks induce dendritic cell motility via activation of the C-type lectin receptor CLEC-2. Immunity 37 (2):276–289

Acton SE, Farrugia AJ, Astarita JL, Mourao-Sa D, Jenkins RP, Nye E, Hooper S, van Blijswijk J, Rogers NC, Snelgrove KJ, Rosewell I, Moita LF, Stamp G, Turley SJ, Sahai E, Reis e Sousa C (2014) Dendritic cells control fibroblastic reticular network tension and lymph node expansion. Nature 514(7523):498–502

Astarita JL, Cremasco V, Fu J, Darnell MC, Peck JR, Nieves-Bonilla JM, Song K, Kondo Y, Woodruff MC, Gogineni A, Onder L, Ludewig B, Weimer RM, Carroll MC, Mooney DJ, Xia L, Turley SJ (2015) The CLEC-2-podoplanin axis controls the contractility of fibroblastic reticular cells and lymph node microarchitecture. Nat Immunol 16(1):75–84

Bender M, May F, Lorenz V, Thielmann I, Hagedorn I, Finney BA, Vogtle T, Remer K, Braun A, Bosl M, Watson SP, Nieswandt B (2013) Combined in vivo depletion of glycoprotein VI and C-type lectin-like receptor 2 severely compromises hemostasis and abrogates arterial thrombosis in mice. Arterioscler Thromb Vasc Biol 33(5):926–934

Benezech C, Nayar S, Finney BA, Withers DR, Lowe K, Desanti GE, Marriott CL, Watson SP, Caamano JH, Buckley CD, Barone F (2014) CLEC-2 is required for development and maintenance of lymph nodes. Blood 123(20):3200–3207

Bergmeier W, Bouvard D, Eble JA, Mokhtari-Nejad R, Schulte V, Zirngibl H, Brakebusch C, Fassler R, Nieswandt B (2001) Rhodocytin (aggretin) activates platelets lacking alpha(2)beta (1) integrin, glycoprotein VI, and the ligand-binding domain of glycoprotein Ibalpha. J Biol Chem 276(27):25121–25126

Bertozzi CC, Schmaier AA, Mericko P, Hess PR, Zou Z, Chen M, Chen CY, Xu B, Lu MM, Zhou D, Sebzda E, Santore MT, Merianos DJ, Stadtfeld M, Flake AW, Graf T, Skoda R, Maltzman JS, Koretzky GA, Kahn ML (2010) Platelets regulate lymphatic vascular development through CLEC-2-SLP-76 signaling. Blood 116(4):661–670

Boulaftali Y, Hess PR, Getz TM, Cholka A, Stolla M, Mackman N, Owens AP 3rd, Ware J, Kahn ML, Bergmeier W (2013) Platelet ITAM signaling is critical for vascular integrity in inflammation. J Clin Invest 123(2):908–916

Chaipan C, Soilleux EJ, Simpson P, Hofmann H, Gramberg T, Marzi A, Geier M, Stewart EA, Eisemann J, Steinkasserer A, Suzuki-Inoue K, Fuller GL, Pearce AC, Watson SP, Hoxie JA, Baribaud F, Pohlmann S (2006) DC-SIGN and CLEC-2 mediate human immunodeficiency virus type 1 capture by platelets. J Virol 80(18):8951–8960

Chang CH, Chung CH, Hsu CC, Huang TY, Huang TF (2010) A novel mechanism of cytokine release in phagocytes induced by aggretin, a snake venom C-type lectin protein, through CLEC-2 ligation. J Thromb Haemost 8(11):2563–2570

Colonna M, Samaridis J, Angman L (2000) Molecular characterization of two novel C-type lectin-like receptors, one of which is selectively expressed in human dendritic cells. Eur J Immunol 30(2):697–704

Finney BA, Schweighoffer E, Navarro-Nunez L, Benezech C, Barone F, Hughes CE, Langan SA, Lowe KL, Pollitt AY, Mourao-Sa D, Sheardown S, Nash GB, Smithers N, Reis ESC, Tybulewicz VL, Watson SP (2011) CLEC-2 and Syk in the megakaryocytic/platelet lineage are essential for development. Blood 119(7):1747–1756

Fu J, Gerhardt H, McDaniel JM, Xia B, Liu X, Ivanciu L, Ny A, Hermans K, Silasi-Mansat R, McGee S, Nye E, Ju T, Ramirez MI, Carmeliet P, Cummings RD, Lupu F, Xia L (2008) Endothelial cell O-glycan deficiency causes blood/lymphatic misconnections and consequent fatty liver disease in mice. J Clin Invest 118(11):3725–3737

Fuller GL, Williams JA, Tomlinson MG, Eble JA, Hanna SL, Pohlmann S, Suzuki-Inoue K, Ozaki Y, Watson SP, Pearce AC (2007) The C-type lectin receptors CLEC-2 and Dectin-1, but not DC-SIGN, signal via a novel YXXL-dependent signaling cascade. J Biol Chem 282 (17):12397–12409

Herzog BH, Fu J, Wilson SJ, Hess PR, Sen A, McDaniel JM, Pan Y, Sheng M, Yago T, Silasi-Mansat R, McGee S, May F, Nieswandt B, Morris AJ, Lupu F, Coughlin SR, McEver RP, Chen H, Kahn ML, Xia L (2013) Podoplanin maintains high endothelial venule integrity by interacting with platelet CLEC-2. Nature 502(7469):105–109

Hess PR, Rawnsley DR, Jakus Z, Yang Y, Sweet DT, Fu J, Herzog B, Lu M, Nieswandt B, Oliver G, Makinen T, Xia L, Kahn ML (2014) Platelets mediate lymphovenous hemostasis to maintain blood-lymphatic separation throughout life. J Clin Invest 124(1):273–284

Huang TF, Liu CZ, Yang SH (1995) Aggretin, a novel platelet-aggregation inducer from snake (Calloselasma rhodostoma) venom, activates phospholipase C by acting as a glycoprotein Ia/IIa agonist. Biochem J 309(Pt 3):1021–1027

Hughes CE, Pollitt AY, Mori J, Eble JA, Tomlinson MG, Hartwig JH, O'Callaghan CA, Futterer K, Watson SP (2010a) CLEC-2 activates Syk through dimerization. Blood 115 (14):2947–2955

Hughes CE, Navarro-Nunez L, Finney BA, Mourao-Sa D, Pollitt AY, Watson SP (2010b) CLEC-2 is not required for platelet aggregation at arteriolar shear. J Thromb Haemost 8(10):2328–2332

Ichise H, Ichise T, Ohtani O, Yoshida N (2009) Phospholipase Cgamma2 is necessary for separation of blood and lymphatic vasculature in mice. Development 136(2):191–195

Inoue K, Ozaki Y, Satoh K, Wu Y, Yatomi Y, Shin Y, Morita T (1999) Signal transduction pathways mediated by glycoprotein Ia/IIa in human platelets: comparison with those of glycoprotein VI. Biochem Biophys Res Commun 256(1):114–120

Katagiri Y, Hayashi Y, Baba I, Suzuki H, Tanoue K, Yamazaki H (1991) Characterization of platelet aggregation induced by the human melanoma cell line HMV-I: roles of heparin, plasma adhesive proteins, and tumor cell membrane proteins. Cancer Res 51(4):1286–1293

Kato Y, Fujita N, Kunita A, Sato S, Kaneko M, Osawa M, Tsuruo T (2003) Molecular identification of Aggrus/T1alpha as a platelet aggregation-inducing factor expressed in colorectal tumors. J Biol Chem 278(51):51599–51605

Kato Y, Sasagawa I, Kaneko M, Osawa M, Fujita N, Tsuruo T (2004) Aggrus: a diagnostic marker that distinguishes seminoma from embryonal carcinoma in testicular germ cell tumors. Oncogene 23(52):8552–8556

Kato Y, Kaneko M, Sata M, Fujita N, Tsuruo T, Osawa M (2005) Enhanced expression of Aggrus (T1alpha/podoplanin), a platelet-aggregation-inducing factor in lung squamous cell carcinoma. Tumour Biol 26(4):195–200

Kato Y, Kaneko MK, Kuno A, Uchiyama N, Amano K, Chiba Y, Hasegawa Y, Hirabayashi J, Narimatsu H, Mishima K, Osawa M (2006) Inhibition of tumor cell-induced platelet aggregation using a novel anti-podoplanin antibody reacting with its platelet-aggregation-stimulating domain. Biochem Biophys Res Commun 349(4):1301–1307

Kato Y, Kaneko MK, Kunita A, Ito H, Kameyama A, Ogasawara S, Matsuura N, Hasegawa Y, Suzuki-Inoue K, Inoue O, Ozaki Y, Narimatsu H (2008) Molecular analysis of the pathophysiological binding of the platelet aggregation-inducing factor podoplanin to the C-type lectin-like receptor CLEC-2. Cancer Sci 99(1):54–61

Kerrigan AM, Dennehy KM, Mourao-Sa D, Faro-Trindade I, Willment JA, Taylor PR, Eble JA, Reis e Sousa C, Brown GD (2009) CLEC-2 is a phagocytic activation receptor expressed on murine peripheral blood neutrophils. J Immunol 182(7):4150–4157

Kitagawa H, Yamamoto N, Yamamoto K, Tanoue K, Kosaki G, Yamazaki H (1989) Involvement of platelet membrane glycoprotein Ib and glycoprotein IIb/IIIa complex in thrombin-dependent and -independent platelet aggregations induced by tumor cells. Cancer Res 49(3):537–541

May F, Hagedorn I, Pleines I, Bender M, Vogtle T, Eble J, Elvers M, Nieswandt B (2009) CLEC-2 is an essential platelet-activating receptor in hemostasis and thrombosis. Blood 114 (16):3464–3472

Michelson AD (2013) Platelets, 3rd edn. Academic Press/Elsevier Science, Amsterdam

Mishima K, Kato Y, Kaneko MK, Nakazawa Y, Kunita A, Fujita N, Tsuruo T, Nishikawa R, Hirose T, Matsutani M (2006a) Podoplanin expression in primary central nervous system germ cell tumors: a useful histological marker for the diagnosis of germinoma. Acta Neuropathol 111(6):563–568

Mishima K, Kato Y, Kaneko MK, Nishikawa R, Hirose T, Matsutani M (2006b) Increased expression of podoplanin in malignant astrocytic tumors as a novel molecular marker of malignant progression. Acta Neuropathol 111(5):483–488

Navdaev A, Clemetson JM, Polgar J, Kehrel BE, Glauner M, Magnenat E, Wells TN, Clemetson KJ (2001) Aggretin, a heterodimeric C-type lectin from Calloselasma rhodostoma (malayan pit viper), stimulates platelets by binding to alpha 2beta 1 integrin and glycoprotein Ib, activating Syk and phospholipase Cgamma 2, but does not involve the glycoprotein VI/Fc receptor gamma chain collagen receptor. J Biol Chem 276(24):20882–20889

Nieswandt B, Hafner M, Echtenacher B, Mannel DN (1999) Lysis of tumor cells by natural killer cells in mice is impeded by platelets. Cancer Res 59(6):1295–1300

Osada M, Inoue O, Ding G, Shirai T, Ichise H, Hirayama K, Takano K, Yatomi Y, Hirashima M, Fujii H, Suzuki-Inoue K, Ozaki Y (2012) Platelet activation receptor CLEC-2 regulates blood/lymphatic vessel separation by inhibiting proliferation, migration, and tube formation of lymphatic endothelial cells. J Biol Chem 287(26):22241–22252

Ozaki Y, Suzuki-Inoue K, Inoue O (2009) Novel interactions in platelet biology: CLEC-2/podoplanin and laminin/GPVI. J Thromb Haemost 7(Suppl 1):191–194

Schacht V, Dadras SS, Johnson LA, Jackson DG, Hong YK, Detmar M (2005) Up-regulation of the lymphatic marker podoplanin, a mucin-type transmembrane glycoprotein, in human squamous cell carcinomas and germ cell tumors. Am J Pathol 166(3):913–921

Severin S, Pollitt AY, Navarro-Nunez L, Nash CA, Mourao-Sa D, Eble JA, Senis YA, Watson SP (2011) Syk dependent phosphorylation of CLEC-2: a novel mechanism of hemitam signalling. J Biol Chem 286(6):4107–4116

Shin Y, Morita T (1998) Rhodocytin, a functional novel platelet agonist belonging to the heterodimeric C-type lectin family, induces platelet aggregation independently of glycoprotein Ib. Biochem Biophys Res Commun 245(3):741–745

Shirai T, Inoue O, Hirayama K, Endo H, Fujii H, Utida-Sato H, K S-I YO (2013) A role of CLEC-2 in tumor growth and metastasis. J Thromb Haemost 11(Suppl 2):139

Spalton JC, Mori J, Pollitt AY, Hughes CE, Eble JA, Watson SP (2009) The novel Syk inhibitor R406 reveals mechanistic differences in the initiation of GPVI and CLEC-2 signaling in platelets. J Thromb Haemost 7(7):1192–1199

Suzuki-Inoue K, Ozaki Y, Kainoh M, Shin Y, Wu Y, Yatomi Y, Ohmori T, Tanaka T, Satoh K, Morita T (2001) Rhodocytin induces platelet aggregation by interacting with glycoprotein Ia/IIa (GPIa/IIa, Integrin alpha 2beta 1). Involvement of GPIa/IIa-associated src and protein tyrosine phosphorylation. J Biol Chem 276(2):1643–1652

Suzuki-Inoue K, Fuller GL, Garcia A, Eble JA, Pohlmann S, Inoue O, Gartner TK, Hughan SC, Pearce AC, Laing GD, Theakston RD, Schweighoffer E, Zitzmann N, Morita T, Tybulewicz VL, Ozaki Y, Watson SP (2006) A novel Syk-dependent mechanism of platelet activation by the C-type lectin receptor CLEC-2. Blood 107(2):542–549

Suzuki-Inoue K, Kato Y, Inoue O, Kaneko MK, Mishima K, Yatomi Y, Yamazaki Y, Narimatsu H, Ozaki Y (2007) Involvement of the snake toxin receptor CLEC-2, in podoplanin-mediated platelet activation, by cancer cells. J Biol Chem 282(36):25993–26001

Suzuki-Inoue K, Inoue O, Ding G, Nishimura S, Hokamura K, Eto K, Kashiwagi H, Tomiyama Y, Yatomi Y, Umemura K, Shin Y, Hirashima M, Ozaki Y (2010) Essential in vivo roles of the C-type lectin receptor CLEC-2: embryonic/neonatal lethality of CLEC-2-deficient mice by blood/lymphatic misconnections and impaired thrombus formation of CLEC-2-deficient platelets. J Biol Chem 285(32):24494–24507

Suzuki-Inoue K, Inoue O, Ozaki Y (2011) Novel platelet activation receptor CLEC-2: from discovery to prospects. J Thromb Haemost 9(Suppl 1):44–55

Tammela T, Alitalo K (2010) Lymphangiogenesis: molecular mechanisms and future promise. Cell 140(4):460–476

Tang T, Li L, Tang J, Li Y, Lin WY, Martin F, Grant D, Solloway M, Parker L, Ye W, Forrest W, Ghilardi N, Oravecz T, Platt KA, Rice DS, Hansen GM, Abuin A, Eberhart DE, Godowski P, Holt KH, Peterson A, Zambrowicz BP, de Sauvage FJ (2010) A mouse knockout library for secreted and transmembrane proteins. Nat Biotechnol 28(7):749–755

Tsuruo T, Fujita N (2008) Platelet aggregation in the formation of tumor metastasis. Proc Jpn Acad Ser B Phys Biol Sci 84(6):189–198

Uhrin P, Zaujec J, Breuss JM, Olcaydu D, Chrenek P, Stockinger H, Fuertbauer E, Moser M, Haiko P, Fassler R, Alitalo K, Binder BR, Kerjaschki D (2010) Novel function for blood platelets and podoplanin in developmental separation of blood and lymphatic circulation. Blood 115(19):3997–4005

Watson SP, Auger JM, McCarty OJ, Pearce AC (2005) GPVI and integrin alphaIIb beta3 signaling in platelets. J Thromb Haemost 3(8):1752–1762

Watson AA, Brown J, Harlos K, Eble JA, Walter TS, O'Callaghan CA (2007) The crystal structure and mutational binding analysis of the extracellular domain of the platelet-activating receptor CLEC-2. J Biol Chem 282(5):3165–3172

Watson AA, Christou CM, James JR, Fenton-May AE, Moncayo GE, Mistry AR, Davis SJ, Gilbert RJ, Chakera A, O'Callaghan CA (2009) The platelet receptor CLEC-2 is active as a dimer. Biochemistry 48(46):10988–10996

Yuan P, Temam S, El-Naggar A, Zhou X, Liu DD, Lee JJ, Mao L (2006) Overexpression of podoplanin in oral cancer and its association with poor clinical outcome. Cancer 107 (3):563–569

Part III
ITIM-Bearing Receptors

Chapter 7
Dendritic Cell Immunoreceptor (DCIR): An ITIM-Harboring C-Type Lectin Receptor

Tomonori Kaifu and Yoichiro Iwakura

Abstract C-type lectin receptors (CLRs) have been recognized as one of the pattern recognition receptors that trigger immune responses against pathogens. Dendritic cell immunoreceptor (DCIR) is a type II membrane protein that contains an extracellular carbohydrate recognition domain (CRD) and a long cytoplasmic tail with an immunoreceptor tyrosine-based inhibition motif (ITIM). Only one molecular species is identified in humans (DCIR), while four family molecules (DCIR1-4) are present in mice. Human DCIR and mouse DCIR1 dampen immune responses through ITIM-mediated reaction. DCIR binds mannose and fucose and also pathogenic organisms, but the structure of the ligand carbohydrates still remains to be determined. DCIR1 is important for the homeostasis of the immune system, and the deficiency causes autoimmune diseases. DCIR also acts as an attachment factor for HIV in dendritic cells and HIV-infected T cells. DCIR1 is also implicated in the pathogenesis of mosquito-transmitted virus and protozoan infections. This chapter highlights the roles of human and mouse DCIR in immune responses and immune homeostasis revealed by in vitro cell-based studies as well as in vivo gene-depleted mouse analyses.

Keywords C-type lectins • HIV • Phosphatases • ITIM • Homeostasis • Bone metabolism

A large array of membrane receptors is expressed on professional antigen-presenting cells (APCs) to sense pathogen invasions for activating innate immunity and to present foreign peptides on major histocompatibility complex II (MHC-II) or MHC-I for activating adaptive immunity. DCs and macrophages express plentiful germline-encoded innate receptors to recognize pathogen-associated molecular patterns (PAMPs) of invading pathogens and conserved molecular structures of pathogens. These receptors are called as pattern recognition receptors (PRRs). Toll-

T. Kaifu (✉) • Y. Iwakura
Research Institute for Biomedical Science, Division of Experimental Animal Immunology, Center for Animal Disease Models, Tokyo University of Science, 2669 Yamazaki, Noda, Chiba, Japan
e-mail: kaifu@rs.tus.ac.jp

S. Yamasaki (ed.), *C-Type Lectin Receptors in Immunity*,
DOI 10.1007/978-4-431-56015-9_7

like receptors (TLRs) and C-type lectin receptors (CLRs) are membrane-embedded PRRs that response to multiple types of PAMPs (van Vliet et al. 2007). Although TLRs are able to stimulate maturation of DCs through recognition of different types of PAMPs, they do not mediate antigen uptake to present peptides on MHC-I or MHC-II. While CLRs detect diverse PAMPs to synthesize pro- and anti-inflammatory mediators, they can incorporate pathogens into cytoplasm to facilitate antigen presentation (Bergman et al. 2004; Hunger et al. 2004).

There is a diverse group of proteins in C-type lectin family. C-type lectins were originally characterized by their abilities to bind carbohydrates. Soluble and membrane-bound forms function as C-type lectin proteins, but they have a common domain, termed the carbohydrate recognition domain (CRD), which contains disulfide bridges between cysteine residues. C type is termed by the property that requires Ca^{2+} for binding carbohydrate structures, but some C-type lectin receptors bind their ligands in a Ca^{2+}-independent manner. Moreover, they recognize multiple types of molecules, proteins, and lipids, other than carbohydrates. Whereas a majority of CLRs recognize carbohydrate structures on pathogens, some are found to interact with self-glycoproteins or endogenous ligands.

A group of immune receptors is consisted of counterbalancing two types of receptors: one mediates upregulation of cellular activities and the other counter-regulates the activation induced by these receptors. The former type of receptors contains an immunoreceptor tyrosine-based activation motif (ITAM) in the cytoplasmic domain. Another type of receptors in this group, with a short cytoplasmic tail and a charged amino acid residue in the transmembrane region, forms a complex with an adaptor protein containing ITAM, such as Fc receptor γ-chain. On the other hand, counterbalanced receptors that are capable of dampening activation signal contain immunoreceptor tyrosine-based inhibitory motif (ITIM), which provides a docking site for phosphatases to dephosphorylate phosphotyrosine residues in signaling molecules or ITAMs. Thus, it is suggested that the balance of signals between ITAM-harboring and ITIM-harboring receptors determines the adequate cellular conditions, and interruption of the delicate signal balance causes harmful havoc, such as excessive inflammation, self-tissue damages, and autoimmune diseases. So far, two CLRs have been identified as an ITIM-harboring receptor, and dendritic cell immunoreceptor (DCIR) is one of the best studied receptors. In this review, we will introduce its expression pattern, intracellular signal pathways, and physiological roles at the levels of cells and animals.

7.1 The *DCIR* Gene Is Located in the CLR Cluster

Cloning of the human and mouse DCIR gene has been reported in four reports. (1) Human DCIR was identified by the homology search for the conserved seven amino acid sequences of the CRD in hepatic asialoglycoprotein receptors (ASGPRs) and macrophage lectin in the Human Genome Sciences databases. Amino acid identity of the deduced CRD domain of human DCIR is 42 %, 35 %, and 37 % for macrophage lectin, ASGPR1, and ASGPR2, respectively. Mouse DCIR1 molecule was also identified in the EST databases, using the deduced protein sequence of human DCIR. The predicted protein sequence of mouse DCIR is 54 % identical to human DCIR (Bates et al. 1999). (2) Human DCIR gene was cloned by subtractive cloning of differentially expressed cDNA from human monocyte-derived DCs, which were treated with or without keyhole limpet hemocyanin. The gene was named lectin-like immunoreceptor (LLIR) gene in that study. The predicted protein sequence of the LLIR is highly similar to ASGPRs in the BLAST search. The LILR gene is located in the chromosome 12p13 locus, close to NK cell receptors clusters that encode multiple lectin-like NK receptors. Two shorter variants were isolated from human myeloid cell line, which lacked the nucleotide sequence encoding the transmembrane region (Huang et al. 2001). (3) Upregulation of *DCIR* mRNA was shown in GM-CSF-treated human neutrophils compared with non-treated neutrophils using differential display polymerase chain reaction method. The gene was assigned as C-type lectin superfamily 6 (*CLECSF6*). A homologous BAC clone in nucleotide sequence databases was assigned in chromosome 12p13.1 (Richard et al. 2002). (4) Yeast signal sequence trap screening identified cDNA of mouse DCIR1 from mouse bone marrow-derived DCs, which were induced maturation by TGF-β1. The predicted amino acid sequence of the CRD region of mouse DCIR1 was 48 % identical to Dectin-2 (Kanazawa et al. 2002). Meanwhile, homology search for the rat *Dcir* to various human and mouse C-type lectin-like receptors identified only one *DCIR* gene located in the chromosome 12 in humans and four genes for mouse DCIR (*Dcir1-4*) in the chromosome 6 at the telomeric region near the NK cell receptor clusters (Kerscher et al. 2013; Flornes et al. 2004). Human DCIR and mouse DCIR1 have 55 % amino acid sequence identity, and mouse DCIR1 is 74 %, 57 %, and 52 % identical to DCIR2, DCIR3, and DCIR4, respectively. In this review, we use mouse DCIR as mouse DCIR1 hereafter.

7.2 *DCIR* Gene Is Expressed in Cells in the Immune System

Human *DCIR* mRNA is highly detected in peripheral blood leukocytes and moderate in bone marrow, spleen, and lymph node, while the *DCIR* mRNA level is low in the thymus, spinal cord, and trachea. *DCIR* mRNA is observed in DCs, macrophages, and granulocytes, which are induced in vitro from $CD34^+$ cord blood progenitors as well as freshly isolated blood monocytes. *DCIR* mRNA is slightly detected in tonsil B cells, but not in blood T cells and NK cells. Peripheral blood neutrophils express a low level of *DCIR* mRNA. In accordance with the presence of *DCIR* mRNA, DCIR protein is detected in peripheral blood $CD14^+$ monocytes, $CD15^+$ granulocytes (including neutrophils, eosinophils, and monocytes/macrophages), and $CD19^+$ B cells, but not in $CD56^+$ NK cells and $CD3^+$ T cells. In immunohistochemical staining, the expression of DCIR in the tonsil is found in the broad area below the epithelium and weakly in the mantle zone of the follicle, but not in the germinal center. DCIR-positive cells have a dendritic shape and are negative for CD1a, which is a maker for Langerhans cells in the epithelium (Bates et al. 1999). Thus, human DCIR is expressed in myeloid cells and APCs including B cells, and the expression pattern is distinct among subsets of DCs.

The expression pattern of mouse *Dcir* is similar to humans. Mouse *Dcir* mRNA is high in the spleen and lymph nodes, in which APCs, such as DCs, macrophages, and B cells, accumulate in the resting and inflammatory conditions. *Dcir* mRNA is detected in bone marrow-derived DCs but not in NK cells. Consistent with the expression profiles of *Dcir* mRNA, mouse DCIR was detected on the surface of splenic B220+ or CD19+ cells, CD11c+ cells, and Mac1+ cells (Kanazawa et al. 2002). Indeed, all the DCIR-expressing cells express MHC class II, indicating that mouse DCIR is exclusively expressed on APCs. However, mouse DCIR expression in neutrophils still remains undetermined. Mouse *Dcir2* mRNA is observed in macrophages, DCs, neutrophils, B cells, and NK cells. At present, mouse DCIR2 is used as a specific marker for CD8– DCs (Dudziak et al. 2007). Mouse *Dcir3* mRNA is detected in macrophages, DCs, neutrophils, and B cells, while mouse *Dcir4* mRNA is in macrophages, DCs, and neutrophils (Flornes et al. 2004). The precise expression patterns of mouse *Dcir3* and mouse *Dcir4* remain yet unclear.

7.3 DCIR Is One of the ITIM-Harboring C-Type Lectin Receptors

The deduced amino acid alignment displays human DCIR as a type II receptor of 237 aa with a single CRD in the extracellular C-terminus portion and ITIM in the intracellular N-terminus portion. Human DCIR carries no signal sequence but a putative hydrophobic signal anchor sequence. A number of conserved cysteine residues remain in the CRD of human DICR, in comparison with ASGPRs and macrophage lectin. Human DCIR forms disulfide bonds with a pair of cysteines at aa positions 106 and 117, 134 and 230, and 203 and 222 (UniPortKB; Q9UMR7). The CRD has the conserved calcium-binding domain consisted of an EPS motif (at position 195–197), glutamate (at position 201), and asparagine and aspartate (at positions 218 and 219) (Bates et al. 1999). The calcium-binding motif is conserved in ASGPRs and macrophage lectin, although they have QPD motif instead of EPS motif. It is noteworthy that the NKG2 receptors, which are Ca2+-independent C-type lectin receptors mainly in NK cells, lack the motif in the amino acid alignment. Crystal structure analysis of CD94 revealed the lack of the corresponding loop at the position of the calcium-binding domain (Boyington et al. 1999). Human DCIR has a potential consensus N-glycosylation site at position 185, which is suggested to have a hindrance effect on binding with DCIR ligands (Bloem et al. 2013).

The amino acid sequence of human DCIR displays the intracellular domain with 48 amino acids, containing ITYAEV at the position from residue 5–10. The sequence matches the canonical ITIM, an inhibitory signal motif whose conserved sequence is S/I/V/LxYxxI/V/L (Ravetch and Lanier 2000). The signal motif has been shown to act as a docking site for phosphatases that counterbalances cellular activities through inhibition of a cascade of tyrosine kinases.

Two mouse homologues have structural features similar to human DCIR. Mouse DCIR and DCIR2 are type II protein with a single CRD, conserved cysteine residues, and the ITIM sequence. They also have potential cysteine residues that are supposed to be involved in disulfide bond formation. Mouse DCIR has conserved EPS motif (at position 197–199), glutamate (at position 203), and asparagine and aspartate (at positions 219 and 220), but DCIR2 has EPN motif, which is thought to be responsible for mannose binding. DCIR has three potential N-glycosylation sites, whereas DCIR2 has four potential N-glycosylation sites. In contrast, DCIR3 and DCIR4 lack the consensus ITIM sequence, although a tyrosine residue remains at the corresponding region. Asparagine in the calcium-binding domain of DCIR3 and DCIR4 is replaced by serine, and aspartate of DCIR4 is substituted for valine, suggesting that they may function in a Ca2+-independent way (Flornes et al. 2004).

Human DCIR is a unique C-type lectin receptor with a long cytoplasmic tail and an ITIM, similarly to Clec12a. The same structural features are conserved in mouse DCIR and DCIR2. In B-cell lymphoma lacking Fc receptors, chimeric proteins consisting of the extracellular region of FcγRIIB and the cytoplasmic of mouse DCIR dampened B-cell receptor (BCR)-mediated Ca2+ flux and total tyrosine phosphorylation. The inhibitory effect of DCIR depends on the tyrosine residue in the ITIM (Kanazawa et al. 2002). The peptide covering the ITIM region of human DCIR that contains phosphorylated tyrosine binds to SHP-1 and SHP-2, but not SHIP, upon incubation with human neutrophil lysates (Richard et al. 2006) (Fig. 7.1).

The expression of DCIR in DCs implies inhibitory roles in maturation, secretion of soluble mediators, and antigen presentation. Human pDCs enriched by anti-BDCA-4 magnetic beads from peripheral blood expressed BDCA-2 and DCIR but not DC-SIGN, mannose receptor, or Dectin-1. Induction of DCIR-mediated signal by cross-linking with anti-DCIR antibody reduced IFN-α production in response to TLR9 ligands, whereas DCIR triggering did not affect the expression levels of co-stimulatory molecules in TLR9-activated pDCs (Meyer-Wentrup et al. 2008). Thus, DCIR in DCs appears not to be involved in the regulation of antigen recognition and co-stimulation in the process of antigen presentation. Many of C-type lectin receptors on DCs are known to function as an endocytic receptor and internalize bound ligands (Figdor et al. 2002). DCIR activation with antibody causes internalization of DCIR in a clathrin-dependent manner. This may explain the mechanism of the inhibitory effect of DCIR on TLR9-induced IFN-α production, which locates in endosomal compartments. However, it still remains to be elucidated whether the ITIM motif of DCIR is required for the downregulation of cytokine production and whether the phosphatases (SHP-1 and SHP-2) are involved

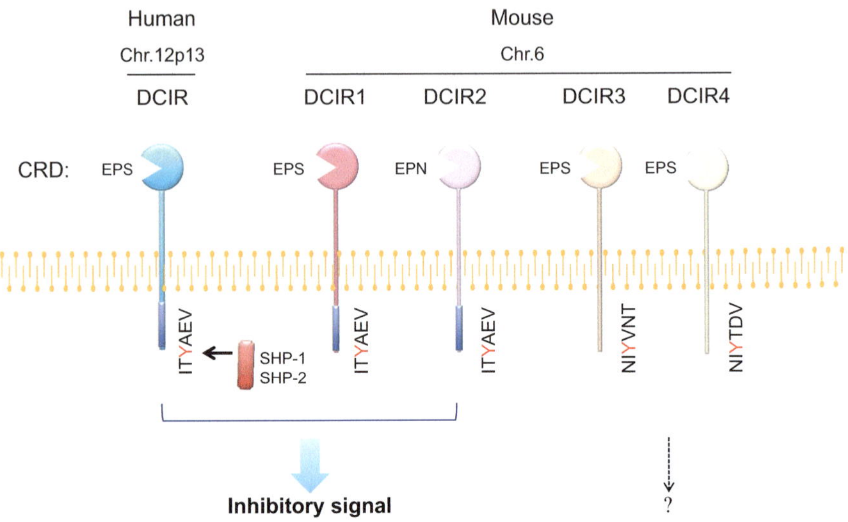

Fig. 7.1 Structures of human DCIR and mouse DCIR1, DCIR2, DCIR3, and DCIR4. Dendritic cell immunoreceptor (DCIR) is a C-type lectin receptor with the ITIM in the cytoplasmic region. The genes for human and mouse DCIR are located in chromosomes 12p13 and 6, respectively. Four genes are identified in mice. The CRD of human DCIR and mouse DCIR1, DCIR3, and DCIR4 has EPS motif that is important for Ca2+-dependent carbohydrate recognition, but mouse DCIR2 has EPN motif. Human DCIR, mouse DCIR1, and mouse DCIR2 have the canonical sequence of ITIM and mediate inhibitory signals through SHP-1 and SHP-2. In contrast, mouse DCIR3 and DCIR4 lack the consensus ITIM sequence. They have a tyrosine residue in the cytoplasmic part, but it is not yet known whether they transduce intracellular signal. CRD: carbohydrate recognition domain. ITIM motif: S/I/V/LxYxxI/V/L. E: glutamic acid, P: proline, N: asparagine, S: serine

in the signaling components downstream of TLR9. In addition, DCIR inhibits TLR8-induced pro-inflammatory cytokine production from human monocyte-derived DCs (moDCs). DCIR triggering by specific antibodies reduced the production of TNF-α and IL-12 from moDCs upon stimulation with a ligand for TLR8 but not ligands for TLR3, TLR4, and TLR2/Dectin-1. This is probably explained by the fact that DCIR is internalized into LAMP-1-positive lysosome compartments and TLR8 and TLR9, but not TLR3, TLR4, or TLR2/Dectin-1, are also present in endosomal/lysosomal compartment (Meyer-Wentrup et al. 2009; Kawai and Akira 2008; Kawasaki and Kawai 2014). There is no evidence of the crosstalk between mouse DCIR and TLRs in mouse APCs. Nonetheless, these results indicate the importance of DCIR-mediated inhibition in limiting and modifying DC-associated immune responses.

7.4 DCIR Interacts with Carbohydrates and Ligands on Pathogens

Cell surface-embedded receptors such as TLR and immunoglobulin-like receptors trigger downstream signaling cascades to exert their functions after engagement with their ligands. CLRs bind not only to sugar moieties of membrane components of invading pathogens but also to endogenous proteins and glycolipids (Zhang et al. 2012; Ahrens et al. 2012; Ishikawa et al. 2009). To understand the molecular basis of DCIR-mediated signals and its function, it is important to elucidate the carbohydrate structures of the ligands. Glycan array data of the Consortium for Functional Glycomics (CFG) shows that human DCIR-Fc protein binds to sulfated lactose (Lac), LacNAc, and alpha-1-acid glycoprotein (AGP), which contains biantennary N-glycans (Hsu et al. 2009). The glycan array data of CLRs, including DCIR, are available from the CFG database. The human DCIR-Fc protein does not bind polysaccharide extract from a fungus, *Ganoderma lucidum*, analyzed by enzyme-linked immunosorbent assay. CLRs containing EPN motif, such as mannose-binding protein (MBP) and macrophage mannose receptor (MMR), are known to bind mannose (Man) and fucose (Fuc). Human DCIR-Fc protein, which has EPS motif in the CRD, binds not only Man and Fuc but also galactose (Gal) and GalNAc, although the binding affinities are lower than those of Man and Fuc (Lee et al. 2011). Human DCIR-Fc protein binds Lewis^b and Man_3 glycans, and truncation of N-glycosyl carbohydrate chains in the CRD increases the binding affinity for these glycans. However, the truncated and non-truncated forms of human DCIR expressed on CHO cells could not bind Lewis^b and Man_3 glycans. These binding assays using Fc-fusion protein reveal that human DCIR has a potential to bind a variety of sugar chains.

Tyrosine phosphorylation of human DCIR was increased when it was expressed in mutant CHO cells that lacked DCIR ligands compared to parent CHO cells, suggesting that *cis* interaction of DCIR with its ligands may affect the phosphorylation status of the ITIM (Bloem et al. 2013). Further analysis showed that human DCIR-Fc protein with truncated form of N-glycans binds Lewis^a and sulfo-Lewis^a, but human DCIR-Fc protein with N-glycan does not bind the glycans. Human DCIR-Fc binds to keratinocytes, gastric cancer cells, colon carcinoma cells, and prostate cancer cells as well as helminths such as *S. mansoni* and *T. spiralis* (Bloem et al. 2014). Moreover, DCIR-Fc fusion protein weakly binds commensal microbiota (Hutter et al. 2014). The structure of the DCIR ligands on microbiota is unclear, but probably a minor population of commensal microbiota expresses DCIR ligands. These data support the idea that DCIR recognizes both endogenous and pathogenic ligands.

Intravenous immunoglobulin (IVIg) therapy is widely used as a therapy for autoimmune diseases and inflammatory diseases. Although the action mechanism of IVIg remains controversial, sialylated immunoglobulins (SA-IgGs) are suggested to mediate the IVIg through SING-R1 (the murine homologue of DC-SIGN) in the treatment of arthritis and thrombocytopenic purpura in mice as well as through CD22 (Siglec-2) in promoting apoptosis of human B cells (Anthony et al. 2008; Kaneko et al. 2006; Seite et al. 2010). SA-IgG also ameliorates ovalbumin-induced airway hyperresponsiveness (AHR) in mice by inducing Foxp3+ Treg cells through induction of tolerogenic DCs (Massoud et al. 2012). In this model, the effect of SA-IgG depends on DCIR in DCs. In immunofluorescence analysis, SA-IgG is fluorescently colocalized with DCIR, and DCIR-expressing CHO cells are stained with SA-IgG, but not with non-SA-IgG, indicating that DCIR serves an important role in SA-IgG-induced inhibition (Massoud et al. 2014). However, the direct binding between DCIR and sialic acid on IgG has not yet been proved.

7.5 DCIR Is an Attachment Factor for HIV and HCV

Human DCIR is implicated in the attachment and transmission of HIV in DCs and CD4+ T cells. Attachment of HIV to DCs is known as one of the entry routes to establish infection to CD4+ T cells. The interaction of viral envelope glycoproteins with CLRs, such as DC-SIGN, MR, and langerin, contributes to the attachment of HIV to DCs and transfers the virus to CD4+ T cells or macrophages (Turville et al. 2001, 2002; de Witte et al. 2007). Then, viral envelope protein binds CD4 protein and coreceptors such as CXCR4 or CCR5 on T cells and macrophages, causing membrane fusion of virus, following penetration into these cells. Knockdown of DCIR expression with siRNA or treatment with antibodies against DCIR in human immature moDCs decreased HIV transmission to autologous CD4+ T cells. Raji-CD4+ T cells that was transfected with *DCIR* gene remarkably augmented the binding and viral production, while Raji-CD4+ T cells expressing neck domain-deficient human DCIR failed to promote HIV binding (Lambert et al. 2008). As the stalk region of CLRs is responsible for multimerization, this result suggests that the formation of dimeric DCIR is necessary for efficient recognition of possible DCIR ligands on HIV (Jin et al. 2014). When the efficiency of HIV attachment and transfer compares that of DC-SIGN which can form tetramer, DCIR is less efficient than DC-SIGN (Jin et al. 2014). Under the physiological conditions, T cells do not express DCIR, but *DCIR* expression is detected in CD4+ T cells and CD8+ T cells of rheumatoid arthritis patients (Eklow et al. 2008). Interestingly, surface expression of DCIR is induced in CD4+ T cells not only by HIV infection but also by soluble factors from HIV-infected cells. The induction of DCIR in CD4+ T cells results in increases of HIV binding, replication, and transmission to uninfected CD4 + T cells (Lambert et al. 2010). The pathological importance of DCIR expression on DCs and CD4+ T cells in HIV pathogenesis is still obscure, but the types of DCIR-expressing cells imply potential roles of DCIR in the pathogenesis. Contribution of ITIM-dependent signaling in HIV infection has also not yet been elucidated completely, but pharmacologic inhibitors and gene silencing with oligonucleotides suggest that phosphatases (SHP-1 and SHP-2) and tyrosine kinases are involved in HIV binding to DCIR-expressing Raji cells (Lambert et al. 2011). Furthermore,

DCIR is suggested to be a potential receptor for HCV. HCV glycoprotein E2, which inhibits IFN-α and IFN-λ production, binds DCIR on pDCs from healthy donors, and treatment with anti-DCIR antibody recovers the cytokine production (Florentin et al. 2012).

7.6 DCIR Maintains Immune Homeostasis by Negatively Regulating DCs

DCIR is an important negative regulator in controlling autoimmune diseases. Mouse *Dcir* has been identified as one of potential pathological genes related to rheumatoid arthritis (Fujikado et al. 2006). The expression of *Dcir* gene was increased in the inflamed joints of two models of rheumatoid arthritis, HTLV-1-Tg mice, and *IL-1rn–/–* mice. To understand the physiological and pathological roles of DCIR, *Dcir–/–* mice have been generated. Aged *Dcir–/–* mice on the C57BL/6 background spontaneously develop several symptoms, such as swelling of joints, enthesitis, and sialadenitis (Fujikado et al. 2008). In addition, young *Dcir–/–* mice are highly susceptible to collagen-induced arthritis compared to wild-type littermates. The development of experimental autoimmune encephalomyelitis, an animal model for multiple sclerosis, is also exacerbated in *Dcir–/–* mice, indicating a protective role for DCIR in autoimmune diseases (Seno et al. 2015). Interestingly, the proportion of CD11c+ DCs in lymph node increases in these mice and *Dcir–/–* mouse-derived BM cells differentiates more efficiently into DCs upon induction with GM-CSF compared with WT mice, suggesting that excess differentiation and expansion of DCs in *Dcir–/–* mice may cause autoimmunity. Thus, DCIR is important for the homeostasis of the immune system by, at least in part, regulating DC differentiation. Consistent with the results obtained in mouse models, polymorphisms of DCIR gene are associated with the susceptibility to rheumatoid arthritis in humans (Eklow et al. 2008; Lorentzen et al. 2007; Guo et al. 2012).

Recently, DCIR has been reported to be involved in the homeostasis of bone. DCIR deficiency increases bone volume and causes aberrant ossification in joints, and *Dcir–/–* mice spontaneously develop ankylosis accompanied by fibrocartilage proliferation and ectopic ossification. Moreover, *Dcir–/–* mice increase IFN-γ+ T cells in peripheral blood and lymph nodes. Splenic CD11c+ DCs isolated from *Dcir–/–* mice have higher potency to induce IFN-γ+ T cells, relative to WT mice. Notably, IFN-γ is identified as an osteogenic factor that promotes chondrogenesis and osteoblastogenesis (Maruhashi et al. 2015). These demonstrate that DCIR deficiency causes an increase of IFN-γ+ T cells, which accelerates mineralization of OBs and proliferation of chondrocytes, leading to the increased bone volume. Thus, DCIR has broad functional activities and is involved in not only the homeostasis of the immune system but also the bone metabolism.

DCIR-Fc protein weakly binds commensal microbiota, and *Dcir*–/– mouse-bone-marrow-derived DCs (031932-UCD) secreted higher concentrations of TNF-α in response to heat-killed microbiota. However, *Dcir* deficiency only marginally influenced the development of DSS-induced colitis, a mouse model for ulcerative colitis, and pro-inflammatory cytokine production (TNF-α, IL-6, and IL-1β) in colon organ culture, indicating a limited role of DCIR in experimental colitis (Hutter et al. 2014). On the other hand, DCIR has a protective role in the pathogenesis of chikungunya virus (CHIKV) infection (Long et al. 2013). CHIKV-induced weight loss and edema were exacerbated in *Dcir*–/– mice but not in DC-SIGN–/– and SIGNR3–/– mice. Cell infiltration and inflammatory tissue damage were enhanced in the fascia of the inoculated foot and the ankle joint of *Dcir*–/– mice. Cytokine production in response to CHIKV in DCIR-defective BMDCs was enhanced even in the absence of viral replication in DCs (Long et al. 2013). There was no evidence that DCIR directly interacts with CHIKV. Thus, DCIR plays a protective role in limiting the CHIKV-induced inflammatory response and subsequent tissue and joint damages, probably by regulating the activation state and pathological potential of the inflammatory cells entering the virally infected tissue.

Interestingly, *Dcir*–/– mice were protected from neurological symptoms of cerebral malaria (CM) during *Plasmodium berghei* infection, although the incidence of parasitemia was similar between *Dcir*–/– and wild-type mice (Maglinao et al. 2013). *Dcir* deficiency diminished infiltration of CD8+ T cell and inflammation in the brain, TNF-α level in sera, and frequency of activated T cells in the spleen. Because these phenotypes are rather unexpected from the function of DCIR in DCs and macrophages, roles of DCIR in the pathogenesis of CM remain unclear.

DCIR can be used as a target molecule on DCs for enhancing antigen presentation to T cells. Targeting of antigens through CLRs on DCs induces efficient antigen presentation to MHC-I-restricted or MHC-II-restricted T cells (Bonifaz et al. 2004; Carter et al. 2006). Similar to the other CLRs, targeting of DCIR with antigen-conjugated anti-DCIR antibody on human pDCs enhanced antigen-specific proliferation of peripheral blood leukocytes, because antigens are internalized together with DCIR after receptor triggering (Meyer-Wentrup et al. 2008). In addition, delivery of antigens by DCIR to human skin-derived DCs and blood-derived DCs induced cross-presentation to CD8+ T cells. TLR7/8 agonists or combination of TLR7/8 agonist and CD40L further augmented the DCIR-mediated cross-presentation (Klechevsky et al. 2010). Anti-DCIR antibody conjugated with antigens may become a strong inducer of cytotoxic CD8+ T cells for virus-infected cells and tumors.

7.7 Conclusion

DCIR is a unique CLR containing the ITIM in the cytoplasmic part. As *Dcir*–/– mice spontaneously develop autoimmunity and bone morphogenic abnormality, this molecule has important roles in not only the immune system but also the bone metabolism. DCIR in DCs negatively regulates their differentiation and functions, and inhibition of DCIR shapes T cells to produce IFN-γ. DCIR is suggested to be involved in the susceptibility to HIV infection, CHIKV-induced pathogenesis, and *Plasmodium berghei*-induced cerebral malaria. In addition, targeting this molecule to deliver antigens to DCs efficiently induces antigen-specific T cell responses. Further investigation is needed to elucidate its functional ligands and the precise mechanisms to regulate the immune and bone system.

References

Ahrens S et al (2012) F-actin is an evolutionarily conserved damage-associated molecular pattern recognized by DNGR-1, a receptor for dead cells. Immunity 36:635–645. doi:10.1016/j.immuni.2012.03.008

Anthony RM, Wermeling F, Karlsson MC, Ravetch JV (2008) Identification of a receptor required for the anti-inflammatory activity of IVIG. Proc Natl Acad Sci U S A 105:19571–19578. doi:10.1073/pnas.0810163105

Bates EE et al (1999) APCs express DCIR, a novel C-type lectin surface receptor containing an immunoreceptor tyrosine-based inhibitory motif. J Immunol 163:1973–1983

Bergman MP et al (2004) Helicobacter pylori modulates the T helper cell 1/T helper cell 2 balance through phase-variable interaction between lipopolysaccharide and DC-SIGN. J Exp Med 200:979–990. doi:10.1084/jem.20041061

Bloem K et al (2013) Ligand binding and signaling of dendritic cell immunoreceptor (DCIR) is modulated by the glycosylation of the carbohydrate recognition domain. PLoS One 8:e66266. doi:10.1371/journal.pone.0066266

Bloem K et al (2014) DCIR interacts with ligands from both endogenous and pathogenic origin. Immunol Lett 158:33–41. doi:10.1016/j.imlet.2013.11.007

Bonifaz LC et al (2004) In vivo targeting of antigens to maturing dendritic cells via the DEC-205 receptor improves T cell vaccination. J Exp Med 199:815–824. doi:10.1084/jem.20032220

Boyington JC et al (1999) Structure of CD94 reveals a novel C-type lectin fold: implications for the NK cell-associated CD94/NKG2 receptors. Immunity 10:75–82

Carter RW, Thompson C, Reid DM, Wong SY, Tough DF (2006) Preferential induction of CD4+ T cell responses through in vivo targeting of antigen to dendritic cell-associated C-type lectin-1. J Immunol 177:2276–2284

de Witte L et al (2007) Langerin is a natural barrier to HIV-1 transmission by Langerhans cells. Nat Med 13:367–371. doi:10.1038/nm1541

Dudziak D et al (2007) Differential antigen processing by dendritic cell subsets in vivo. Science 315:107–111. doi:10.1126/science.1136080

Eklow C et al (2008) Cellular distribution of the C-type II lectin dendritic cell immunoreceptor (DCIR) and its expression in the rheumatic joint: identification of a subpopulation of DCIR+ T cells. Ann Rheum Dis 67:1742–1749. doi:10.1136/ard.2007.076976

Figdor CG, van Kooyk Y, Adema GJ (2002) C-type lectin receptors on dendritic cells and Langerhans cells. Nat Rev Immunol 2:77–84. doi:10.1038/nri723

Florentin J et al (2012) HCV glycoprotein E2 is a novel BDCA-2 ligand and acts as an inhibitor of IFN production by plasmacytoid dendritic cells. Blood 120:4544–4551. doi:10.1182/blood-2012-02-413286

Flornes LM et al (2004) Identification of lectin-like receptors expressed by antigen presenting cells and neutrophils and their mapping to a novel gene complex. Immunogenetics 56:506–517. doi:10.1007/s00251-004-0714-x

Fujikado N, Saijo S, Iwakura Y (2006) Identification of arthritis-related gene clusters by microarray analysis of two independent mouse models for rheumatoid arthritis. Arthritis Res Ther 8: R100. doi:10.1186/ar1985

Fujikado N et al (2008) Dcir deficiency causes development of autoimmune diseases in mice due to excess expansion of dendritic cells. Nat Med 14:176–180. doi:10.1038/nm1697

Guo J et al (2012) A replication study confirms the association of dendritic cell immunoreceptor (DCIR) polymorphisms with ACPA – negative RA in a large Asian cohort. PLoS One 7: e41228. doi:10.1371/journal.pone.0041228

Hsu TL et al (2009) Profiling carbohydrate-receptor interaction with recombinant innate immunity receptor-Fc fusion proteins. J Biol Chem 284:34479–34489. doi:10.1074/jbc.M109.065961

Huang X et al (2001) Cloning and characterization of a novel ITIM containing lectin-like immunoreceptor LLIR and its two transmembrane region deletion variants. Biochem Biophys Res Commun 281:131–140. doi:10.1006/bbrc.2001.4322

Hunger RE et al (2004) Langerhans cells utilize CD1a and langerin to efficiently present nonpeptide antigens to T cells. J Clin Invest 113:701–708. doi:10.1172/JCI19655

Hutter J et al (2014) Role of the C-type lectin receptors MCL and DCIR in experimental colitis. PLoS One 9:e103281. doi:10.1371/journal.pone.0103281

Ishikawa E et al (2009) Direct recognition of the mycobacterial glycolipid, trehalose dimycolate, by C-type lectin Mincle. J Exp Med 206:2879–2888. doi:10.1084/jem.20091750

Jin W et al (2014) DC-SIGN plays a stronger role than DCIR in mediating HIV-1 capture and transfer. Virology 458–459:83–92. doi:10.1016/j.virol.2014.04.016

Kanazawa N et al (2002) DCIR acts as an inhibitory receptor depending on its immunoreceptor tyrosine-based inhibitory motif. J Investig Dermatol 118:261–266. doi:10.1046/j.0022-202x.2001.01633.x

Kaneko Y, Nimmerjahn F, Ravetch JV (2006) Anti-inflammatory activity of immunoglobulin G resulting from Fc sialylation. Science 313:670–673. doi:10.1126/science.1129594

Kawai T, Akira S (2008) Toll-like receptor and RIG-I-like receptor signaling. Ann N Y Acad Sci 1143:1–20. doi:10.1196/annals.1443.020

Kawasaki T, Kawai T (2014) Toll-like receptor signaling pathways. Front Immunol 5:461. doi:10.3389/fimmu.2014.00461

Kerscher B, Willment JA, Brown GD (2013) The Dectin-2 family of C-type lectin-like receptors: an update. Int Immunol 25:271–277. doi:10.1093/intimm/dxt006

Klechevsky E et al (2010) Cross-priming CD8+ T cells by targeting antigens to human dendritic cells through DCIR. Blood 116:1685–1697. doi:10.1182/blood-2010-01-264960

Lambert AA, Gilbert C, Richard M, Beaulieu AD, Tremblay MJ (2008) The C-type lectin surface receptor DCIR acts as a new attachment factor for HIV-1 in dendritic cells and contributes to trans- and cis-infection pathways. Blood 112:1299–1307. doi:10.1182/blood-2008-01-136473

Lambert AA, Imbeault M, Gilbert C, Tremblay MJ (2010) HIV-1 induces DCIR expression in CD4+ T cells. PLoS Pathog 6:e1001188. doi:10.1371/journal.ppat.1001188

Lambert AA, Barabe F, Gilbert C, Tremblay MJ (2011) DCIR-mediated enhancement of HIV-1 infection requires the ITIM-associated signal transduction pathway. Blood 117:6589–6599. doi:10.1182/blood-2011-01-331363

Lee RT et al (2011) Survey of immune-related, mannose/fucose-binding C-type lectin receptors reveals widely divergent sugar-binding specificities. Glycobiology 21:512–520. doi:10.1093/glycob/cwq193

Long KM et al (2013) Dendritic cell immunoreceptor regulates Chikungunya virus pathogenesis in mice. J Virol 87:5697–5706. doi:10.1128/JVI.01611-12

Lorentzen JC et al (2007) Association of arthritis with a gene complex encoding C-type lectin-like receptors. Arthritis Rheum 56:2620–2632. doi:10.1002/art.22813

Maglinao M, Klopfleisch R, Seeberger PH, Lepenies B (2013) The C-type lectin receptor DCIR is crucial for the development of experimental cerebral malaria. J Immunol 191:2551–2559. doi:10.4049/jimmunol.1203451

Maruhashi T et al (2015) DCIR maintains bone homeostasis by regulating IFN-gamma production in T cells. J Immunol 194:5681–5691. doi:10.4049/jimmunol.1500273

Massoud AH et al (2012) Intravenous immunoglobulin attenuates airway inflammation through induction of forkhead box protein 3-positive regulatory T cells. J Allergy Clin Immunol 129:1656–1665. doi:10.1016/j.jaci.2012.02.050, e1653

Massoud AH et al (2014) Dendritic cell immunoreceptor: a novel receptor for intravenous immunoglobulin mediates induction of regulatory T cells. J Allergy Clin Immunol 133:853–863. doi:10.1016/j.jaci.2013.09.029, e855

Meyer-Wentrup F et al (2008) Targeting DCIR on human plasmacytoid dendritic cells results in antigen presentation and inhibits IFN-alpha production. Blood 111:4245–4253. doi:10.1182/blood-2007-03-081398

Meyer-Wentrup F et al (2009) DCIR is endocytosed into human dendritic cells and inhibits TLR8-mediated cytokine production. J Leukoc Biol 85:518–525. doi:10.1189/jlb.0608352

Ravetch JV, Lanier LL (2000) Immune inhibitory receptors. Science 290:84–89

Richard M, Veilleux P, Rouleau M, Paquin R, Beaulieu AD (2002) The expression pattern of the ITIM-bearing lectin CLECSF6 in neutrophils suggests a key role in the control of inflammation. J Leukoc Biol 71:871–880

Richard M, Thibault N, Veilleux P, Gareau-Page G, Beaulieu AD (2006) Granulocyte macrophage-colony stimulating factor reduces the affinity of SHP-2 for the ITIM of CLECSF6 in neutrophils: a new mechanism of action for SHP-2. Mol Immunol 43:1716–1721. doi:10.1016/j.molimm.2005.10.006

Seite JF et al (2010) IVIg modulates BCR signaling through CD22 and promotes apoptosis in mature human B lymphocytes. Blood 116:1698–1704. doi:10.1182/blood-2009-12-261461

Seno A et al (2015) Exacerbation of experimental autoimmune encephalomyelitis in mice deficient for DCIR, an inhibitory C-type lectin receptor. Exp Anim/Japan Assoc Lab Anim Sci 64(2):109–19

Turville SG et al (2001) HIV gp120 receptors on human dendritic cells. Blood 98:2482–2488

Turville SG et al (2002) Diversity of receptors binding HIV on dendritic cell subsets. Nat Immunol 3:975–983. doi:10.1038/ni841

van Vliet SJ, den Dunnen J, Gringhuis SI, Geijtenbeek TB, van Kooyk Y (2007) Innate signaling and regulation of dendritic cell immunity. Curr Opin Immunol 19:435–440. doi:10.1016/j.coi.2007.05.006

Zhang JG et al (2012) The dendritic cell receptor Clec9A binds damaged cells via exposed actin filaments. Immunity 36:646–657. doi:10.1016/j.immuni.2012.03.009

Chapter 8
Regulation and Function of the Inhibitory C-Type Lectin Clec12A/MICL

Konstantin Neumann and Jürgen Ruland

Abstract The C-type lectin domain family 12 member A (Clec12A/MICL) is an inhibitory receptor encoded in the Dectin-1 cluster. It is widely expressed in myeloid cells and was identified as a specific marker for cancer stem cells in acute myeloid leukemia. Clec12A possesses an immunoreceptor tyrosine-based inhibition motif (ITIM), which can counteract activating signals from immunoreceptor tyrosine-based activation motifs (ITAMs). The receptor can sense necrotic cell death and limit ITAM-coupled receptor-induced inflammation in response to cell death or tissue damage. One Clec12A agonist released from dead cells was identified as uric acid in its crystallized form. Clec12A limits ITAM-dependent respiratory burst and IL-8 release from neutrophils in response to uric acid crystal binding but does not interfere with crystal-induced inflammasome activation. This review discusses recent insights into the regulation and biological functions of Clec12A.

Keywords Clec12A • MICL • CLL1 • CLL-1 • DCAL2 • DCAL-2 • KLRL1 • MSU • Sodium urate • Uric acid • Crystals • ITAM • ITIM • Sterile inflammation

8.1 Structure and Expression Pattern of Clec12A

The C-type lectin domain family 12 member A (Clec12A) is alternatively called myeloid inhibitory C-type lectin-like receptor (MICL) (Marshall et al. 2004), C-type lectin-like molecule 1 (CLL-1) (Bakker et al. 2004), dendritic cell-associated lectin 2 (DCAL-2) (Chen et al. 2006), or KLRL1 (Han et al. 2004). The Clec12A gene is located in the Dectin-1 cluster of C-type lectin receptors (CLRs). Similar to other members encoded in this cluster, it is a type II transmembrane protein with an extracellular C-type lectin domain (CTLD), which lacks the amino acid motif of classical CTLDs required for Ca^{2+} complexation and carbohydrate binding. Clec12A and the closely related orphan CLR Clec12B (Hoffmann

K. Neumann (✉) • J. Ruland (✉)
Institut für Klinische Chemie und Pathobiochemie, Klinikum rechts der Isar, Technische Universität München, 81675 Munich, Germany
e-mail: konstantin.neumann@lrz.tu-muenchen.de; j.ruland@tum.de

S. Yamasaki (ed.), *C-Type Lectin Receptors in Immunity*,
DOI 10.1007/978-4-431-56015-9_8

et al. 2007) are the only inhibitory receptors encoded in the Dectin-1 cluster. Their intracellular domains contain an immunoreceptor tyrosine-based inhibition motif (ITIM). Receptors harboring ITIMs generally antagonize activating immune receptors harboring immunoreceptor tyrosine-based activation motifs (ITAMs) (Reth 1989). While ITAM-coupled receptors signal via tyrosine kinases like Syk, ITIMs upon phosphorylation recruit the lipid phosphatase SHIP or protein tyrosine phosphatases SHP-1 or SHP-2 (Lanier 2003). Clec12A and B were shown to associate both with SHP-1 and SHP-2 (Han et al. 2004; Marshall et al. 2004; Pyz et al. 2008; Hoffmann et al. 2007). Several isoforms of Clec12A exist (Gerhard et al. 2004; Marshall et al. 2004). Notably, one isoform has an additional tyrosine-based signaling motif that might transduce ITIM-independent signals. Another isoform lacks the transmembrane domain, which potentially gives rise to an intracellular protein, as Clec12A is a type II transmembrane protein. A functional relevance for these isoforms is, however, unknown to date.

Clec12A is mainly expressed on myeloid cells including monocytes, granulocytes (neutrophils, eosinophils, basophils), both myeloid and plasmacytoid dendritic cells, macrophages, and nearly absent in lymphocytes and NK cells (Marshall et al. 2004, 2006; Bakker et al. 2004). Intriguingly, while the expression level of Clec12A does not change during the differentiation of human monocytes into macrophages, the amount of Clec12A glycosylation was shown to increase during macrophage differentiation (Marshall et al. 2006). However, similar effects have not been reported for murine Clec12A. Whether the differential glycosylation of Clec12a has functional significance, e.g., by changing the affinity to its ligands or to other receptors on the cell surface, remains to be determined.

Human Clec12A was transcriptionally downregulated after TLR stimulation in vitro, and human granulocytes and monocytes recruited to the site of acute inflammation had reduced expression of Clec12A (Marshall et al. 2006). Similarly, mouse Clec12A was downregulated on myeloid cells after TLR stimulation (Pyz et al. 2008) and on cells recruited to the peritoneum during peritonitis induced by uric acid crystals or thioglycolate broth (Heng et al. 2008). These findings indicate a role for Clec12A in limiting immune responses in the absence of microbes or danger. However, Clec12A-deficient mice do not develop spontaneous autoimmune or autoinflammatory syndromes indicating that Clec12A is largely dispensable for immune homeostasis (Neumann et al. 2014).

8.2 Clec12A as a Dead Cell Receptor That Recognizes Uric Acid Crystals

Most of the activating CLRs encoded in the Dectin-1 and the Dectin-2 clusters are pattern recognition receptors (PRRs) that either recognize pathogen-associated molecular patterns (PAMPs), damage-associated molecular patterns (DAMPs), or both (Iborra and Sancho 2014). ITIM-containing immunoreceptors often recognize endogenous ligands to prevent autoimmunity, e.g., inhibitory NK cell receptors recognize MHC class I (MHC I) molecules to prevent NK cell activation by healthy host cells. Upon virus-induced downregulation of MHC I, ITAM-coupled activating NK cell receptors are no longer blocked by MHC I-specific ITIM receptors and the host cell is attacked (Long 2008). Given the homology of Clec12A to inhibitory NK cell killer cell lectin-like receptors, it was initially speculated that Clec12A could be an inhibitory NK cell receptor (Han et al. 2004). Yet, NK cells do not seem to express Clec12A under homeostatic conditions. Recombinant mouse Clec12A (Clec12A-Fc fusion protein) still bound to cells from diverse primary mouse tissues indicating that Clec12A can sense endogenous ligands. Furthermore, reporter cells expressing mouse Clec12A fused to the intracellular signaling domain of CD3ζ responded to stimulation with mouse tissues (Pyz et al. 2008) further supporting the notion that Clec12A binds to endogenous agonists. By searching for cell types and conditions that expose the ligand of Clec12A, we observed that in all tested tissues, murine or human Clec12A selectively bound to dead cells that had lost the integrity of the plasma membrane. In addition, reporter cells expressing either human or mouse Clec12A fused to the intracellular domain of CD3ζ did not respond to viable cells. However, when certain cell types were killed by freeze-thaw cycles, they were able to activate Clec12A reporter cells (Neumann et al. 2014). These findings indicate that Clec12A is a sensor of cell death suggesting that Clec12A could potentially inhibit cell death-induced immune cell activation by ITAM-coupled PRRs like Mincle (Clec4E), which recognizes the intracellular protein SAP130 (Yamasaki et al. 2008), and DNGR-1 (Clec9A), which recognizes filamentous (F) actin (Zhang et al. 2012; Ahrens et al. 2012; Sancho et al. 2009). Indeed, we found enhanced sterile inflammatory responses in Clec12A-deficient mice in a model of low-dose X-ray irradiation, in which inflammation is sensitive to an antibody that blocks Mincle (Clec4E) and MCL (Clec4D) (Yamasaki et al. 2008; Miyake et al. 2013) indicating that Clec12A can indeed inhibit Mincle-induced inflammation in vivo. Whether Clec12A also regulates DNGR-1-dependent immune responses is unclear at this point.

Since Clec12A reporter cell activation by dead cells was sensitive to pretreatment of the killed cells with proteases, we anticipated that Clec12A recognizes proteinaceous ligands. However, we have so far been unable to isolate a proteinaceous ligand from dead cells using recombinant Clec12A, while we could easily purify actin with recombinant DNGR-1 and SAP130 with recombinant Mincle (K. Neumann, unpublished). Therefore, we additionally searched for other danger-associated molecular patterns released from dead cells, including uric acid (Shi et al. 2003). Uric acid is the end product of purine catabolism. The degradation of nucleic acids after cell death leads to a local increase in uric acid concentration that favors its crystallization (Kono et al. 2010).

Allopurinol is a drug, which is used to lower uric acid levels in gout patients by inhibiting the conversion of xanthine to uric acid by xanthine dehydrogenase (XDH). Interestingly, we found that allopurinol inhibited Clec12A-reporter cell activation by dead cells. Moreover, both human and mouse recombinant Clec12A specifically bound to uric acid crystals in vitro. In addition, reporter cells expressing human or mouse Clec12A were specifically activated by uric acid crystals but not by the Dectin-1 ligands zymosan or curdlan or by other crystalline structures like silica, calcium pyrophosphate (CPPD), or polystyrol beads (Neumann et al. 2014). Together, these findings indicate that Clec12A is a uric acid crystal-specific receptor.

Dead cell-mediated activation of Clec12A reporter cells or binding of Clec12A to dead cells was not completely blocked by allopurinol or uricase treatment, respectively (Neumann et al. 2014). Therefore, we speculate that Clec12A may sense additional agonists exposed or released by dead cells. Clec12A may bind to parts of the protein complex containing the Mincle ligand SAP130. Given the high expression of both Clec12A and DNGR-1 on $CD8^+$ dendritic cells in the mouse (Heng et al. 2008; Kasahara and Clark 2012; Lahoud et al. 2009), it is also conceivable that certain forms of cell death generate Clec12A binding sites associated with the DNGR-1 ligand F-actin. Alternatively, ions released from the cytoplasm of dead cells may crystallize with calcium ions in the extracellular space, creating other crystalline structures that may be recognized by Clec12A. These possibilities need to be further explored.

While we identified Clec12A as the first mammalian crystal-recognition receptor, it is interesting to note that C-type lectin-like domains had been identified as crystal-binding domains before. They constitute one of three classes of antifreeze proteins found in cold-water fish that prevent ice crystal formation, thereby lowering the serum freezing temperature (Gronwald et al. 1998; Zelensky and Gready 2005). Thus, the C-type lectin domain is in principle well suited to bind to crystals. The future will show if there are other mammalian C-type lectin domains or CLRs that recognize other crystals like bone mineral or cholesterol.

8.3 Clec12A Regulates Inflammatory Responses

The inflammatory properties of uric acid crystals (in the form of monosodium urate, MSU) have been intensively investigated since they were discovered as the cause of gout (Shi et al. 2010; Mccarty and Hollander 1961). It is now widely accepted that uric acid has inflammatory properties only in its crystalline state and that crystallization is facilitated in vivo by crystal-specific antibodies (Kanevets et al. 2009). The crystals activate the complement cascade leading to the release of inflammatory breakdown products (e.g., C5a) (Hasselbacher 1979; Russell et al. 1982), and complement-deficient animals show reduced inflammation in response to uric acid crystals (Tramontini et al. 2004). In myeloid cells, the crystals activate intracellular NLRP3 inflammasomes, which leads to maturation of IL-1β and a form of inflammatory cell death called pyroptosis. NLRP3-deficient mice have a severely diminished inflammatory response to crystals (Martinon et al. 2006). It has further been shown that in human whole blood, C5a release by complement activation is required to prime the NLRP3 inflammasome (An et al. 2014). Whether specific recognition of the uric acid crystals by myeloid cell surface receptors is required for

NLRP3 inflammasome activation is unknown. It is becoming clear though that most crystals or crystal-like structures activate the NLRP3 inflammasome, indicating that probably there is not one specific crystal-recognition receptor linking to the NLRP3 inflammasome. Still, interaction of crystalline structures with the cell membrane is essential for inflammasome activation (Hari et al. 2014).

Such interactions may occur via recognition of crystal-attached proteins like complement or antibodies that are bound by complement receptors or Fc receptors, respectively, but their contribution has not been investigated. The activation of human neutrophils by pure unopsonized crystals depends on CD11b and/or CD16, as antibodies targeting these receptors block crystal-induced neutrophil activation (Barabe et al. 1998; Ryckman et al. 2004). Whether one of these receptors directly binds to the uric acid crystals is currently unknown. Another receptor implicated in the recognition of uric acid crystals is CD14, a co-receptor for TLR4. CD14 was shown to bind to uric acid crystals and facilitate activation of macrophages by uric acid crystals in the absence of serum, but CD14 is considered to neither have sufficient affinity nor specificity in the presence of other proteins (Scott et al. 2006).

Last, it was shown that uric acid crystals directly interact with lipids in cellular plasma membranes. The binding to cholesterol leads to membrane reorganization (lipid sorting) that induces activation of Syk, presumably by cross-linking ITAM-coupled receptors independent of their specificity (Ng et al. 2008). Alum crystals were also shown to activate Syk in a similar manner, which seems to be required for its adjuvanticity (Flach et al. 2011). Together, there seem to be various direct and indirect modes of recognition for uric acid crystals by innate immune cells. All of them depend on ITAM-coupled receptors and the downstream tyrosine kinase Syk. Therefore, independent of how myeloid cells recognize uric acid crystals in a given situation, their activation by the ITAM-coupled kinase Syk should be sensitive to ITIM receptor-mediated inhibition by Clec12A. In line with this notion, neutrophils deficient of Clec12A showed enhanced Syk-dependent reactive oxygen species (ROS) production in response to uric acid crystals, which was accompanied by enhanced phosphorylation of the NADPH oxidase (Neumann et al. 2014). Similarly, antibody-mediated downregulation of Clec12A on human neutrophils or siRNA-mediated downregulation of Clec12A on a human neutrophil cell line led to enhanced IL-8 production in response to uric acid crystals (Gagne et al. 2013). Most importantly, Clec12A has nonredundant functions in regulating the inflammatory response to uric acid crystals in vivo, as Clec12A-deficient mice showed increased neutrophil influx in response to intraperitoneal injection of uric acid crystal (Neumann et al. 2014). Although crystal-induced activation of the NLRP3 inflammasome and IL-1β secretion is dependent on Syk (Gross et al. 2009; Hara et al. 2013; Shio et al. 2009), blocking of Clec12A on neutrophils did not enhance IL-1β secretion in response to uric acid crystals (Gagne et al. 2013). This suggests that Syk activity may not be the limiting factor for inflammasome activation. The role of Clec12A in regulation of neutrophil activation in response to uric acid crystals is summarized in Fig. 8.1.

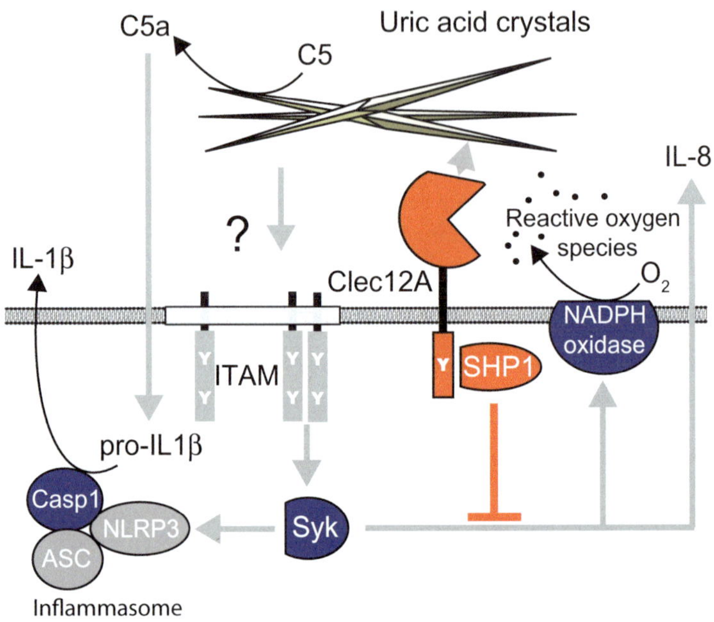

Fig. 8.1 Clec12A limits uric acid crystal-induced respiratory burst (ROS production) and IL-8 but not IL-1β secretion. Uric acid crystals induce activation of the complement system, which leads to cleavage of C5 to C5a and C5b. C5a induces priming of the inflammasome by upregulating transcription of pro-IL-1β. Direct or indirect interaction of the crystals with the plasma membrane induces ITAM-dependent activation of Syk, which is required for NLRP3 inflammasome activation and IL-1β secretion. Syk also activates a signaling cascade leading to NADPH oxidase activation that produces reactive oxygen species and to production of the chemotactic cytokine IL-8. Upon cross-linking to the activated ITAMs, Clec12A is phosphorylated and recruits via its ITIM protein phosphatase SHP-1 or SHP-2, which counteract Syk-induced NADPH oxidase activation and IL-8 production

8.4 Function of Clec12A on Dendritic Cells

While Clec12A-mediated suppression of neutrophil inflammatory responses is genetically established, the role in other cell types is unclear. Clec12A is strongly expressed on various dendritic cell subsets (Lahoud et al. 2009; Kasahara and Clark 2012). As dendritic cells are the major antigen-presenting cells that initiate and shape T cell responses, Clec12A may also regulate adaptive immunity. In this context, it was shown that antibody-mediated targeting of antigen to Clec12A for antigen delivery to dendritic cells enhances humoral immune responses (Lahoud et al. 2009). In the original study, Lahoud et al. used a monoclonal IgG antibody against Clec12A, which probably does not extensively crosslink Clec12A. Since this Clec12A antibody did not lead to activation of dendritic cells, co-injection of adjuvant (LPS or CpG) was required to induce strong humoral immune responses to Clec12A-targeted antigen (Lahoud et al. 2009). In a similar approach, in which the activating receptor DNGR-1 was targeted for antigen delivery, an adaptive humoral immune response was achieved even in the absence of adjuvant (Lahoud et al. 2011). Thus, targeting of an antigen to Clec12A does not activate dendritic cells but seems to be sufficient for antigen internalization, processing, and presentation. When uric acid crystals were co-injected with an antigen in vivo, these crystals significantly enhanced the generation $CD8^+$ T cell responses (Shi et al. 2003), which requires antigen cross-presentation. Since ROS production favors antigen cross-presentation (Hari et al. 2015; Savina et al. 2006) and since Clec12A can in principle regulate ROS production, this process might also be sensitive to Clec12A inhibition.

It should be noted that cross-linking of Clec12A on human dendritic cells with a monoclonal IgM antibody specific for Clec12A induced activation of the mitogen-activated protein kinase (MAPK) pathway and CCR7 upregulation. This signal synergized with the CD40 signaling pathway to enhance cytokine expression (Chen et al. 2006). Whether this is physiologically relevant or a property of this specific cross-linking antibody remains to be determined.

Together, these findings along with the expression of Clec12A on dendritic cells indicate that Clec12A might modulate certain adaptive immune responses during sterile injury or crystal recognition, but additional experiments are required to address these questions.

8.5 Clec12A Expression on Acute Myeloid Leukemia Cells

Clec12A was independently identified in a screen for novel antibody targets for acute myeloid leukemia (AML) (Bakker et al. 2004). Novel therapies are required for this malignancy as tumor cell progenitors, or leukemic stem cells confer resistance to chemotherapy (Dick 2005). Therefore, antibodies targeting surface markers of leukemic stem cells could be useful to eradicate these cells. A preclinical study using the surface marker CD44 already showed a promising outcome (Jin et al. 2006). Interestingly, Clec12A is expressed in the malignant $CD34^+CD38^-$ stem cell compartment in the majority of $CD34^+$ AML patients, while it is absent on normal $CD34^+CD38^-$ resting bone marrow cells (van Rhenen et al. 2007). Indeed, it was recently shown that Clec12A could serve as a diagnostic marker to quantify minimal residual disease (Roug et al. 2014; Larsen et al. 2012). Research on the possible use of Clec12A as a therapeutic target for immunotherapies against AML is currently conducted both in academia and industry (Zhao et al. 2010; Hangalapura et al. 2014; Noordhuis et al. 2010; Lu et al. 2014). Whether Clec12A is simply a marker for AML cells or whether this receptor has also a functional role in AML biology is currently unclear. Yet, it is conceivable that an inhibitory signal delivered by Clec12A within the right microenvironment could keep leukemic stem cells in a quiescent state.

8.6 Conclusion

Work over the last few years identified Clec12A as an important inhibitory pattern recognition receptor that regulates inflammatory responses to cell death. Such reactions are important for the maintenance of homeostasis and tissue regeneration, but inflammation can also cause further tissue damage that can even lead to organ failure (Arslan et al. 2010; Imaeda et al. 2009; Jiang et al. 2005). Because cell damage is often induced during infection, CLRs that sense dead cells may also play critical roles in immune response against pathogens that have few or successfully hide their PAMPs. The activating dead cell receptor DNGR-1 has already been shown to be essential for protective immune responses against certain viruses (Iborra et al. 2012; Zelenay et al. 2012). The circumstances in which Clec12A is essential to either prevent excessive inflammation that is harmful to the host or regulate protective immune responses against pathogens are currently unknown. The further analysis of Clec12A-deficient mice will identify these circumstances and broaden our understanding of the immunological consequences of cell death recognition. During production of this article a recent study has shown that Clec12A indeed regulates immune responses against pathogens, as Clec12A-deficient mice were shown to be more susceptible to *Salmonella* infection (Begun et al. 2015).

Conflict of Interest The authors declare no conflict of interest.

Acknowledgments We thank Paul-Albert Koenig for helpful discussions. This work was supported by SFB grants from Deutsche Forschungsgemeinschaft and an ERC Advanced Grant to J.R.

References

Ahrens S, Zelenay S, Sancho D, Hanc P, Kjaer S, Feest C, Fletcher G, Durkin C, Postigo A, Skehel M, Batista F, Thompson B, Way M, Reis e Sousa C, Schulz O (2012) F-actin is an evolutionarily conserved damage-associated molecular pattern recognized by DNGR-1, a receptor for dead cells. Immunity 36(4):635–645

An LL, Mehta P, Xu L, Turman S, Reimer T, Naiman B, Connor J, Sanjuan M, Kolbeck R, Fung M (2014) Complement C5a potentiates uric acid crystal-induced IL-1beta production. Eur J Immunol 44(12):3669–3679. doi:10.1002/eji.201444560

Arslan F, Smeets MB, O'Neill LAJ, Keogh B, McGuirk P, Timmers L, Tersteeg C, Hoefer IE, Doevendans PA, Pasterkamp G, de Kleijn DPV (2010) Myocardial ischemia/reperfusion injury is mediated by leukocytic toll-like receptor-2 and reduced by systemic administration of a novel anti-toll-like receptor-2 antibody. Circulation 121(1):80–90. doi:10.1161/Circulationaha.109.880187

Bakker AB, van den Oudenrijn S, Bakker AQ, Feller N, van Meijer M, Bia JA, Jongeneelen MA, Visser TJ, Bijl N, Geuijen CA, Marissen WE, Radosevic K, Throsby M, Schuurhuis GJ, Ossenkoppele GJ, de Kruif J, Goudsmit J, Kruisbeek AM (2004) C-type lectin-like molecule-1: a novel myeloid cell surface marker associated with acute myeloid leukemia. Cancer Res 64(22):8443–8450. doi:10.1158/0008-5472.CAN-04-1659

Barabe F, Gilbert C, Liao N, Bourgoin SG, Naccache PH (1998) Crystal-induced neutrophil activation VI. Involvement of FcgammaRIIIB (CD16) and CD11b in response to inflammatory microcrystals. FASEB J 12(2):209–220

Begun J, Lassen KG, Jijon HB, Baxt LA, Goel G, Heath RJ, Ng A, Tam JM, Kuo SY, Villablanca EJ, Fagbami L, Oosting M, Kumar V, Schenone M, Carr SA, Joosten LA, Vyas JM, Daly MJ, Netea MG, Brown GD, Wijmenga C, Xavier RJ (2015) Integrated genomics of Crohn's disease risk variant identifies a role for CLEC12A in antibacterial autophagy. Cell reports 11 (12):1905–1918. doi:10.1016/j.celrep.2015.05.045

Chen CH, Floyd H, Olson NE, Magaletti D, Li C, Draves K, Clark EA (2006) Dendritic-cell-associated C-type lectin 2 (DCAL-2) alters dendritic-cell maturation and cytokine production. Blood 107(4):1459–1467. doi:10.1182/blood-2005-08-3264

Dick JE (2005) Acute myeloid leukemia stem cells. Ann N Y Acad Sci 1044:1–5. doi:10.1196/annals.1349.001

Flach TL, Ng G, Hari A, Desrosiers MD, Zhang P, Ward SM, Seamone ME, Vilaysane A, Mucsi AD, Fong Y, Prenner E, Ling CC, Tschopp J, Muruve DA, Amrein MW, Shi Y (2011) Alum interaction with dendritic cell membrane lipids is essential for its adjuvanticity. Nat Med 17 (4):479–U121. doi:10.1038/Nm.2306

Gagne V, Marois L, Levesque JM, Galarneau H, Lahoud MH, Caminschi I, Naccache PH, Tessier P, Fernandes MJ (2013) Modulation of monosodium urate crystal-induced responses in neutrophils by the myeloid inhibitory C-type lectin-like receptor: potential therapeutic implications. Arthritis Res Ther 15(4):R73. doi:10.1186/ar4250

Gerhard DS, Wagner L, Feingold EA, Shenmen CM, Grouse LH, Schuler G, Project Team MGC et al (2004) The status, quality, and expansion of the NIH full-length cDNA project: the Mammalian Gene Collection (MGC). Genome Res 14(10B):2121–2127. doi:10.1101/gr.2596504

Gronwald W, Loewen MC, Lix B, Daugulis AJ, Sonnichsen FD, Davies PL, Sykes BD (1998) The solution structure of type II antifreeze protein reveals a new member of the lectin family. Biochemistry 37(14):4712–4721. doi:10.1021/bi972788c

Gross O, Poeck H, Bscheider M, Dostert C, Hannesschlager N, Endres S, Hartmann G, Tardivel A, Schweighoffer E, Tybulewicz V, Mocsai A, Tschopp J, Ruland J (2009) Syk kinase signalling couples to the Nlrp3 inflammasome for anti-fungal host defence. Nature 459(7245):433–U149. doi:10.1038/Nature07965

Han Y, Zhang M, Li N, Chen T, Zhang Y, Wan T, Cao X (2004) KLRL1, a novel killer cell lectin like receptor, inhibits natural killer cell cytotoxicity. Blood 104(9):2858–2866. doi:10.1182/blood-2004-03-0878

Hangalapura BN, van Loo PF, Leenders M, de Kruif J, Bakker AB, Dolstra H (2014) A novel Clec12axcd3 bispecific antibody efficiently induces T-cell mediated lysis of Clec12a+Aml blasts. Haematologica 99:294

Hara H, Tsuchiya K, Kawamura I, Fang RD, Hernandez-Cuellar E, Shen YN, Mizuguchi J, Schweighoffer E, Tybulewicz V, Mitsuyama M (2013) Phosphorylation of the adaptor ASC acts as a molecular switch that controls the formation of speck-like aggregates and inflammasome activity. Nat Immunol 14(12):1247. doi:10.1038/Ni.2749

Hari A, Zhang Y, Tu Z, Detampel P, Stenner M, Ganguly A, Shi Y (2014) Activation of NLRP3 inflammasome by crystalline structures via cell surface contact. Sci Rep 4:7281. doi:10.1038/srep07281

Hari A, Ganguly A, Mu LB, Davis SP, Stenner MD, Lam R, Munro F, Namet I, Alghamdi E, Furstenhaupt T, Dong W, Detampel P, Shen LJ, Amrein MW, Yates RM, Shi Y (2015) Redirecting soluble antigen for MHC class I cross-presentation during phagocytosis. Eur J Immunol 45(2):383–395. doi:10.1002/Eji.201445156

Hasselbacher P (1979) C3-activation by monosodium urate monohydrate and other crystalline material. Arthritis Rheum 22(6):571–578. doi:10.1002/Art.1780220603

Heng TS, Painter MW, Immunological Genome Project C (2008) The immunological genome project: networks of gene expression in immune cells. Nat Immunol 9(10):1091–1094. doi:10.1038/ni1008-1091

Hoffmann SC, Schellack C, Textor S, Konold S, Schmitz D, Cerwenka A, Pflanz S, Watzl C (2007) Identification of CLEC12B, an inhibitory receptor on myeloid cells. J Biol Chem 282(31):22370–22375. doi:10.1074/jbc.M704250200

Iborra S, Sancho D (2014) Signalling versatility following self and non-self sensing by myeloid C-type lectin receptors. Immunobiology. doi:10.1016/j.imbio.2014.09.013

Iborra S, Izquierdo HM, Martinez-Lopez M, Blanco-Menendez N, Reis e Sousa C, Sancho D (2012) The DC receptor DNGR-1 mediates cross-priming of CTLs during vaccinia virus infection in mice. J Clin Invest 122(5):1628–1643. doi:10.1172/JCI60660

Imaeda AB, Watanabe A, Sohail MA, Mahmood S, Mohamadnejad M, Sutterwala FS, Flavell RA, Mehall WZ (2009) Acetaminophen-induced hepatotoxicity in mice is dependent on Tlr9 and the Nalp3 inflammasome. J Clin Invest 119(2):305–314. doi:10.1172/Jci35958

Jiang DH, Liang JR, Fan J, Yu S, Chen SP, Luo Y, Prestwich GD, Mascarenhas MM, Garg HG, Quinn DA, Homer RJ, Goldstein DR, Bucala R, Lee PJ, Medzhitov R, Noble PW (2005) Regulation of lung injury and repair by toll-like receptors and hyaluronan. Nat Med 11(11):1173–1179. doi:10.1038/Nm1315

Jin L, Hope KJ, Zhai Q, Smadja-Joffe F, Dick JE (2006) Targeting of CD44 eradicates human acute myeloid leukemic stem cells. Nat Med 12(10):1167–1174. doi:10.1038/nm1483

Kanevets U, Sharma K, Dresser K, Shi Y (2009) A role of IgM antibodies in monosodium urate crystal formation and associated adjuvanticity. J Immunol 182(4):1912–1918. doi:10.4049/Jimmunol.0803777

Kasahara S, Clark EA (2012) Dendritic cell-associated lectin 2 (DCAL2) defines a distinct CD8alpha- dendritic cell subset. J Leukoc Biol 91(3):437–448. doi:10.1189/jlb.0711384

Kono H, Chen CJ, Ontiveros F, Rock KL (2010) Uric acid promotes an acute inflammatory response to sterile cell death in mice. J Clin Invest 120(6):1939–1949

Lahoud MH, Proietto AI, Ahmet F, Kitsoulis S, Eidsmo L, Wu L, Sathe P, Pietersz S, Chang HW, Walker ID, Maraskovsky E, Braley H, Lew AM, Wright MD, Heath WR, Shortman K, Caminschi I (2009) The C-type lectin Clec12A present on mouse and human dendritic cells can serve as a target for antigen delivery and enhancement of antibody responses. J Immunol 182(12):7587–7594. doi:10.4049/Jimmunol.0900464

Lahoud MH, Ahmet F, Kitsoulis S, Wan SS, Vremec D, Lee CN, Phipson B, Shi W, Smyth GK, Lew AM, Kato Y, Mueller SN, Davey GM, Heath WR, Shortman K, Caminschi I (2011) Targeting antigen to mouse dendritic cells via Clec9A induces potent CD4 T cell responses biased toward a follicular helper phenotype. J Immunol 187(2):842–850. doi:10.4049/Jimmunol.1101176

Lanier LL (2003) Natural killer cell receptor signaling. Curr Opin Immunol 15(3):308–314

Larsen HO, Roug AS, Just T, Brown GD, Hokland P (2012) Expression of the hMICL in acute myeloid leukemia-a highly reliable disease marker at diagnosis and during follow-up. Cytometry B Clin Cytom 82(1):3–8. doi:10.1002/cyto.b.20614

Long EO (2008) Negative signaling by inhibitory receptors: the NK cell paradigm. Immunol Rev 224:70–84. doi:10.1111/j.1600-065X.2008.00660.x

Lu H, Zhou Q, Deshmukh V, Phull H, Ma J, Tardif V, Naik RR, Bouvard C, Zhang Y, Choi S, Lawson BR, Zhu S, Kim CH, Schultz PG (2014) Targeting human C-type lectin-like molecule-1 (CLL1) with a bispecific antibody for immunotherapy of acute myeloid leukemia. Angew Chemie 53(37):9841–9845. doi:10.1002/anie.201405353

Marshall AS, Willment JA, Lin HH, Williams DL, Gordon S, Brown GD (2004) Identification and characterization of a novel human myeloid inhibitory C-type lectin-like receptor (MICL) that is predominantly expressed on granulocytes and monocytes. J Biol Chem 279(15):14792–14802. doi:10.1074/jbc.M313127200

Marshall ASJ, Willment JA, Pyz E, Dennehy KM, Reid DM, Dri P, Gordon S, Wong SYC, Brown GD (2006) Human MICL (CLEC12A) is differentially glycosylated and is down-regulated following cellular activation. Eur J Immunol 36(8):2159–2169. doi:10.1002/Eji.200535628

Martinon F, Petrilli V, Mayor A, Tardivel A, Tschopp J (2006) Gout-associated uric acid crystals activate the NALP3 inflammasome. Nature 440(7081):237–241. doi:10.1038/nature04516

Mccarty DJ, Hollander JL (1961) Identification of urate crystals in gouty synovial fluid. Ann Intern Med 54(3):452

Miyake Y, Toyonaga K, Mori D, Kakuta S, Hoshino Y, Oyamada A, Yamada H, Ono KI, Suyama M, Iwakura Y, Yoshikai Y, Yamasaki S (2013) C-type lectin MCL is an FcR gamma-coupled receptor that mediates the adjuvanticity of mycobacterial cord factor. Immunity 38(5):1050–1062. doi:10.1016/J.Immuni.2013.03.010

Neumann K, Castineiras-Vilarino M, Hockendorf U, Hannesschlager N, Lemeer S, Kupka D, Meyermann S, Lech M, Anders HJ, Kuster B, Busch DH, Gewies A, Naumann R, Gross O, Ruland J (2014) Clec12a is an inhibitory receptor for uric acid crystals that regulates inflammation in response to cell death. Immunity 40(3):389–399. doi:10.1016/J.Immuni.2013.12.015

Ng G, Sharma K, Ward SM, Desrosiers MD, Stephens LA, Schoel WM, Li T, Lowell CA, Ling CC, Amrein MW, Shi Y (2008) Receptor-independent, direct membrane binding leads to cell-surface lipid sorting and Syk kinase activation in dendritic cells. Immunity 29(5):807–818. doi:10.1016/j.immuni.2008.09.013

Noordhuis P, Terwijn M, Rutten AP, Smit L, Ossenkoppele GJ, Schuurhuis GJ (2010) Targeting of CLEC12A in acute myeloid leukemia by antibody-drug-conjugates and bispecific CLL-1 x CD3 BiTE antibody. Blood 116(21):1191

Pyz E, Huysamen C, Marshall ASJ, Gordon S, Taylor PR, Brown GD (2008) Characterisation of murine MICL (CLEC12A) and evidence for an endogenous ligand. Eur J Immunol 38 (4):1157–1163. doi:10.1002/Eji.200738057

Reth M (1989) Antigen receptor tail clue. Nature 338(6214):383–384. doi:10.1038/338383b0

Roug AS, Larsen HO, Nederby L, Just T, Brown G, Nyvold CG, Ommen HB, Hokland P (2014) hMICL and CD123 in combination with a CD45/CD34/CD117 backbone – a universal marker combination for the detection of minimal residual disease in acute myeloid leukaemia. Br J Haematol 164(2):212–222. doi:10.1111/bjh.12614

Russell IJ, Mansen C, Kolb LM, Kolb WP (1982) Activation of the 5th component of human-complement (C5) induced by monosodium urate crystals – C5 convertase assembly on the crystal-surface. Clin Immunol Immunopathol 24(2):239–250. doi:10.1016/0090-1229(82)90235-5

Ryckman C, Gilbert C, de Medicis R, Lussier A, Vandal K, Tessier PA (2004) Monosodium urate monohydrate crystals induce the release of the proinflammatory protein S100A8/A9 from neutrophils. J Leukoc Biol 76(2):433–440. doi:10.1189/jlb.0603294

Sancho D, Joffre OP, Keller AM, Rogers NC, Martinez D, Hernanz-Falcon P, Rosewell I, Reis e Sousa C (2009) Identification of a dendritic cell receptor that couples sensing of necrosis to immunity. Nature 458(7240):899–903

Savina A, Jancic C, Hugues S, Guermonprez P, Vargas P, Moura IC, Lennon-Dumenil AM, Seabra MC, Raposo G, Amigorena S (2006) NOX2 controls phagosomal pH to regulate antigen processing during crosspresentation by dendritic cells. Cell 126(1):205–218. doi:10.1016/J.Cell.2006.05.035

Scott P, Ma H, Viriyakosol S, Terkeltaub R, Liu-Bryan R (2006) Engagement of CD14 mediates the inflammatory potential of monosodium urate crystals. J Immunol 177(9):6370–6378

Shi Y, Evans JE, Rock KL (2003) Molecular identification of a danger signal that alerts the immune system to dying cells. Nature 425(6957):516–521

Shi Y, Mucsi AD, Ng G (2010) Monosodium urate crystals in inflammation and immunity. Immunol Rev 233(1):203–217. doi:10.1111/j.0105-2896.2009.00851.x

Shio MT, Eisenbarth SC, Savaria M, Vinet AF, Bellemare MJ, Harder KW, Sutterwala FS, Bohle DS, Descoteaux A, Flavell RA, Olivier M (2009) Malarial hemozoin activates the NLRP3 inflammasome through Lyn and Syk Kinases. PLos Pathog 5(8). doi:10.1371/Journal.Ppat.1000559

Tramontini N, Huber C, Liu-Bryan R, Terkeltaub RA, Kilgore KS (2004) Central role of complement membrane attack complex in monosodium urate crystal-induced neutrophilic rabbit knee synovitis. Arthritis Rheum 50(8):2633–2639. doi:10.1002/art.20386

van Rhenen A, van Dongen GA, Kelder A, Rombouts EJ, Feller N, Moshaver B, Stigter-van Walsum M, Zweegman S, Ossenkoppele GJ, Jan Schuurhuis G (2007) The novel AML stem cell associated antigen CLL-1 aids in discrimination between normal and leukemic stem cells. Blood 110(7):2659–2666. doi:10.1182/blood-2007-03-083048

Yamasaki S, Ishikawa E, Sakuma M, Hara H, Ogata K, Saito T (2008) Mincle is an ITAM-coupled activating receptor that senses damaged cells. Nat Immunol 9(10):1179–1188

Zelenay S, Keller AM, Whitney PG, Schraml BU, Deddouche S, Rogers NC, Schulz O, Sancho D, Reis e Sousa C (2012) The dendritic cell receptor DNGR-1 controls endocytic handling of necrotic cell antigens to favor cross-priming of CTLs in virus-infected mice. J Clin Invest 122 (5):1615–1627. doi:10.1172/JCI60644

Zelensky AN, Gready JE (2005) The C-type lectin-like domain superfamily. FEBS J 272 (24):6179–6217. doi:10.1111/j.1742-4658.2005.05031.x

Zhang JG, Czabotar PE, Policheni AN, Caminschi I, Wan SS, Kitsoulis S, Tullett KM, Robin AY, Brammananth R, van Delft MF, Lu J, O'Reilly LA, Josefsson EC, Kile BT, Chin WJ, Mintern JD, Olshina MA, Wong W, Baum J, Wright MD, Huang DC, Mohandas N, Coppel RL, Colman PM, Nicola NA, Shortman K, Lahoud MH (2012) The dendritic cell receptor Clec9A binds damaged cells via exposed actin filaments. Immunity 36(4):646–657

Zhao X, Singh S, Pardoux C, Zhao J, Hsi ED, Abo A, Korver W (2010) Targeting C-type lectin-like molecule-1 for antibody-mediated immunotherapy in acute myeloid leukemia. Haematologica 95(1):71–78. doi:10.3324/haematol.2009.009811

Part IV
Other Receptors and Related Topics

Chapter 9
DC-SIGN in Infection and Immunity

Joris K. Sprokholt, Ronald J. Overmars, and Teunis B.H. Geijtenbeek

Abstract Dendritic cells (DCs) play a central role in the immune system by patrolling peripheral tissues to sample antigens to induce antigen-specific adaptive immune responses in lymphoid tissues. DCs express pattern recognition receptors such as toll-like receptors (TLRs), RIG-I-like receptors (RLRs), and C-type lectin receptors (CLRs) to interact with pathogens for antigen presentation and immune activation. One of the CLRs involved in different processes of DC function is DC-specific intercellular adhesion molecule-3-grabbing non-integrin (DC-SIGN). DC-SIGN recognition of pathogens leads to efficient internalization and processing of antigen for MHC class I and II presentation. In addition, triggering of DC-SIGN induces intracellular signaling that affects immune responses. Although DC-SIGN signaling by itself does not lead to activation of transcription factors such as NFκB, it greatly modifies signaling pathways induced by other receptors, including TLRs, RLRs, and interferon receptors. Modulation of signaling pathways by DC-SIGN tailors adaptive immune responses to different pathogens by driving specific T-helper cell responses. Intriguingly, DC-SIGN signaling depends on the carbohydrate structures present on pathogens as mannose structures induce very different signaling cascades than fucose structures, providing DCs with the plasticity to tailor immune responses to a diverse range of pathogens. Several pathogens however have evolved to subvert DC-SIGN functions for effective infection of DCs and efficient transmission to target cells. In this chapter we will discuss DC-SIGN structure, expression, and DC-SIGN functionality in shaping adaptive immune responses and immunopathogenesis.

Keywords Dendritic cell-specific intercellular adhesion molecule-3-grabbing non-integrin (DC-SIGN) • Pathogen-recognition receptor (PRR) • Adhesion receptor • Antigen receptor • Adaptive immune responses • Innate signaling

---

J.K. Sprokholt • R.J. Overmars • T.B.H. Geijtenbeek (✉)
Academic Medical Center, University of Amsterdam, Amsterdam, The Netherlands
e-mail: t.b.geijtenbeek@amc.uva.nl

S. Yamasaki (ed.), *C-Type Lectin Receptors in Immunity*,
DOI 10.1007/978-4-431-56015-9_9

9.1 Introduction

The primary function of dendritic cells (DCs) is patrolling peripheral tissues, sampling the environment for antigens, presentation of antigen to T cells, and shaping T-cell differentiation for effective immune responses and long-lasting immunity. DCs express an array of pattern recognition receptors (PRRs), such as toll-like receptors (TLRs), RIG-I-like receptors (RLRs), NOD-like receptors (NLRs), and C-type lectin receptors (CLRs) which recognize conserved molecular structures of pathogens. DCs use these receptors for pathogen binding, internalization, antigen presentation, and immune activation. In addition to pathogens, PRRs can recognize endogenous ligands to support key functions of DCs, including cellular contact with other (immune) cells. DC-specific intercellular adhesion molecule-3-grabbing non-integrin (DC-SIGN) (Geijtenbeek and Gringhuis 2009; Švajger et al. 2010), first described in 1992 in placenta (Curtis et al. 1992) and identified in 2000 as a DC-specific receptor (Geijtenbeek et al. 2000a), is a multifaceted CLR involved in primary functions of DCs.

DC-SIGN functions as an adhesion receptor involved in DC migration as well as DC-T-cell interactions. DC-SIGN mediates endothelial rolling for DC emigration from blood into peripheral tissue or secondary lymphoid organs (Geijtenbeek et al. 2000c). Once inside lymphoid tissues, DC-SIGN establishes the initial contact between DCs and T cells in order to scan MHC-peptide complexes by T-cell receptors (TCR) (Geijtenbeek et al. 2000a). Interestingly, DC-SIGN not only recognizes self-ligands but also interacts with different pathogens; and binding by DC-SIGN leads to internalization for antigen processing and presentation on MHC molecules (Engering et al. 2002a; Moris et al. 2004; Smith et al. 2007; Cambi et al. 2009). Pathogen recognition by DC-SIGN shapes adaptive immune responses by modulating signaling pathways induced by other receptors to tailor adaptive immune responses to different pathogens (Hovius et al. 2008; Geijtenbeek et al. 2009; Geijtenbeek and Gringhuis 2009; Gringhuis et al. 2009b, 2014a, b). However, DC-SIGN is also used by a diverse range of pathogens to subvert host immune mechanisms and establish productive infection (Geijtenbeek et al. 2000b; Trumpfheller et al. 2003; Ludwig et al. 2004; Mesman et al. 2014). These aspects of DC-SIGN functionality will be discussed in detail in this chapter.

9.2 DC-SIGN Structure Dictates Function

The molecular structure of DC-SIGN contains the blueprint for ligand specificity, endocytic potential, and signaling capacities. DC-SIGN is a type II transmembrane CLR with an intracellular N-terminus and an extracellular C-terminus (Cambi et al. 2009), similar to dectin-1 and mincle (Matsumoto et al. 1999; Ariizumi et al. 2000). The overall structure of DC-SIGN comprises a cytoplasmic tail followed by a membrane domain, neck region, and carbohydrate recognition domain (CRD).

The cytoplasmic domain contains a di-leucine motif, which facilitates internalization (Engering et al. 2002a; Azad et al. 2008), and a tyrosine residue, which serves as a docking site for adapter proteins (Hodges et al. 2007; Gringhuis et al. 2009b). Several proteins have been proposed as adapter proteins for DC-SIGN, including leukemia-associated Rho guanine nucleotide exchange factor (LARG) and Ras homolog gene family member A (RhoA), but only leukocyte-specific protein 1 (LSP1) directly interacts with the cytoplasmic domain of DC-SIGN and serves as a docking site for additional proteins (Gringhuis et al. 2009b, 2014a; Hodges et al. 2007). The cytoplasmic domain is followed by a transmembrane domain of 18 amino acids and a neck domain, which contains 7.5 repeats of 23 amino acids that form α-helicases. These α-helicases contain hydrophobic residues which direct the CRD away from the cell surface and mediate the spontaneous formation of tetramers, thereby stabilizing DC-SIGN oligomers (Frisont et al. 2003). Additionally, the formation of tetramers enhances DC-SIGN specificity and avidity for repetitive structures (Frisont et al. 2003). Despite the formation of stable tetramers, sufficient flexibility is retained in the structure of DC-SIGN to adapt the CRDs for optimal binding of spatially distributed ligands (Leckband et al. 2011), which could explain why DC-SIGN is capable of binding a wide variety of ligands.

The CRD of DC-SIGN requires Ca^{2+} for structural stability as well as ligand binding (Geijtenbeek et al. 2000a; Cambi et al. 2009) and contains a conserved EPN motif that is essential for ligand binding (Geijtenbeek et al. 2002b; van Die et al. 2003). DC-SIGN is capable of binding mannose and fucose structures and to a lesser extent N-acetylglucosamine (GlcNAc) structures (Appelmelk et al. 2003; van Die et al. 2003; Meyer et al. 2005; Steeghs et al. 2006; van Liempt et al. 2006). Site-directed mutagenesis has revealed that mannose and fucose binding depends on the EPN motif and a valine residue in close proximity of the EPN motif (Geijtenbeek et al. 2002b; van Die et al. 2003). For GlcNAc structures, little is known about the amino acid residues of DC-SIGN involved in binding. In general, DC-SIGN affinity for GlcNAc structures is lower compared to fucose or mannose structures. It has been suggested that DC-SIGN binding to GlcNAc depends on the oligosaccharide backbone on which GlcNAc is presented, which could increase DC-SIGN avidity for GlcNAc structures (Steeghs et al. 2006).

In summary, the cytoplasmic domain of DC-SIGN is important for internalization (i.e., antigen uptake), signaling to shape immune responses, and intracellular trafficking for antigen processing; the neck domain is vital for protein stability and tetramer formation, and the CRD determines ligand specificity.

9.3 Cellular Expression of DC-SIGN Specifies Its Function

Cell-specific expression of DC-SIGN specifies its function, either by peripheral location, expression of relevant (signaling) proteins, or cell-intrinsic characteristics. DC-SIGN is expressed by distinct DC subsets in peripheral and lymphoid tissues but can also be found on certain macrophages (Geijtenbeek et al. 2000a, 2002a; Soilleux et al. 2001; Engering et al. 2002b, 2004; Granelli-Piperno et al. 2012).

Immature monocyte-derived DCs, differentiated using IL-4 and GM-CSF, express high levels of DC-SIGN, and this model has been used extensively to study DC-SIGN (Geijtenbeek et al. 2000a; Kwon et al. 2002; Engering et al. 2002a; Gringhuis et al. 2009b, 2010). However, DC-SIGN$^+$ DCs can also be found in peripheral tissues, including the skin, small and large intestine, and blood. Human skin harbors three distinct DC subsets: conventional CD1c$^+$ DCs, cross-presenting CD141$^+$ DCs, and monocyte-derived CD14$^+$ DCs (Haniffa et al. 2012; Mcgovern et al. 2014). Only CD14$^+$ DCs express DC-SIGN, and targeting this subset in human tissue using DC-SIGN-specific ligands has been investigated as an approach to induce tumor-specific immunity (Joshi et al. 2012; Unger et al. 2012). Although CD14$^+$ DCs are not classified as bona fide cross-presenting DCs, DC-SIGN-mediated endocytosis does lead to cross presentation on MHC class I molecules and activation of CD8$^+$ T cells (Unger et al. 2012; Fehres et al. 2015), highlighting that general statements about cell-intrinsic capacities do not necessarily hold true in the light of receptor-specific processes.

DC-SIGN$^+$ DCs in the large intestine are of clinical relevance in the sexual transmission of human immunodeficiency virus (HIV)-1, and possibly other pathogens, as HIV-1 targets DC-SIGN for efficient transmission to CD4$^+$ T cells (Geijtenbeek et al. 2000c; Kwon et al. 2002; Trumpfheller et al. 2003; Gurney et al. 2005). The same holds true for DC-SIGN$^+$ myeloid DCs in blood, which efficiently transmit HIV-1 to CD4$^+$ T cells (Engering et al. 2002b). In addition, DC-SIGN expression on blood DCs is important for endothelial rolling and DC migration into tissues by interacting with ICAM-2 on endothelial cells (Geijtenbeek et al. 2000c).

DC-SIGN$^+$ DCs in lymph nodes are likely to represent migrated DCs from peripheral tissues as well as resident DCs (Engering et al. 2004). DC-SIGN expression in lymph nodes is involved in establishing initial DC-T-cell interactions via DC-SIGN and intercellular adhesion molecule (ICAM)-3 expressed on naïve T cells (Geijtenbeek et al. 2000a).

In addition to DCs, DC-SIGN is expressed by decidual macrophages and Hofbauer cells in the placenta, where it could be involved in vertical transmission of HIV from mother to newborn (Geijtenbeek et al. 2001; Soilleux et al. 2001). DC-SIGN is also expressed by alveolar macrophages, which could be important during mycobacteria infections of the lung as DC-SIGN binds the mycobacterial cell wall component ManLAM (Soilleux et al. 2002; Geijtenbeek et al. 2003). However, little is known about the role of DC-SIGN in endocytosis, antigen processing, and signaling in macrophages. On a genomic level, DC-SIGN is encoded by *CD209*, located on chromosome 19p13 and regulated by transcription factor PU.1 in combination with either MYB or RUNX3, which confine DC-SIGN expression to DCs and macrophages (Domínguez-Soto et al. 2005). Expression of DC-SIGN is enhanced by IL-4 via IL-4 receptor-mediated JAK-STAT6 signaling (Relloso et al. 2002). This signaling cascade is blocked by type I and II interferons (IFN) as they inhibit phosphorylation and nuclear translocation of STAT6 (Dickensheets et al. 1999; Svajger et al. 2010). This indicates that STAT6 could be an additional transcription factor for *CD209* or that STAT6 indirectly influences PU.1-MYB/RUNX3 transactivation activity. After translation, DC-SIGN is shuttled to the cell surface where it clusters into micro domains of 100–200 nm in diameter. These clusters are important for binding and internalization of virus particles (Cambi et al. 2005), probably by increasing avidity or by creating signaling scaffolds.

In conclusion, DC-SIGN is encoded by *CD209*, regulated by transcription factors PU.1, MYB, RUNX3, and possibly STAT6 and is expressed in micro domains on the cell membrane of different DC as well as macrophage subsets. The majority of studies investigating DC-SIGN are based on DCs, and it is therefore difficult to address whether DC-SIGN function is similar in macrophages without functional studies.

9.4 DC-SIGN Functionality Exposed

9.4.1 DC-SIGN Takes the Lead in Adhesion

DCs are professional antigen-presenting cells, which orchestrate adaptive immune responses by sampling antigens in the periphery and presenting peptides on MHC molecules to $CD4^+$ and $CD8^+$ T cells. One DC can present a vast number of different peptides, and initial DC-T-cell contact allows scanning of MHC-peptide complexes by TCRs. The initial contact is mediated by adhesion receptors, and DC-SIGN facilitates scanning by T cells by binding ICAM-3 on T cells, which is a N-linked glycosylated protein with high mannose oligosaccharides (Fig. 9.1b) (Bleijs et al. 2001; Geijtenbeek et al. 2000a). This transient contact is further stabilized via LFA-1-ICAM-1 interactions facilitating the formation of the immunological synapse for induction of T cell proliferation upon peptide recognition (Geijtenbeek et al. 2000a).

DC progenitors originate in the bone marrow and circulate the blood before migrating into peripheral tissues, where they sample antigen and home to secondary lymphoid organs. The migration from blood to tissue is a complex process involving leukocyte rolling, adhesion to endothelial cells, and transendothelial migration (Springer 1995). Endothelial cells lining blood and lymphatic vessels constitutively express the glycoprotein ICAM-2, which is crucial in leukocyte transendothelial migration. DC-SIGN functions as a DC-specific rolling receptor for ICAM-2 and mediates adhesion of DCs to the endothelium and subsequent transendothelial migration (Fig. 9.1a) (Geijtenbeek et al. 2000c, 2002a). In addition to ICAM molecules,

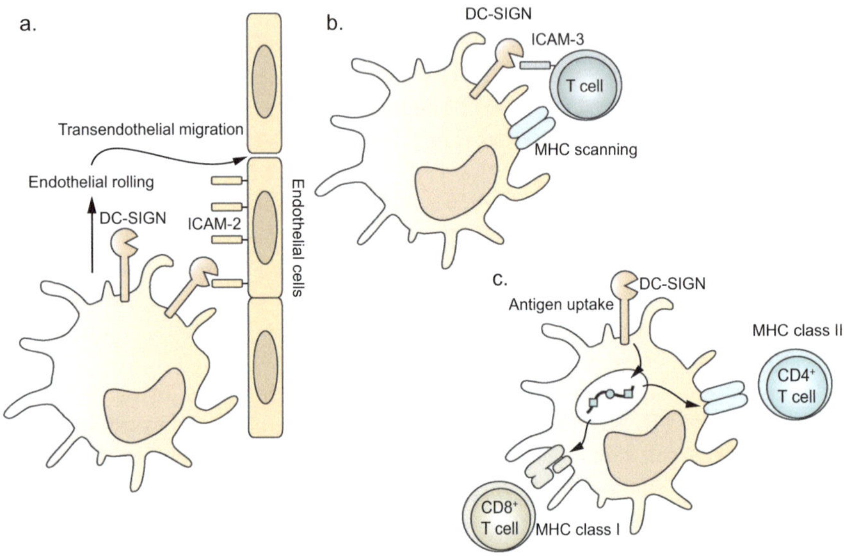

Fig. 9.1 DC-SIGN function in DC migration, T cell responses and antigen processing. (**a**) DC progenitors originate in the bone marrow and circulate the blood before migrating into peripheral tissues to sample antigens. DC migration from blood into tissue is receptor-dependent and requires interactions which can resist the shearing forces of blood circulation (Springer 1995). Endothelial cells express the glycoprotein ICAM-2 and adherence of DC-SIGN to ICAM-2 mediates DC rolling over endothelial surfaces. Once DCs have adhered to endothelial cells via LFA-1, DC-SIGN-ICAM-2 interactions mediate transendothelial migration of DCs to peripheral tissues. After sampling antigen, DCs migrate to secondary lymphoid structures for antigen presentation to T cells. (**b**) In secondary lymphoid structures, DCs present antigens on MHC class I or II molecules to T cells for the induction of adaptive immune responses. Initial DC-T-cell contact is antigen independent to allow scanning of MHC-peptide complexes by TCRs. DC-SIGN mediates this process by adhering to ICAM-3 on T cells. (**c**) DCs are professional antigen-presenting cells, which requires internalization and processing of antigens. Targeting DC-SIGN with either single molecules, pathogens, or carbohydrate-coated liposomes results in rapid internalization into endosomes. The endosomal cargo is then further routed to lysosomal compartments where antigen is processed for MHC class II presentation to $CD4^+$ T cells. Although the precise mechanism is unclear, endosomal cargo is probably also transported to the cytosol where it is loaded on MHC class I molecules for $CD8^+$ T cell presentation

DC-SIGN specifically interacts with β-integrin Mac-1 expressed by neutrophils due to neutrophil-specific glycosylation of Mac-1. This interaction facilitates DC-neutrophil clustering and DC maturation by activated neutrophils (van Gisbergen et al. 2005). Hence, DC-SIGN plays a central role in establishing cell-cell contact between DCs and other cells such as T cells, endothelial cells, and neutrophils to support DC migration, maturation, and induction of adaptive immune responses.

9.4.2 DC-SIGN Has Excellent Presentation Skills

A key characteristic of DCs is endocytosis of antigens and subsequent presentation of peptides on MHC class I molecules for $CD8^+$ cytotoxic T cell responses or on MHC class II molecules for the induction of $CD4^+$ T cell responses. Soluble antigens, intact pathogens, and individual molecules are captured by DC-SIGN and internalized via clathrin-coated pits (Cambi et al. 2009; Engering et al. 2002a). Endocytosis of DC-SIGN depends on the di-leucine motif in the cytoplasmic domain of DC-SIGN (Engering et al. 2002a; Azad et al. 2008). Upon internalization, $DC\text{-}SIGN^+$ endosomes are routed to the endo-lysosomal pathway for antigen processing and subsequent loading of peptides on MHC class II molecules for presentation to $CD4^+$ T cells (Fig. 9.1c) (Geijtenbeek et al. 2002a; Engering et al. 2002a; Schjetne et al. 2002). However, several pathogens that are captured by DC-SIGN are not targeted to lysosomal compartments. Instead, they are retained in (early) endosomes and remain infectious for extended periods of time, as has been shown for HIV and HCV (Geijtenbeek et al. 2000b; Cormier et al. 2004; Ludwig et al. 2004). This indicates that DC-SIGN-dependent internalization can lead to different intracellular routing pathways depending on the ligand. Multivalency of the carbohydrates or carbohydrate structures might be involved in the different internalization routes.

In addition to MHC class II presentation, DCs have the unique capacity to internalize and process antigen for cross-presentation on MHC class I molecules, which can occur via two mechanisms. Either by (1) transporting antigen from endosomes to the cytosol for proteasomal degradation and loading on MHC class I molecules in the ER or (2) by processing and loading of antigen on MHC class I molecules in endo-lysosomal compartments (Joffre et al. 2012). Targeting antigens to DC-SIGN leads to efficient cross presentation and induction of $CD8^+$ T cell responses using either monocyte-derived DCs and $CD14^+$ dermal DCs in human skin explants or in vivo using mice (Tacken et al. 2012; Unger et al. 2012; Fehres et al. 2015). Although it is unclear what the underlying mechanisms are for DC-SIGN-dependent cross presentation, these studies indicate that internalization of pathogens by DC-SIGN could lead to MHC class I presentation and $CD8^+$ T cell responses. Indeed, HIV-1 capture by DC-SIGN can lead to MHC class I presentation under specific circumstances (Moris et al. 2004).

Receptor-mediated endocytosis is a fundamental component in inducing antigen-specific immune responses and forms the basis for antigen uptake, processing, and presentation. DC-SIGN binds with high affinity to a wide variety of pathogens and mediates internalization and processing of pathogens for peptide presentation on MHC molecules. Therefore, DC-SIGN plays a central role in antigen-specific immune responses against numerous pathogens, which could be harnessed in vaccine development.

9.4.3 DC-SIGN Signaling Tailors Immune Responses

DC-SIGN is a multivalent receptor that interacts with mannose, fucose, and GlcNAc structures present on a diverse range of pathogens, including HIV-1, *Mycobacterium tuberculosis*, *Candida albicans*, and *Schistosoma mansoni*. Although DC-SIGN signaling does not directly lead to induction of immune responses, it greatly affects signaling induced by other PRRs such as TLRs and RLRs (Geijtenbeek et al. 2009; Geijtenbeek and Gringhuis 2009; Gringhuis et al. 2007, 2009b; Mesman et al. 2014). Interestingly, DC-SIGN signaling depends on the nature of the pathogen as mannose structures induce very different signaling cascades from fucose structures, which critically alters adaptive T cell responses (Geijtenbeek et al. 2003; Gringhuis et al. 2009b, 2014a, b; Geijtenbeek and Gringhuis 2009).

9.4.3.1 Mannose Signaling

Under homeostatic conditions, DC-SIGN is constitutively associated with a signaling complex consisting of LSP1, kinase suppressor of Ras (KSR)-1, connector enhancer of KSR (CNK), and serine/threonine kinase Raf-1 (Gringhuis et al. 2009b). Activation of DC-SIGN by mannose-containing structures or pathogens, including HIV-1, measles virus (MV), *Candida albicans*, and mycobacteria recruits the additional proteins LARG, RhoA, and GTP-Ras (Gringhuis et al. 2009b). These proteins induce a complex chain of events that ultimately leads to the phosphorylation, and thereby activation, of Raf-1 at Ser338 and Tyr340-341 by p21-activated kinases (PAKs) and Src kinases, respectively (Gringhuis et al. 2007). Activation of Raf-1 subsequently leads to the phosphorylation of NFκB subunit p65 at Ser276 (Gringhuis et al. 2007). A prerequisite for p65 phosphorylation by Raf-1 is prior activation of NFκB by others PRRs, such as TLR4, as DC-SIGN signaling by itself does not induce NFκB activation. Phosphorylated p65 by Raf-1 facilitates complex formation of p65 with CREB-binding protein (CBP) and p300, leading to acetylation of p65 (Chen et al. 2005; Gringhuis et al. 2007). Acetylation of p65 increases its DNA-binding affinity and transcriptional rate and prolongs nuclear activity (Chen et al. 2002). This results in increased transcription of *Il10*, *Il12a*, *Il12b*, and *Il6* genes, which are critical cytokines for skewing T_H differentiation (Fig. 9.2a) (Gringhuis et al. 2009b).

Raf-1 is known for its function in the Raf-MEK-ERK pathway. This pathway is involved in cell fate decisions such as cell growth, differentiation, and survival. However, binding of mannose structures to DC-SIGN leads to specific Raf-1 activation, without triggering ERK1/2 or MEK-1/2 (Wellbrock et al. 2004; Gringhuis et al. 2007). It is unclear why DC-SIGN-dependent Raf-1 activation does not lead to canonical ERK1/2-MEK1/2 signaling, but this is probably caused by the cellular location of Raf-1 during DC-SIGN signaling and the DC-SIGN-specific signaling complex that leads to Raf-1 activation. However, the lack of ERK1/2-MEK1/2 signaling by mannose structures could also be ligand-specific as other ligands of DC-SIGN, such as the tick saliva protein Salp15, induce MEK1/2

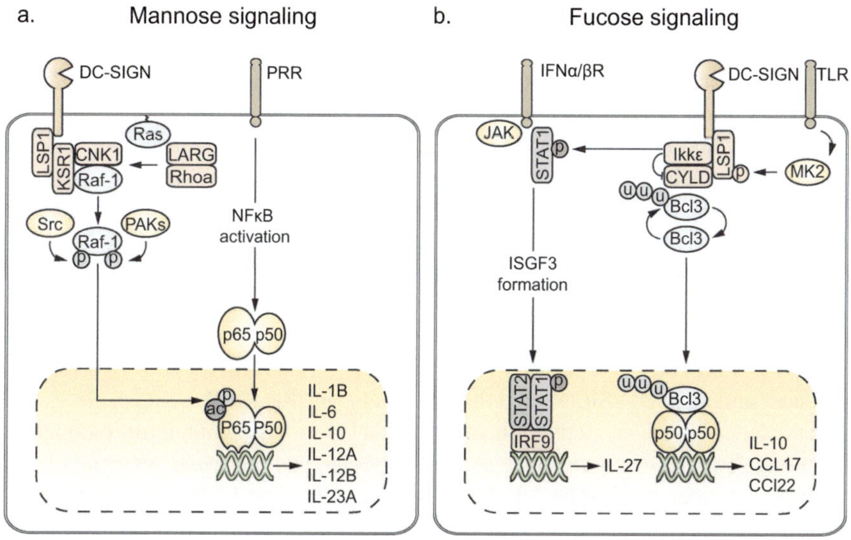

Fig. 9.2 DC-SIGN modulates intracellular signaling pathways. DC-SIGN is unable to induce gene expression on its own, but DC-SIGN signaling greatly affects signaling pathways induced by other receptors with a decisive outcome on adaptive immune responses. (**a**) DC-SIGN is under homeostatic conditions constitutively associated with the adapter molecule LSP1 in combination with KSR1, CNK1, and the kinase Raf-1. Triggering of DC-SIGN by mannose-containing pathogens recruits LARG, Rhoa, and GTP-Ras, which ultimately leads to the phosphorylation and activation of Raf-1 by PAKs and Src kinases. DC-SIGN-activated Raf-1 modulates NFκB signaling of other PRRs by inducing the phosphorylation and acetylation of NFκB subunit p65. This enhances the transcription of *IL-1B*, *IL-6*, *IL-10*, *IL-12A*, *IL-12B*, and *IL-23A* genes. These cytokines are crucial for the induction of specific T cell responses. (**b**) DC-SIGN triggering by fucose-containing pathogens leads to disassociation of KSR1-CNK1-Raf-1 from DC-SIGN-LSP1 and probably provides structural clearance for MK2-dependent phosphorylation of LSP1. MK2 is activated by TLR signaling, including TLR2, TLR3, and TLR4. Phosphorylation of LSP1 recruits the kinase IKKε and the deubiquitinase CYLD, which constantly removes ubiquitin chains from the noncanonical NFκB subunit Bcl3. IKKε and CYLD recruitment to LSP1 results in CYLD phosphorylation, decreasing its deubiquitinase activity and activating Bcl3. Ubiquitinated Bcl3 induces the formation of NFκB subunit p50 homodimers which decreases *IL1B*, *IL-6*, *IL-12A*, *IL-12B*, and *IL-23A* gene transcription and increases *IL-10*, *CCL17*, and *CCL22* transcription, thereby shifting the cytokine responses from a T_H1-skewing profile to a T_H2-inducing profile. In addition to CYLD-Bcl3, IKKε activation modulates IFNR JAK-STAT signaling, which is triggered by TLR-induced type I IFN. IKKε phosphorylates STAT1 at Ser708, leading to the formation of STAT1-STAT2 heterodimers in complex with IRF9 (ISGF3). ISGF3 binds to ISRE-containing genes including IL-27, a crucial cytokine in T_{FH} cell responses

but not ERK1/2 activation (Hovius et al. 2008); and triggering DC-SIGN using recombinant Hepatitis C protein E2 leads to both MEK and ERK activation (Zhao et al. 2013). In addition to distinct DC-SIGN signaling pathways induced by Salp15 and E2, MEK/ERK activation by these proteins could also be the results of triggering additional receptors. Indeed, both E2 and Salpt15 are known to interact with other receptors on DCs (Ogden and Tang 2015; Garg et al. 2010).

Thus, DC-SIGN is under homeostatic conditions in complex with LSP1-KSR1-CNK-Raf-1, and DC-SIGN activation by mannose-containing pathogens leads to the recruitment of LARG and RhoA, which activates kinase Raf-1, leading to p65 phosphorylation and acetylation. Acetylated p65 increases the transcription of *Il10*, *Il12a*, *Il12b*, and *Il6* genes, which are pivotal in adaptive T cell responses. Hence, DC-SIGN signaling plays a central role in innate and adaptive immune responses against mannose-containing pathogens. Notably, fucose binding results in a different composition of the LSP-1 signalosome, which will be discussed below.

9.4.3.2 Fucose Signaling

Mannose-induced DC-SIGN signaling depends on the LSP1-KSR1-CNK-Raf-1 signalosome. However, activation of DC-SIGN by fucose-containing pathogens, including *Schistosoma mansoni*, *Fasciola hepatica,* and *Helicobacter pylori* leads to disassociation of KSR1-CNK-Raf1 without affecting LSP1-DC-SIGN interaction. This allows mitogen-activated protein kinase-activated protein (MAPKAP) kinase 2 (MK2) to phosphorylate LPS1 at Ser204 and Ser252 (Gringhuis et al. 2014a). MK2 activation depends on MAP kinase p38, which is not induced by DC-SIGN signaling and requires additional PRR triggering. In particular, Gringhuis et al. (2014a) have shown that TLR2, TLR3, and TLR4 signaling activates p38-MK2 to phosphorylate LSP1 after fucose triggering of DC-SIGN. Although this study investigated TLRs, numerous cellular processes can activate p38-MK2, including cytokine receptor signaling, osmotic stress, and chemical stress (Roux and Blenis 2004). Whether these processes also lead to LSP1 phosphorylation by fucose-triggered DC-SIGN has not been investigated.

Phosphorylation of LSP1 at Ser252 results in recruitment of IkappaB kinase ε (IKKε) and the deubiquitinase CYLD to DC-SIGN. Under homeostatic conditions, CYLD prevents noncanonical NFκB subunit Bcl3 translocation to the nucleus by continuous removal of K63-linked polyubiquitin chains (Massoumi et al. 2006). However, fucose-DC-SIGN-induced complex formation of LSP1-IKKε-CYLD results in IKKε phosphorylation at Ser172 and subsequent phosphorylation of CYLD at Ser418. Phosphorylation of CYLD decreases its deubiquitinase activity, allowing Bcl3 activation and nuclear translocation (Fig. 9.2b) (Gringhuis et al. 2014a).

TLR4 activation by LPS normally leads to NFκB p50-p65 dimers, which bind to promoter sites in *IL6*, *IL10*, *IL12A*, *IL12B*, and *IL23A* genes (Ghosh and Hayden 2008). However, simultaneous triggering of DC-SIGN by fucose structures leads to Bcl3-induced p50 dimers with decreased transcriptional activity at *IL6*, *IL12A*, *IL12B*, and *IL23A* genes and similar transcriptional activity at *IL10* compared to p50-p65 heterodimers. In addition, Bcl3-p50-p50 complexes induce the expression of T_H2 attracting chemokines CCL17 and CCL22 (Gringhuis et al. 2014a). Therefore, cross talk between TLRs and DC-SIGN shifts the cytokine responses from a T_H1-skewing profile to a T_H2-inducing profile, which is crucial in clearing extracellular parasites like *Schistosoma mansoni* (Gringhuis et al. 2014a).

In addition to IKKε-CYLD-dependent Bcl3 activation by DC-SIGN, IKKε alters IFNα/β receptor (IFNR) signaling. TLR-induced IFN-β production leads to autocrine IFNR activation and JAK-STAT signaling, resulting in the formation of STAT1 homodimers, which bind to IFN-γ-activated sites (GAS) in specific genes (Decker et al. 2005). However, IKKε, activated by fucose-triggered DC-SIGN, phosphorylates STAT1 at Ser708, which leads to the formation of IFN-stimulated gene factor 3 (ISGF3), a protein complex consisting of STAT-1-STAT2 heterodimers and the DNA-binding unit IRF9 (González-Navajas et al. 2012; Gringhuis et al. 2014b). ISGF3 binds to IFN-stimulated response elements (ISRE) instead of GAS elements and thereby induces a different set of genes compared to STAT1 homodimers (Decker et al. 2005). In particular, ISGF3 activation leads to pronounced and prolonged expression of IFN-β as well as specific ISRE-dependent genes. ISRE-containing genes comprise important antiviral genes, including 29,59-oligoadenylate synthetase 1 and myxovirus resistance protein A (Takaoka and Yanai 2006), and cytokines such as IL-27, which is essential for follicular T-helper (T_{FH}) cell formation (Batten et al. 2010; Gringhuis et al. 2014b).

Fucose-specific DC-SIGN signaling not only modulates TLR signaling but also IFNR signaling to mount T_H2 and T_{FH} responses important for antibody-mediated immunity. This suggests that T_H2 cells are closely linked to T_{FH} responses as both cell types are induced simultaneously. As humoral immune responses are critical against viruses, parasites, and bacteria (Tangye et al. 2013), fucose-based strategies targeting DC-SIGN have the potential to be harnessed for vaccine development.

9.4.4 DC-SIGN Fine Tunes Adaptive Immunity

For the innate immune system, pathogens are basically a collection of pathogen-associated molecular patterns (PAMPs). Tailored immune responses require activation of different PRRs by these PAMPs to specify the invading pathogen. This demands cross talk of different PRR-induced signaling pathways to induce tailored adaptive immune responses by DCs. For example, *M. tuberculosis* infection induces T_H1 and T_H17 responses, which rely on IL-12 and IL-23 secretion by DCs (Khader et al. 2007). IL-12 is essential in inducing T_H1 differentiation and stimulation of DCs with *M. tuberculosis* leads to robust IL-12 secretion (Gringhuis et al. 2009b). Interestingly, blocking DC-SIGN or inhibition of Raf-1 decreases IL-12 production by *M. tuberculosis*-stimulated DCs (Gringhuis et al. 2009b), resulting in a shift from protective T_H1 to T_H2 (Gringhuis et al. 2009a). Furthermore, inhibiting DC-SIGN-Raf-1 signaling in *M. tuberculosis*-stimulated DCs also reduces IL-23 production (Gringhuis et al. 2009b), which is critical for T_H17-mediated *M. tuberculosis* responses (Khader et al. 2007). Although the direct effect of DC-SIGN-Raf-1 signaling on T_H17 responses has not been investigated, increased IL-23 production probably promotes T_H17 responses against *M. tuberculosis* (Fig. 9.3a). This indicates that DC-SIGN is crucial for *M. tuberculosis*-induced immune responses and possibly other mannose-containing pathogens.

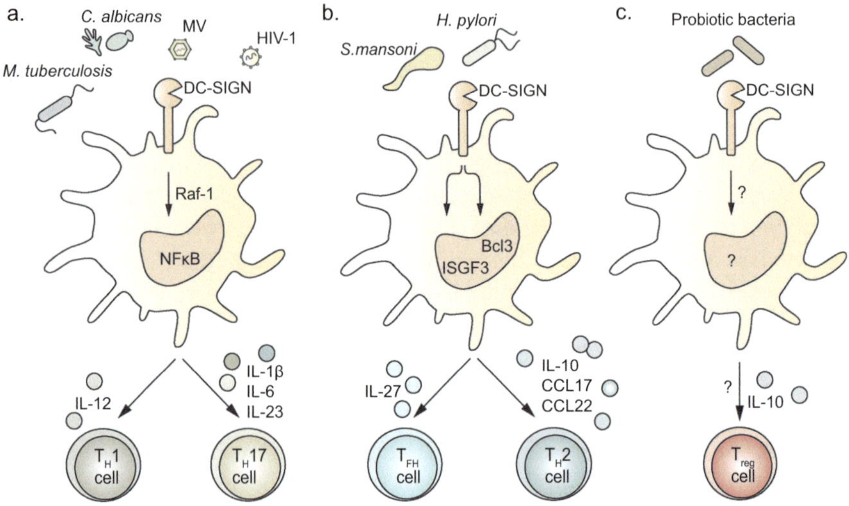

Fig. 9.3 DC-SIGN tailors adaptive immune responses. (**a**) Mannose-containing pathogens including MV, HIV-1, *C. albicans*, and *M. tuberculosis* induce DC-SIGN-dependent Raf-1 activation leading to phosphorylation and acetylation of NFκB subunit p65. This enhances the production of TLR-induced IL-12, which is crucial for the induction of T_H1 responses. In addition, Raf-1 activation leads to increased secretion of IL-1β, IL-6, and IL-23 which drive the formation of T_H17 responses. (**b**) Fucose-containing pathogens such as the parasite *S. mansoni* and the bacterium *H. pylori* induce DC-SIGN-dependent T_H2 and T_{FH} cell responses. Fucose-specific DC-SIGN signaling in combination with TLR activation induces Bcl3-dependent decrease of IL-1β, IL-6, IL-12, and IL-23 secretion while enhancing IL-10 secretion and production of T_H2-attracting chemokines CCL17 and CCL22. This drives the formation of T $_H2$ cells. Furthermore, fucose-specific DC-SIGN signaling modulates IFNR signaling to induce the production of IL-27, which is pivotal in T_{FH} formation. **c**. Probiotic bacteria like *L. reuteri* and *L. casei* are internalized by DCs via DC-SIGN. This leads to the secretion of IL-10 and the formation of T_{reg} cells. Although the formation of T_{reg} depends on DC-SIGN, it is unclear if *L. reuteri* and *L. casei* induce DC-SIGN-specific signaling pathways that differ from mannose or fucose induced signaling pathways

The clearance of extracellular pathogens depends on robust T_H2 responses (Allen and Sutherland 2014). The extracellular bacterium *H. pylori* induces T_H1 or T_H2 responses depending on the phase-variable expression of fucose structures (Bergman et al. 2004). Interestingly, only fucose-expressing *H. pylori* binds to DC-SIGN and induce T_H2 responses (Bergman et al. 2004) by inducing DC-SIGN-dependent Bcl3 activation (Gringhuis et al. 2014a). Activation of Bcl3 by DC-SIGN lowers pro-inflammatory cytokine secretion and increases IL-10 secretion in combination with increased production of T_H2-attracting chemokines CCL17 and CCL22 (Fig. 9.3b) (Gringhuis et al. 2014a). Whether other molecules important for T_H2 responses, such as OX40, are also induced by DC-SIGN-dependent Bcl3 activation has not been investigated (Gringhuis et al. 2014a). These studies indicate that the sole expression of fucose structures can completely shift the paradigm from T_H1 to T_H2 dominated immune responses, highlighting the importance of carbohydrates and CLRs in adaptive immunity.

The induction of regulatory T (T_{reg}) cells by probiotic bacteria, including *Lactobacillus reuteri* and *Lactobacillus casei*, has been linked to DC-SIGN-mediated uptake by DCs (Fig. 9.3c) (Smits et al. 2005). Although the presence of carbohydrate structures on *L. reuteri* and *L. casei* has not been investigated, the closely related bacterium *L. plantarum* expresses glycoproteins on the surface containing GlcNAcs (Fredriksen et al. 2012). DC-SIGN is known to interact with GlcNAcs, but it is unclear if this triggers DC-SIGN signaling (Steeghs et al. 2006; Zhang et al. 2006). Furthermore, the probiotic bacteria used by Federiksen et al. (2012) did not trigger any TLRs, which could also explain the induction of T_{reg} cells independent of DC-SIGN.

In addition to T-cell-driven immunity, humoral immune responses are pivotal in protection against many diseases. Humoral immune responses are induced in specific areas in secondary lymphoid organs called germinal centers (GC), where B cells differentiate into memory B cells and long-lived plasma cells. T_{FH} in GC play critical roles in regulating B cell differentiation and antibody isotype class switching. The differentiation and maintenance of T_{FH} by DCs is not fully understood but involves IL-6, IL-21, and IL-27 (Vogelzang et al. 2008; Batten et al. 2010; Nurieva et al. 2010; Gringhuis et al. 2014b). Interestingly, fucose-specific DC-SIGN signaling, in combination with TLR signaling, induces the production of IL-27 by DCs, which drives the formation of T_{FH} cells (Fig. 9.3b) (Gringhuis et al. 2014b). Furthermore, DC-SIGN-dependent IL-27-generated T_{FH} cells produce IL-21 and induce B cell class switching from IgM to IgG (Gringhuis et al. 2014b), indicating that DC-SIGN signaling by fucose-containing pathogens is crucial for humoral immune responses.

Hence, DC-SIGN shapes adaptive immune responses by directing T cell differentiation towards T_H1/T_H17 or T_H2/T_{FH} dominated responses, depending on the carbohydrate profile of pathogens and cross talk with other receptors, greatly enhancing the plasticity of DCs to tailor immune responses to a diverse range of pathogens.

9.5 DC-SIGN Tale from the Murine Perspective

Animal models have proven valuable in determining the overall outcome of molecular or cellular processes studied in vitro. The human DC-SIGN family consists of two receptors: DC-SIGN and L-SIGN, whereas mice have eight DC-SIGN homologs (SIGNR 1–8). However, all mouse homologs differ in glycan specificity, internalization capacity, innate signaling, and cell-specific expression, making mice unsuitable to study DC-SIGN functionality (Garcia-Vallejo and van Kooyk 2013; Soilleux et al. 2000; Bashirova et al. 2001; Powlesland et al. 2006; Park et al. 2001; Takahara et al. 2004; Tanne et al. 2009).

A transgenic mouse model has been created by expressing DC-SIGN under transcriptional control of the murine CD11c promoter to limit expression to DCs (Schaefer et al. 2008). Transgenic hSIGN mice express low levels of DC-SIGN, are protected against *M. tuberculosis* infections compared to wild type mice, and in vivo targeting of DC-SIGN using antibody-antigen complexes enhances T-cell-mediated immune responses (Schaefer et al. 2008; Hesse et al. 2013). However, since cross talk with other PRRs is essential for DC-SIGN signaling, the question remains if hSIGN transgenic mice are suitable to study the full potential of DC-SIGN (Schaefer et al. 2008; Garcia-Vallejo and van Kooyk 2013).

9.6 DC-SIGN Picks Up the Wrong Hitchhikers

Pathogens are under constant pressure of the immune system for survival and have evolved intriguing ways to escape or prevent immune responses. Some pathogens have even hijacked immune components for efficient infection and dissemination in the host. In particular, HIV-1 uses DC-SIGN for efficient transmission to T cells. HIV-1 primarily infects T cells via CD4-CCR5/CXR4, which requires dissemination of the virus from the primary side of infection to secondary lymphoid tissues rich in $CD4^+$ T cells. As DCs constantly sample antigens in peripheral tissues for presentation to T cells, DCs form an ideal transport vehicle for HIV-1. Indeed, DC-SIGN binding to HIV-1 glycoprotein gp120 leads to effective internalization of HIV-1 by DCs (Geijtenbeek et al. 2000b; Turville et al. 2001; Engering et al. 2002b; Smith et al. 2007). Notably, HIV-1 is not routed for lysosomal degradation after DC-SIGN-mediated uptake but is retained in endosomes for several days (Geijtenbeek et al. 2000b). While stored, HIV-1 remains infectious and is released upon DC-T-cell interaction to infect $CD4^+$ T cells (Geijtenbeek et al. 2000b). In addition, DC-SIGN signaling is essential for productive infection of DCs by HIV-1. Release of HIV-1 ssRNA into the cytoplasm depends on HIV-1 binding to CD4 in combination with either CCR5 or CXR4, ultimately resulting in the integration of HIV-1 DNA into the DC genome. However, HIV-1 integration does not lead to productive transcription of HIV-1 DNA and requires TLR8 activation by HIV-1 ssRNA and DC-SIGN activation by gp120. HIV-1-triggered DC-SIGN signaling activates Raf-1, leading to the acetylation of p65 – after NFκB activation by TLR8 – which is essential for full-length transcription of HIV-1 (Gringhuis et al. 2010). Hence, HIV-1 hijacks DC-SIGN for productive infection of DCs and efficient transmission to T cells.

Induction of protective antiviral responses mediated by type I IFNs is paramount to limit viral infections. However, viruses have evolved ways to subvert host innate immunity by shielding replication complexes from detection, degrading essential host molecules, and inhibiting innate detection by activating specific signaling cascades (Ye et al. 2013). Measles virus (MV) is a highly contagious pathogen which infects $CD150^+$ DCs and T and B lymphocytes (Lemon et al. 2011; De Vries et al. 2012). MV is a negative stranded RNA virus, which replicates in the cytoplasm of infected cells and is prone for detection by cytosolic RIG-I-like receptors (RLRs). RLR activation is tightly regulated by continuous phosphorylation of caspase recruitment domains, and these require dephosphorylation by PP1 phosphatases for type I IFN induction (Wies et al. 2013). Intriguingly, MV induces DC-SIGN-dependent activation of Raf-1, which phosphorylates PP1 inhibitor I-1 and thereby prevents dephosphorylation of RLRs and type I IFN production, leading to enhanced infection of DCs (Mesman et al. 2014). Furthermore, the main target cells of dengue virus are DCs and multiple studies have shown that dengue virus depends on DC-SIGN for productive infection of DCs (Tassaneetrithep et al. 2003; Lozach et al. 2005). Whether dengue virus also activates DC-SIGN signaling to inhibit type I IFN induction in DCs has not been investigated.

These studies highlight that pathogens take advantage of the binding capacity of DC-SIGN in combination with specific DC-SIGN signaling for effective infection of DCs and transmission to target cells. Together with the migratory capacity of DCs, this activity makes DC-SIGN$^+$ DCs the ideal vehicle for viral dissemination.

9.7 Clinical Relevance and Therapeutic Potential

Single nucleotide polymorphisms (SNPs) can affect protein expression and function and can therefore be valuable tools to study the clinical relevance of molecular processes, which would otherwise be impossible in humans. Certain SNPs in CD209, the gene coding for DC-SIGN, have been identified that alter the pathogeneses of several diseases. For instance, DCSIGN1-366G is a variant of DC-SIGN with a SNP in the promoter sequence of CD209, which results in lower expression levels at the cell membrane. Individuals carrying the DCSIGN1-366G variant have lower incidence of dengue fever, which is in concordance with in vitro studies (Sakuntabhai et al. 2005; Lozach et al. 2005). In addition, DCSIGN1-366G is associated with protection against *M. tuberculosis*, as individuals carrying this variant have a lower incidence of lung cavitation (Vannberg et al. 2008). SNPs located in other noncoding parts of *CD209* have been implicated in the vertical transmission of HIV-1 from mother to child (Boily-Larouche et al. 2012; da Silva et al. 2012). SNP variants occurring in the neck region of DC-SIGN have been associated with increased vertical transmission. Molecular studies revealed that these variants increase HIV-1 binding and transmission to T cells, which could explain enhanced vertical transmission (Boily-Larouche et al. 2012). These studies emphasize the important role of DC-SIGN in HIV-1, *M. tuberculosis*, and dengue virus pathogenesis, which has spiked the interest to design prophylactics or carbohydrate-based therapies directed against DC-SIGN to prevent or treat disease (Alen et al. 2011; Varga et al. 2014). However, certain precaution should be considered before using DC-SIGN-inhibiting therapies as DC-SIGN is involved in basic functions of DCs.

DC-based vaccination is under intense interest to induce effective immunity against cancer and infectious diseases. Currently used DC therapies in the clinic are complex and require the isolation of monocytes (for monocyte-derived DCs) or DCs from blood, which are loaded with antigen ex vivo and injected back into patients (Kreutz et al. 2013). The capacity of DC-SIGN to internalize antigen for

MHC presentation and simultaneous modulation of cytokine profiles to direct T cell differentiation makes DC-SIGN an ideal receptor for vaccination strategies without the need to isolate DCs. For example, carbohydrate-coated liposomes filled with specific antigens targeting DC-SIGN induce strong $CD4^+$ and $CD8^+$ T cell responses directed against tumor antigens (Hesse et al. 2013; van Kooyk et al. 2013). Differential coating of liposomes with either mannose or fucose structures adds another level of refinement to these strategies to direct adaptive immune responses to the desired outcome. Whether similar successes can be achieved in humans needs to be addressed, but the enormous potential of these therapies requires quick action to investigate the efficiency of DC-SIGN-based therapies, particularly because these therapies can also be used as treatment for autoimmune diseases and infectious diseases (Kreutz et al. 2013).

9.8 Concluding Remarks

From the identification of DC-SIGN as a DC-specific receptor to the clarification of DC-SIGN-induced signaling cascades, DC-SIGN has revealed many aspects of DC functionality and the importance of CLRs and carbohydrates in immune responses. DC-SIGN contributes to basic functions of DCs such as DC migration from blood into tissues and establishing cellular interaction with other immune cells (Geijtenbeek et al. 2000a, b; Bleijs et al. 2001; van Gisbergen et al. 2005). Unraveling DC-SIGN signaling has greatly contributed to our understanding of cross talk between different receptors on a molecular level with great effects on adaptive immune responses. DC-SIGN cross talk with other receptors provides DCs with the plasticity to mount specific immune responses to a wide variety of pathogens (Gringhuis et al. 2007, 2009b, 2014a, b; Hovius et al. 2008; Geijtenbeek and Gringhuis 2009). However, there is still much to discover. Although DC-SIGN signaling in the light of mannose and fucose structures is becoming clear, GlcNAc structures might induce yet another signaling cascade, which could affect T_{reg} responses induced by probiotic bacteria (Smits et al. 2005). In addition to IL-27, fucose-specific DC-SIGN formation of ISGF3 induces prolonged expression of type I interferon and antiviral interferon-stimulated genes (Gringhuis et al. 2014b), thereby inducing a highly antiviral state in DCs. Whether this indeed limits viral replication in DCs remains to be investigated, but this mechanism could be used to decrease viral infections in patients. The antigen routing properties of DC-SIGN together with its strong affinity for different antigens makes it an ideal receptor for targeted vaccines strategies to induce long-lasting immunity (Engering et al. 2002a; Cambi et al. 2009; Hesse et al. 2013; van Kooyk et al. 2013). This, in combination with the modulating effects of DC-SIGN signaling on adaptive immune responses, provides researchers with an extended toolbox to develop effective vaccines against multiple diseases, including cancer, autoimmune diseases, and infectious diseases.

References

Alen MMF, de Burghgraeve T, Kaptein SJF et al (2011) Broad antiviral activity of carbohydrate-binding agents against the four serotypes of dengue virus in monocyte-derived dendritic cells. PLoS One. doi:10.1371/journal.pone.0021658

Allen JE, Sutherland TE (2014) Host protective roles of type 2 immunity: parasite killing and tissue repair, flip sides of the same coin. Semin Immunol 26:329–340. doi:10.1016/j.smim.2014.06.003

Appelmelk BJ, van Die I, van Vliet SJ et al (2003) Cutting edge: carbohydrate profiling identifies new pathogens that interact with dendritic cell-specific ICAM-3-grabbing nonintegrin on dendritic cells. J Immunol 170:1635–1639. doi:10.4049/jimmunol.170.4.1635

Ariizumi K, Shen GL, Shikano S et al (2000) Identification of a novel, dendritic cell-associated molecule, dectin-1, by subtractive cDNA cloning. J Biol Chem 275:20157–20167. doi:10.1074/jbc.M909512199

Azad AK, Torrelles JB, Schlesinger LS (2008) Mutation in the DC-SIGN cytoplasmic triacidic cluster motif markedly attenuates receptor activity for phagocytosis and endocytosis of mannose-containing ligands by human myeloid cells. J Leukoc Biol 84:1594–1603. doi:10.1189/jlb.0308192

Bashirova AA, Geijtenbeek TB, van Duijnhoven GC et al (2001) A dendritic cell-specific intercellular adhesion molecule 3-grabbing nonintegrin (DC-SIGN)-related protein is highly expressed on human liver sinusoidal endothelial cells and promotes HIV-1 infection. J Exp Med 193:671–678. doi:10.1084/jem.193.6.671

Batten M, Ramamoorthi N, Kljavin NM et al (2010) IL-27 supports germinal center function by enhancing IL-21 production and the function of T follicular helper cells. J Exp Med 207:2895–2906. doi:10.1084/jem.20100064

Bergman MP, Engering A, Smits HH et al (2004) Helicobacter pylori modulates the T helper cell 1/T helper cell 2 balance through phase-variable interaction between lipopolysaccharide and DC-SIGN. J Exp Med 200:979–990. doi:10.1084/jem.20041061

Bleijs DA, Geijtenbeek TBH, Figdor CG, Van Kooyk Y (2001) DC-SIGN and LFA-1: a battle for ligand. Trends Immunol 22:457–463. doi:10.1016/S1471-4906(01)01974-3

Boily-Larouche G, Milev MP, Zijenah LS et al (2012) Naturally-occurring genetic variants in human DC-SIGN increase HIV-1 capture, cell-transfer and risk of mother-to-child transmission. PLoS One. doi:10.1371/journal.pone.0040706

Cambi A, Koopman M, Figdor C (2005) How C-type lectins detect pathogens. Cell Microbiol 7:481–488. doi:10.1111/j.1462-5822.2005.00506.x

Cambi A, Beeren I, Joosten B et al (2009) The C-type lectin DC-SIGN internalizes soluble antigens and HIV-1 virions via a clathrin-dependent mechanism. Eur J Immunol 39:1923–1928. doi:10.1002/eji.200939351

Chen LF, Mu Y, Greene WC (2002) Acetylation of RelA at discrete sites regulates distinct nuclear functions of NF-??B. EMBO J 21:6539–6548. doi:10.1093/emboj/cdf660

Chen L, Williams SA, Mu Y et al (2005) NF- κ B RelA phosphorylation regulates RelA acetylation NF- □ B RelA phosphorylation regulates RelA acetylation †. Mol Cell Biol 25:7966–7975. doi:10.1128/MCB.25.18.7966

Cormier EG, Durso RJ, Tsamis F et al (2004) L-SIGN (CD209L) and DC-SIGN (CD209) mediate transinfection of liver cells by hepatitis C virus Emmanuel. Proc Natl Acad Sci U S A 101:14067–14072. doi:10.1073pnas.0405695101

Curtis BM, Scharnowske S, Watson AJ (1992) Sequence and expression of a membrane-associated C-type lectin that exhibits CD4-independent binding of human immunodeficiency virus envelope glycoprotein gp120. Proc Natl Acad Sci U S A 89:8356–8360. doi:10.1073/pnas.89.17.8356

Da Silva RC, Segat L, Zanin V et al (2012) Polymorphisms in DC-SIGN and L-SIGN genes are associated with HIV-1 vertical transmission in a Northeastern Brazilian population. Hum Immunol 73:1159–1165. doi:10.1016/j.humimm.2012.07.338

De Vries RD, Mesman AW, Geijtenbeek TBH et al (2012) The pathogenesis of measles. Curr Opin Virol 2:248–255. doi:10.1016/j.coviro.2012.03.005

Decker T, Müller M, Stockinger S (2005) The yin and yang of type I interferon activity in bacterial infection. Nat Rev Immunol 5:675–687. doi:10.1038/nri1684

Dickensheets HL, Venkataraman C, Schindler U, Donnelly RP (1999) Interferons inhibit activation of STAT6 by interleukin 4 in human monocytes by inducing SOCS-1 gene expression. Proc Natl Acad Sci U S A 96:10800–10805. doi:10.1073/pnas.96.19.10800

Domínguez-Soto A, Puig-Kröger A, Vega MA, Corbí AL (2005) PU.1 regulates the tissue-specific expression of dendritic cell-specific intercellular adhesion molecule (ICAM)-3-grabbing nonintegrin. J Biol Chem 280:33123–33131. doi:10.1074/jbc.M503401200

Engering A, Geijtenbeek TBH, van Vliet SJ et al (2002a) The dendritic cell-specific adhesion receptor DC-SIGN internalizes antigen for presentation to T cells. J Immunol 168:2118–2126. doi:10.4049/jimmunol.168.5.2118

Engering A, Van Vliet SJ, Geijtenbeek TBH, Van Kooyk Y (2002b) Subset of DC-SIGN+ dendritic cells in human blood transmits HIV-1 to T lymphocytes. Blood 100:1780–1786. doi:10.1182/blood-2001-12-0179

Engering A, van Vliet SJ, Hebeda K et al (2004) Dynamic populations of dendritic cell-specific ICAM-3 grabbing nonintegrin-positive immature dendritic cells and liver/lymph node-specific ICAM-3 grabbing nonintegrin-positive endothelial cells in the outer zones of the paracortex of human lymph nodes. Am J Pathol 164:1587–1595. doi:10.1016/S0002-9440(10)63717-0

Fehres CM, van Beelen AJ, Bruijns SCM et al (2015) In-situ delivery of antigen to DC-SIGN+ CD14+ dermal dendritic cells results in enhanced CD8+ T cell responses. J Invest Dermatol. doi:10.1038/jid.2015.152

Fredriksen L, Mathiesen G, Moen A et al (2012) The major autolysin Acm2 from Lactobacillus plantarum undergoes cytoplasmic O-glycosylation. J Bacteriol 194:325–333. doi:10.1128/JB.06314-11

Frisont N, Taylor ME, Soilleux E et al (2003) Oligolysine-based oligosaccharide clusters: selective recognition and endocytosis by the mannose receptor and dendritic cell-specific intercellular adhesion molecule 3 (ICAM-3)-grabbing nonintegrin. J Biol Chem 278:23922–23929. doi:10.1074/jbc.M302483200

Garcia-Vallejo JJ, van Kooyk Y (2013) The physiological role of DC-SIGN: a tale of mice and men. Trends Immunol 34:482–486. doi:10.1016/j.it.2013.03.001

Garg R, Juncadella IJ, Ramamoorthi N et al (2010) Cutting edge: CD4 is the receptor for the tick saliva immunosuppressor, Salp15. J Immunol 9:1–14. doi:10.1111/j.1600-6143.2008.02497.x.Plasma

Geijtenbeek TBH, Gringhuis SI (2009) Signalling through C-type lectin receptors: shaping immune responses. Nat Rev Immunol 9:465–479. doi:10.1038/nri2569

Geijtenbeek TB, Torensma R, van Vliet SJ et al (2000a) Identification of DC-SIGN, a novel dendritic cell-specific ICAM-3 receptor that supports primary immune responses. Cell 100:575–585

Geijtenbeek TBH, Kwon DS, Torensma R et al (2000b) DC-SIGN, a dendritic cell – specific HIV-1-binding protein that enhances trans -infection of T cells. Cell 100:587–597

Geijtenbeek TB, Krooshoop DJ, Bleijs DA et al (2000c) DC-SIGN-ICAM-2 interaction mediates dendritic cell trafficking. Nat Immunol 1:353–357. doi:10.1038/79815

Geijtenbeek TBH, Van Vliet SJ, Van Duijnhoven GCF et al (2001) DC-SIGN, a dentritic cell-specific HIV-1 receptor present in placenta that infects T cells in trans – a review. Placenta 22:19–23. doi:10.1053/plac.2001.0674

Geijtenbeek TBH, Engering A, Van Kooyk Y (2002a) DC-SIGN, a C-type lectin on dendritic cells that unveils many aspects of dendritic cell biology. J Leukoc Biol 71:921–931

Geijtenbeek TBH, Van Duijnhoven GCF, Van Vliet SJ et al (2002b) Identification of different binding sites in the dendritic cell-specific receptor DC-SIGN for intercellular adhesion molecule 3 and HIV-1. J Biol Chem 277:11314–11320. doi:10.1074/jbc.M111532200

Geijtenbeek TBH, Van Vliet SJ, Koppel EA et al (2003) Mycobacteria target DC-SIGN to suppress dendritic cell function. J Exp Med 197:7–17. doi:10.1084/jem.20021229
Geijtenbeek TBH, den Dunnen J, Gringhuis SI (2009) Pathogen recognition by DC-SIGN shapes adaptive immunity. Future Microbiol 4:879–890. doi:10.2217/fmb.09.51
Ghosh S, Hayden MS (2008) New regulators of NF-kappaB in inflammation. Nat Rev Immunol 8: 837–848. doi:10.1038/nri2423
González-Navajas JM, Lee J, David M, Raz E (2012) Immunomodulatory functions of type I interferons. Changes 29:997–1003. doi:10.1016/j.biotechadv.2011.08.021.Secreted
Granelli-Piperno A, Pritsker A, Pack M et al (2012) Dendritic cell-specific intercellular adhesion molecule 3-grabbing nonintegrin/CD209 is abundant on macrophages in the normal human lymph node and is not required for dendritic cell stimulation of the mixed leukocyte reaction. J Immunol 127:358–366. doi:10.1016/j.jsbmb.2011.07.002.Identification
Gringhuis SI, den Dunnen J, Litjens M et al (2007) C-type lectin DC-SIGN modulates toll-like receptor signaling via Raf-1 kinase-dependent acetylation of transcription factor NF-κB. Immunity 26:605–616. doi:10.1016/j.immuni.2007.03.012
Gringhuis SI, den Dunnen J, Litjens et al (2009a) Dectin-1 directs T helper cell differentiation by controlling noncanonical NF-kappaB activation through Raf-1 and Syk. Nat Immunol 10(2): 203–213, http://doi.org/10.1038/ni.1692
Gringhuis SI, den Dunnen J, Litjens M et al (2009b) Carbohydrate-specific signaling through the DC-SIGN signalosome tailors immunity to mycobacterium tuberculosis, HIV-1 and helicobacter pylori. Nat Immunol 10:1081–1088. doi:10.1038/ni.1778
Gringhuis SI, van der Vlist M, van den Berg LM et al (2010) HIV-1 exploits innate signaling by TLR8 and DC-SIGN for productive infection of dendritic cells. Nat Immunol 11:419–426. doi:10.1038/ni.1858
Gringhuis SI, Kaptein TM, Wevers BA et al (2014a) Fucose-specific DC-SIGN signalling directs T helper cell type-2 responses via IKKε- and CYLD-dependent Bcl3 activation. Nat Commun 5:3898. doi:10.1038/ncomms4898
Gringhuis SI, Kaptein TM, Wevers BA et al (2014b) Fucose-based PAMPs prime dendritic cells for follicular T helper cell polarization via DC-SIGN-dependent IL-27 production. Nat Commun 5:5074. doi:10.1038/ncomms6074
Gurney KB, Elliott J, Nassanian H et al (2005) Binding and transfer of human immunodeficiency virus by DC-SIGN + cells in human rectal mucosa. J Virol 79:5762–5773. doi:10.1128/JVI.79.9.5762
Haniffa M, Shin A, Bigley V et al (2012) Human tissues contain CD141hi cross-presenting dendritic cells with functional homology to mouse CD103+ nonlymphoid dendritic cells. Immunity 37:60–73. doi:10.1016/j.immuni.2012.04.012
Hesse C, Ginter W, Förg T et al (2013) In vivo targeting of human DC-SIGN drastically enhances CD8+ T-cell-mediated protective immunity. Eur J Immunol 43:2543–2553. doi:10.1002/eji.201343429
Hodges A, Sharrocks K, Edelmann M et al (2007) Activation of the lectin DC-SIGN induces an immature dendritic cell phenotype triggering Rho-GTPase activity required for HIV-1 replication. Nat Immunol 8:569–577. doi:10.1038/ni1470
Hovius JWR, De Jong M a WP, Den Dunnen J et al (2008) Salp15 binding to DC-SIGN inhibits cytokine expression by impairing both nucleosome remodeling and mRNA stabilization. PLoS Pathog. doi:10.1371/journal.ppat.0040031
Joffre OP, Segura E, Savina A, Amigorena S (2012) Cross-presentation by dendritic cells. Nat Rev Immunol 12:557–569. doi:10.1038/nri3254
Joshi MD, Unger WJ, Storm G et al (2012) Targeting tumor antigens to dendritic cells using particulate carriers. J Control Release 161:25–37. doi:10.1016/j.jconrel.2012.05.010
Khader SA, Bell GK, Pearl JE et al (2007) IL-23 and IL-17 in the establishment of protective pulmonary CD4+ T cell responses after vaccination and during mycobacterium tuberculosis challenge. Nat Immunol 8:369–377. doi:10.1038/ni1449

Kreutz M, Tacken PJ, Figdor CG (2013) Targeting dendritic cells – why bother? Blood 121: 2836–2844. doi:10.1182/blood-2012-09-452078

Kwon DS, Gregorio G, Bitton N et al (2002) DC-SIGN-mediated internalization of HIV is required for trans-enhancement of T cell infection. Immunity 16:135–144. doi:10.1016/S1074-7613(02)00259-5

Leckband DE, Menon S, Rosenberg K et al (2011) Geometry and adhesion of extracellular domains of DC-SIGNR neck length variants analyzed by force-distance measurements. Biochemistry 50:6125–6132. doi:10.1021/bi2003444

Lemon K, de Vries RD, Mesman AW et al (2011) Early target cells of measles virus after aerosol infection of non-human primates. PLoS Pathog. doi:10.1371/journal.ppat.1001263

Lozach P-Y, Burleigh L, Staropoli I et al (2005) Dendritic cell-specific intercellular adhesion molecule 3-grabbing non-integrin (DC-SIGN)-mediated enhancement of dengue virus infection is independent of DC-SIGN internalization signals. J Biol Chem 280:23698–23708. doi:10.1074/jbc.M504337200

Ludwig IS, Lekkerkerker AN, Depla E et al (2004) Hepatitis C virus targets DC-SIGN and L-SIGN to escape lysosomal degradation. J Virol 78:8322–8332. doi:10.1128/JVI.78.15.8322-8332.2004

Massoumi R, Chmielarska K, Hennecke K et al (2006) Cyld inhibits tumor cell proliferation by blocking Bcl-3-dependent NF-κB signaling. Cell 125:665–677. doi:10.1016/j.cell.2006.03.041

Matsumoto M, Tanaka T, Kaisho T et al (1999) A novel LPS-inducible C-type lectin is a transcriptional target of NF-IL6 in macrophages. J Immunol 163:5039–5048, doi: ji_v163n9p5039 [pii]

Mcgovern N, Schlitzer A, Gunawan M et al (2014) Human dermal CD14 + cells are a transient population of monocyte-derived macrophages. Immunity 41:465–477. doi:10.1016/j.immuni.2014.08.006

Mesman AW, Zijlstra-Willems EM, Kaptein TM et al (2014) Measles virus suppresses RIG-I-like receptor activation in dendritic cells via DC-SIGN-mediated inhibition of PP1 phosphatases. Cell Host Microbe 16:31–42. doi:10.1016/j.chom.2014.06.008

Meyer S, Van Liempt E, Imberty A et al (2005) DC-SIGN mediates binding of dendritic cells to authentic pseudo-Lewis Y glycolipids of Schistosoma mansoni cercariae, the first parasite-specific ligand of DC-SIGN. J Biol Chem 280:37349–37359. doi:10.1074/jbc.M507100200

Moris A, Nobile C, Buseyne F et al (2004) DC-SIGN promotes exogenous MHC-I-restricted HIV-1 antigen presentation. Blood 103:2648–2654. doi:10.1182/blood-2003-07-2532

Nurieva RI, Chung Y, Martinez GJ, Yang XO, Tanaka S, Matskevitch TD, Wang Y-H, Dong C (2010) Bcl6 mediates the development of t follicular helper cells. Science 9:1–14. doi:10.1111/j.1600-6143.2008.02497.x.Plasma

Ogden SC, Tang H (2015) The missing pieces of the HCV entry puzzle. Futur Virol 10:415–428. doi:10.2217/FVL.15.12

Park CG, Takahara K, Umemoto E et al (2001) Five mouse homologues of the human dendritic cell C-type lectin, DC-SIGN. Int Immunol 13:1283–1290. doi:10.1093/intimm/13.10.1283

Powlesland AS, Ward EM, Sadhu SK et al (2006) Widely divergent biochemical properties of the complete set of mouse DC-SIGN-related proteins. J Biol Chem 281:20440–20449. doi:10.1074/jbc.M601925200

Relloso M, Puig-Kröger A, Pello OM et al (2002) DC-SIGN (CD209) expression is IL-4 dependent and is negatively regulated by IFN, TGF-beta, and anti-inflammatory agents. J Immunol 168:2634–2643. doi:10.4049/jimmunol.168.6.2634

Roux PP, Blenis J (2004) ERK and p38 MAPK-activated protein kinases : a family of protein kinases with diverse biological functions ERK and p38 MAPK-activated protein kinases: a family of protein kinases with diverse biological functions. Microbiol Mol Biol Rev 68: 320–344. doi:10.1128/MMBR.68.2.320

Sakuntabhai A, Turbpaiboon C, Casadémont I et al (2005) A variant in the CD209 promoter is associated with severity of dengue disease. Nat Genet 37:507–513. doi:10.1038/ng1550

Schaefer M, Reiling N, Fessler C et al (2008) Decreased pathology and prolonged survival of human DC-SIGN transgenic mice during mycobacterial infection. J Immunol 180:6836–6845, doi: 180/10/6836 [pii]

Schjetne KW, Thompson KM, Aarvak T et al (2002) A mouse C k -speci ® c T cell clone indicates that DC-SIGN is an ef ® cient target for antibody-mediated delivery of T cell epitopes for MHC class II presentation. Int Immunol 14:1423–1430

Smith AL, Ganesh L, Leung K et al (2007) Leukocyte-specific protein 1 interacts with DC-SIGN and mediates transport of HIV to the proteasome in dendritic cells. J Exp Med 204:421–430. doi:10.1084/jem.20061604

Smits HH, Engering A, Van Der Kleij D et al (2005) Selective probiotic bacteria induce IL-10-producing regulatory T cells in vitro by modulating dendritic cell function through dendritic cell-specific intercellular adhesion molecule 3-grabbing nonintegrin. J Allergy Clin Immunol 115:1260–1267. doi:10.1016/j.jaci.2005.03.036

Soilleux EJ, Barten R, Trowsdale J (2000) DC-SIGN; a related gene, DC-SIGNR; and CD23 form a cluster on 19p13. J Immunol 165:2937–2942. doi:10.4049/jimmunol.165.6.2937

Soilleux EJ, Morris LS, Lee B et al (2001) Placental expression of DC-SIGN may mediate intrauterine vertical transmission of HIV. J Pathol 195:586–592. doi:10.1002/path.1026

Soilleux EJ, Morris LS, Leslie G et al (2002) Constitutive and induced expression of DC-SIGN on dendritic cell and macrophage subpopulations in situ and in vitro. J Leukoc Biol 71:445–457

Springer TA (1995) Traffic signals on endothelium for lymphocyte recirculation and leukocyte emigration. Annu Rev Physiol 57:827–872

Steeghs L, van Vliet SJ, Uronen-Hansson H et al (2006) Neisseria meningitidis expressing IgtB lipopolysaccharide targets DC-SIGN and modulates dendritic cell function. Cell Microbiol 8: 316–325. doi:10.1111/j.1462-5822.2005.00623.x

Svajger U, Anderluh M, Jeras M, Obermajer N (2010) C-type lectin DC-SIGN: an adhesion, signalling and antigen-uptake molecule that guides dendritic cells in immunity. Cell Signal 22: 1397–1405. doi:10.1016/j.cellsig.2010.03.018

Tacken PJ, De Vries IJM, Gijzen K et al (2012) Effective induction of naive and recall T-cell responses by targeting antigen to human dendritic cells via a humanized anti – DC-SIGN antibody effective induction of naive and recall T-cell responses by targeting antigen to human dendritic cells via a hum. Pharmacia 106:1278–1285. doi:10.1182/blood-2005-01-0318

Takahara K, Yashima Y, Omatsu Y, Yoshida H, Kimura Y, Kang YS, Steinman RM, Park CG, Inaba K (2004) Functional comparison of the mouse DC-SIGN, SIGNR1, SIGNR3 and Langerin, C-type lectins. Int Immunol 16(6):819–829. doi:10.1093/intimm/dxh084

Takaoka A, Yanai H (2006) Interferon signalling network in innate defence. Cell Microbiol 8: 907–922. doi:10.1111/j.1462-5822.2006.00716.x

Tangye SG, Ma CS, Brink R, Deenick EK (2013) The good, the bad and the ugly – TFH cells in human health and disease. Nat Rev Immunol 13:412–426. doi:10.1038/nri3447

Tanne A, Ma B, Boudou F, Tailleux L, Botella H, Badell E, Levillain F, Taylor ME, Drickamer K, Nigou J, Dobos KM, Puzo G, Vestweber D, Wild MK, Marcinko M, Sobieszczuk P, Stewart L, Lebus D, Gicquel B, Neyrolles O (2009) A murine DC-SIGN homologue contributes to early host defense against Mycobacterium tuberculosis. J Exp Med 206(10):2205–2220. doi:10.1084/jem.20090188

Tassaneetrithep B, Burgess TH, Granelli-Piperno A et al (2003) DC-SIGN (CD209) mediates dengue virus infection of human dendritic cells. J Exp Med 197:823–829. doi:10.1084/jem. 20021840

Trumpfheller C, Park CG, Finke J et al (2003) Cell type-dependent retention and transmission of HIV-1 by DC-SIGN. Int Immunol 15:289–298. doi:10.1093/intimm/dxg030

Turville SG, Arthos J, Donald KM et al (2001) HIV gp120 receptors on human dendritic cells. Blood 98:2482–2488. doi:10.1182/blood.V98.8.2482

Unger WWJ, Van Beelen AJ, Bruijns SC et al (2012) Glycan-modified liposomes boost CD4 + and CD8 + T-cell responses by targeting DC-SIGN on dendritic cells. J Control Release 160:88–95. doi:10.1016/j.jconrel.2012.02.007

Van Die I, van Vliet SJ, Nyame AK et al (2003) The dendritic cell-specific C-type lectin DC-SIGN is a receptor for Schistosoma mansoni egg antigens and recognizes the glycan antigen Lewis x. Glycobiology 13:471–478. doi:10.1093/glycob/cwg052

Van Gisbergen KPJM, Sanchez-Hernandez M, Geijtenbeek TBH, van Kooyk Y (2005) Neutrophils mediate immune modulation of dendritic cells through glycosylation-dependent interactions between Mac-1 and DC-SIGN. J Exp Med 201:1281–1292. doi:10.1084/jem.20041276

Van Kooyk Y, Unger WWJ, Fehres CM et al (2013) Glycan-based DC-SIGN targeting vaccines to enhance antigen cross-presentation. Mol Immunol 55:143–145. doi:10.1016/j.molimm.2012.10.031

Van Liempt E, Bank CMC, Mehta P et al (2006) Specificity of DC-SIGN for mannose- and fucose-containing glycans. FEBS Lett 580:6123–6131. doi:10.1016/j.febslet.2006.10.009

Vannberg FO, Chapman SJ, Khor CC et al (2008) CD209 genetic polymorphism and tuberculosis disease. PLoS One 3:1–6. doi:10.1371/journal.pone.0001388

Varga N, Sutkeviciute I, Ribeiro-Viana R et al (2014) A multivalent inhibitor of the DC-SIGN dependent uptake of HIV-1 and Dengue virus. Biomaterials 35:4175–4184. doi:10.1016/j.biomaterials.2014.01.014

Vogelzang A, McGuire HM, Yu D et al (2008) A fundamental role for interleukin-21 in the generation of T follicular helper cells. Immunity 29:127–137. doi:10.1016/j.immuni.2008.06.001

Wellbrock C, Karasarides M, Marais R (2004) The RAF proteins take centre stage. Nat Rev Mol Cell Biol 5:875–885. doi:10.1038/nrm1498

Wies E, Wang MK, Maharaj NP et al (2013) Dephosphorylation of the RNA sensors RIG-I and MDA5 by the phosphatase PP1 is essential for innate immune signaling. Immunity 38:437–449. doi:10.1016/j.immuni.2012.11.018

Ye J, Zhu B, Fu ZF et al (2013) Immune evasion strategies of flaviviruses. Vaccine 31:461–471. doi:10.1016/j.vaccine.2012.11.015

Zhang P, Snyder S, Feng P et al (2006) Role of N-acetylglucosamine within core lipopolysaccharide of several species of gram-negative bacteria in targeting the DC-SIGN (CD209). J Immunol 177:4002–4011. doi:10.4049/jimmunol.177.6.4002

Zhao LJ, Wang W, Ren H, Qi ZT (2013) ERK signaling is triggered by hepatitis C virus E2 protein through DC-SIGN. Cell Stress Chaperones 18:495–501. doi:10.1007/s12192-013-0405-3

Chapter 10
LOX-1 and Immunity

Anja Hofmann, Tatsuya Sawamura, and Henning Morawietz

Abstract An interesting C-type lectin-like receptor is LOX-1. LOX-1 is a membrane glycoprotein with a cytoplasmic domain, a transmembrane domain, and an oxidized low-density lipoprotein (oxLDL)-binding extracellular lectin-like domain. LOX-1 is involved in a variety of physiological and pathophysiological processes. The LOX-1 receptor is able to mediate the uptake of minimally and maximally oxidized LDL. In addition, LOX-1 plays a role in the phagocytosis of aged and apoptotic cells, the uptake of advanced glycation end products, thrombocyte adhesion, and the interaction between bacterial proteins and endothelial cells in sepsis. It is considered as a therapeutic target in atherosclerosis and cardiovascular disease. Recently, LOX-1 has been described in response to different danger signals, rheumatic diseases, preeclampsia, and bone diseases. This review focuses on the increasing evidence supporting a novel role of LOX-1 in immunity. LOX-1 and other pattern recognition receptors are expressed on the surface of myeloid cells like macrophages, dendritic cells, or neutrophils. They have an important role in the immediate innate immune response and also in the regulation of the adaptive immune response.

Keywords C-type lectin-like receptor • LOX-1 • Lipoprotein • Atherosclerosis • Dendritic cells • Immunity

A. Hofmann • H. Morawietz, Ph.D., F.A.H.A. (✉)
Division of Vascular Endothelium and Microcirculation, Department of Medicine III, University Hospital Carl Gustav Carus, Technische Universität Dresden, Fetscherstr. 74, 01307 Dresden, Germany
e-mail: Henning.Morawietz@tu-dresden.de

T. Sawamura
Department of Physiology, Shinshu University School of Medicine, 3-1-1, Asahi, Matsumoto 390-8621, Japan

S. Yamasaki (ed.), *C-Type Lectin Receptors in Immunity*,
DOI 10.1007/978-4-431-56015-9_10

10.1 The C-Type Lectin-Like Receptor LOX-1

Lipids are essential for cellular function and metabolism. Lipoproteins are important lipid carriers in the blood stream. An increased plasma level of low-density lipoprotein (LDL) is also a well-known risk factor of metabolic and cardiovascular diseases. Especially under conditions of oxidative stress, LDL can be oxidized (Muller and Morawietz 2009). Classical receptors of maximally oxidized LDL are scavenger receptor A (SR-A), scavenger receptor B (CLA-1/SR-BI), cluster of differentiation (CD) 36 and CD68, and scavenger receptor that binds phosphatidylserine and oxidized lipoprotein (SR-PSOX)/CXC chemokine ligand 16 (CXCL16), while minimally oxidized LDL can interact with, e.g., CD14/toll-like receptor 4 (TLR) 4 (Shashkin et al. 2005).

An interesting novel receptor for oxLDL is the C-type lectin-like oxLDL receptor LOX-1 (Sawamura et al. 1997). LOX-1 is also known as the first member of the CLEC8A family of C-type lectin-like receptors. LOX-1 is a membrane glycoprotein of 50 kDa. The protein has three important functional domains: a short cytoplasmic domain, a transmembrane domain, and an oxLDL-binding extracellular lectin-like domain (Chen et al. 2002). Three LOX-1 homodimers are necessary to bind an oxLDL molecule (Xie et al. 2004). Proteolytic cleavage of the extracellular domain can lead to two almost identical soluble LOX-1 forms (Murase et al. 2000). Whether these soluble forms correlate with the formation of membrane-bound LOX-1 or represent antagonistic forms is under discussion (Mehta et al. 2006; Sawamura et al. 2012).

Several ligands have been described for LOX-1 (Fig. 10.1). The LOX-1 receptor is able to mediate the uptake of minimally and maximally oxidized LDL (Tanigawa et al. 2006). Interestingly, LOX-1 is also involved in the phagocytosis of aged and apoptotic cells (Oka et al. 1998). In addition, a role of LOX-1 in thrombocyte adhesion and in the interaction between bacterial proteins and endothelial cells in sepsis has been described (Chen et al. 2002). LOX-1 could also act as a receptor of advanced glycation end products (AGEs) (Chen et al. 2008). These data suggest a role of LOX-1 in a variety of physiological and pathophysiological processes.

Human LOX-1 is encoded by the oxidized low-density lipoprotein receptor (OLR1) gene. LOX-1 is a single-copy gene located on the short arm of chromosome 12. Genes encoding natural killer cell (NK) receptors (NKG2A, C, D, E, CD94, CD69, and NKR-P1A) are mapped in this chromosomal region as well. LOX-1 is clustered within this "natural killer gene complex" (Aoyama et al. 1999; Yamanaka et al. 1998). The LOX-1 gene has six exons (Yamanaka et al. 1998). The LOX-1 gene structure suggests a relationship between the exon-intron architecture and the resulting protein structure (Yamanaka et al. 1998). Exon 1 encodes a 5′ untranslated region (5′ UTR) and the intracellular cytoplasmic domain. The second exon encodes the transmembrane domain and the third exon the extracellular neck domain. Exons 4–6 encode the carbohydrate recognition lectin-like domain and exon 6 contains the 3′-untranslated region (Yamanaka et al. 1998; Aoyama et al. 1999). The architecture of LOX-1 gene is similar to other C-type lectin

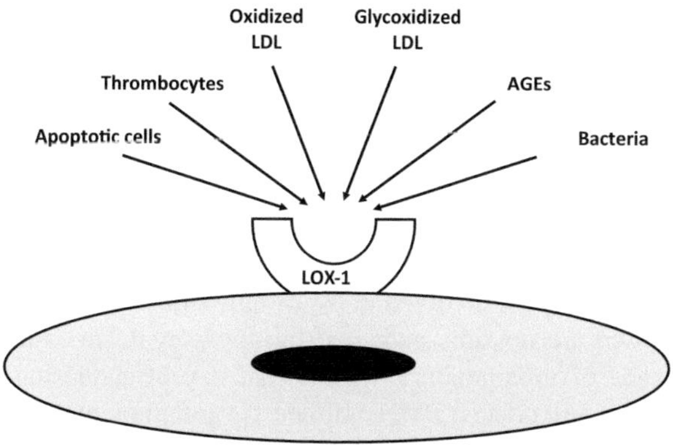

Fig. 10.1 The lectin-like oxLDL receptor LOX-1 mediates the uptake of oxidized or glycoxidized low-density lipoprotein (LDL) and advanced glycation end products (AGEs). In addition, it plays a role in endothelial phagocytosis of aged and apoptotic cells, thrombocyte adhesion, and the interaction between bacterial proteins and endothelial cells in sepsis

receptors expressed on NK cells. The exon-intron junctions in the carbohydrate recognition domain are well conserved between LOX-1 and other NK receptors (Yamanaka et al. 1998). The LOX-1 gene is located 200 kb telomeric of CD94. LOX-1 forms together with CLEC-1/CLEC1A, CLEC-2/CLEC1B, and DECTIN-1 a subfamily of C-type lectin domain receptor genes closely related to the NK receptor genes in the NK complex (Bull et al. 2000; Sobanov et al. 2001). DECTIN-1 shows the highest similarity with LOX-1, whereas CLEC-1/CLEC1A and CLEC-2/CLEC1B are more distantly related (Sobanov et al. 2001). C-type lectin receptors recognize endogenous proteinaceous ligands and are part of detection of virally infected and transformed cells. LOX-1 and other pattern recognition receptors (PRR) are expressed on the surface of myeloid cells like macrophages, dendritic cells (DCs), or neutrophils. PPRs are important for immediate innate immune response and also in regulation of adaptive immune response (Huysamen and Brown 2009; Sattler et al. 2012). Initially, C-type lectin receptors were defined by their carbohydrate-binding function, which is a calcium-dependent mechanism. In later studies, it became obvious that not all of these receptors bind exclusively carbohydrates, sometimes also independent of calcium (Sattler et al. 2012).

LOX-1 is considered to be a therapeutic target in atherosclerosis and cardiovascular disease (Kita 1999; Mehta et al. 2011; Morawietz 2010; Morawietz et al. 1999; Sawamura et al. 2015). Recently, novel roles of LOX-1 have been described in response to different danger signals (Sawamura et al. 2012), rheumatic diseases (Ishikawa et al. 2012), preeclampsia (English et al. 2013; Morton et al. 2012), and bone diseases (Nakayachi et al. 2015). The role of LOX-1 in immunity is less well understood. Therefore, this review article will focus on LOX-1 and immunity.

10.2 LOX-1, Atherosclerosis, and Immunity

The innate immune system is the first line of host defense against different pathogens and comprises eosinophil, neutrophil and basophil granulocytes, mast cells (MCs), monocytes, and macrophages as well as dendritic cells (DCs) and natural killer cells (NKs). These cells trigger inflammatory reactions, which are characteristic for specific diseases. Atherosclerosis is also an inflammatory disease, and innate immune cells are involved in the progression of plaque formation. MCs express PRR which are possible activated by microorganisms and allergens and trigger the release of inflammatory signals that can affect lesion development. Polymorphonuclear leukocytes (PMN) initiate the inflammatory response and are the first line of innate immunity. These cells produce reactive oxygen species (ROS), myeloperoxidase, proteolytic enzymes, and leukotrienes and thereby eliminate microbial pathogens and promote tissue destruction. Most of their functions originate from signal mediated through toll-like receptors (TLR) or C-type lectin receptors. NKs play an important role in response to viruses and tumors. Upon activation by affected cells, perforin-induced cytotoxicity and secretion of pro-inflammatory cytokines increase thus leading to elimination, induction of inflammation, and polarization of adaptive immune response. Production of IFN-γ is thought to be a key mechanism in lesion formation. Monocytes play an important role during progression of atherosclerosis. They mediate the recruitment and adhesion of monocytes to activated endothelial cells. Monocytes are thought to continuously repopulate macrophage or DCs and produce pro-inflammatory cytokines like TNF-α, IL-1β, and IL-6. Macrophages derive from monocytes, produce high amounts of ROS and pro-inflammatory cytokines like TNF-α or IL-1β, and participate in Th1 polarization (Chavez-Sanchez et al. 2014). Scavenger receptors (SR-AI/II, CD36, SR-BI, macrosialin/CD68, LOX-1) expressed on macrophages mediate oxLDL uptake and foam cell formation (Xu et al. 2013). DCs are also present in atherosclerotic lesions. During their maturation, they increase MHC molecules, chemokine receptors, and cytokine production and induce different subtypes of $CD4^+$ T cells as well as naive and memory B cells or NK cells (Chavez-Sanchez et al. 2014). LOX-1 expression was found on human monocytes, macrophages, peripheral blood myeloid DC, and B cells, whereas mature $CD83^+$ DCs and T lymphocytes do not express LOX-1 (Delneste et al. 2002; Delneste 2004). Human monocyte-derived macrophages express higher LOX-1 levels than monocytes. LOX-1 expression on primary human monocytes is very low, but differentiation into macrophages increases LOX-1 (Yoshida et al. 1998; Moriwaki et al. 1998). Especially, the role of LOX-1 in monocytes and macrophages was studied intensively in the last years. Several in vitro studies showed an inducible expression of LOX-1 by pro-inflammatory cytokines, which are involved in immune response. Upon recognition of antigen-MHC complexes on the surface of APCs, $CD4^+$ T cells differentiate into effector and regulatory subsets. These cells exert a different cytokine expression profile and belong to the adaptive immune response. Th1 cells mainly secrete IFN-γ, TNF-α, and IL-2, whereas Th2 cells secrete IL-4. Th1 cells enhance cellular immune response and differentiate from naive T cells in response to IL-12 and IL-18. Differentiation of naive T cells toward Th17 cells, a subpopulation required during fungi and bacterial infections, depends on TGF-β, IL-6, and IL-1β (Ait-Oufella et al. 2014; Lintermans et al. 2014).

Tumor necrosis factor-alpha (TNF-α) and phorbol 12-myristate 13-acetate (PMA) increased LOX-1 expression and oxLDL uptake in endothelial cells (Kume et al. 1998), TNF-α increased LOX-1 expression in macrophages (Moriwaki et al. 1998; Hashizume and Mihara 2012) and vascular smooth muscle cells (Hofnagel et al. 2004). C-reactive protein (CRP), an acute-phase protein associated with progression of cardiovascular outcomes, can actively bind to LOX-1 and thereby activating the complement system via the classical C1q pathway. Blocking of LOX-1 by antibody reduced CRP-induced leukocyte infiltration (Fujita et al. 2011). Several other studies confirmed CRP-induced LOX-1 expression in macrophages (Zhao et al. 2011) and endothelial cells (Li et al. 2004). CRP stimulates oxLDL uptake into endothelial cells, an effect restored by a LOX-1-blocking antibody. LOX-1 expression was also enhanced by interleukin-6 (IL-6) in endothelial cells (Li et al. 2004; Lubrano and Balzan 2013) and transforming growth factor-beta 1 (TGF-β1) in macrophages (Draude and Lorenz 2000; Minami et al. 2000), endothelial cells, and smooth muscle cells (Minami et al. 2000). Several other cytokines are able to induce LOX-1 as well: IL-1α and IL-1β in vascular smooth muscle cells (Hofnagel et al. 2004), IL-4 in macrophages (Higuchi et al. 2001), and IL-18 in TNF-α-activated endothelial cells (Mitsuoka et al. 2009). Histamine is a key player in allergic reactions and is released from dense granules of mast cells and basophiles. Histamine-releasing cells are located in connective tissue, mucosal surface of the lung and the gastrointestinal tract, and dermis of the skin. LOX-1 expression was increased by histamine in monocytes and macrophages, underlining the connection between immune cells in atherogenesis (Higuchi et al. 2001; Tanimoto et al. 2001; Alanne-Kinnunen et al. 2014). Mast cells are histamine-releasing cells, but they also secrete IL-1α, IL-1β, TNF-α, TGF-β, or IFN-γ and link inflammation to atherogenesis. A recently published study underlines a novel role of LOX-1 in macrophages and mast cells. Media captured from activated primary human mast cells induced LOX-1 expression in human monocyte-derived macrophages. Cell culture media of mast cells contained histamine, TNF-α, TGF-β1, and IL-1α. Histamine, TNF-α, and TGF-β1 were able to induce LOX-1 expression also in a synergistic way. However, cellular uptake of oxLDL was not affected by secretory mast cell products. Mast cells neighboring macrophages may stimulate LOX-1 expression and thereby promoting a paracrine modulatory effect on their function. Therefore, LOX-1 could be a novel link between mast cell and macrophages in regulation of innate immune reactions in atherosclerosis (Alanne-Kinnunen et al. 2014). In vitro studies revealed that lipopolysaccharide (LPS) increases LOX-1 expression, augments cellular oxLDL uptake, and promotes foam cell formation. Treatment with an LOX-1 antibody reduced the cellular oxLDL uptake. It seems that LOX-1 promotes cellular oxLDL uptake in activated macrophages and contributes to cardiovascular diseases (Hossain et al. 2015).

A link between LOX-1 and toll-like receptors (TLRs) was also shown in macrophages. TLRs are pattern recognition receptors expressed on macrophages, dendritic cells, or other innate immune cells (Lee et al. 2008). TLRs are signaling receptors, involved in cell activation after contact with pathogens (Jeannin et al. 2005). Activation of TLR9 increased LOX-1 expression in macrophages and enhanced foam cell formation. TLR9 blockade reduced LOX-1 expression and foam cell formation. In contrast, LOX-1 induction and foam cell formation were found after activation of TLR1/2, TLR2, TLR5, and TLR6. TLR4 induced the strongest effect. Therefore, TLRs are able to induce foam cell formation, a mechanism that involves LOX-1 and supports its importance in atherogenesis (Lee et al. 2008).

10.3 LOX-1, Dendritic Cells, and Immunity

Monocytes are able to differentiate into dendritic cells (DCs), a specific subtype of leukocytes. DCs play an important role in the innate and adaptive immune response (Pirillo et al. 2013). Increasing evidence suggests that DCs are involved in the process of atherosclerosis by activating T cells and stimulation of vascular inflammation (Huang et al. 2012). DCs are antigen-presenting cells (APCs). Exogenous antigens are endocytosed by APC and mainly loaded into MHC class II molecules for recognition by $CD4^{+}$ T cells. In contrast, endogenous antigens are presented by MHC class I molecules to be recognized by $CD8^{+}$ T cells. Only DCs are able to present exogenous antigens in MHC class I molecules, a process called cross-presentation. In this case, DCs are able to prime both $CD4^{+}$ and $CD8^{+}$ T cells for generating cytotoxic T cell responses. The initiation of an immune response by DCs requires the capture and internalization of antigens by endocytotic receptors (Delneste 2004; Delneste et al. 2002). The first reports about a role of LOX-1 in dendritic cell-mediated antigen cross-presentation were published in 2002. This study showed that LOX-1 is a binding structure for heat-shock protein 70 (Hsp70) on DCs. An anti-LOX-1 monoclonal antibody partly prevented binding of Hsp70 to LOX-1 in DCs and macrophages. HSPs are chaperones that control the folding and prevent the aggregation of proteins. Tumor-derived HSPs initiate a protective and tumor-specific cytolytic T cell response (Delneste 2004). HSPs form complexes with non-covalent peptides derived from tumor antigens and bind to DCs and macrophages. The internalization requires mainly scavenger receptors and is followed by co-localization with MHC class I molecules (Delneste 2004). Targeting LOX-1 reduced the Hsp70-induced IL-2 production in vitro, suggesting a role of LOX-1 in antigen cross-presentation in MHC class I molecules. In addition, injection of a LOX-1 antibody triggered an antitumor response in a murine tumor model (Delneste et al. 2002). Further studies addressed the role of HSPs as ligands for LOX-1 in DCs. Hsp60 stimulates intracellular signaling molecules that serve as danger signals of stressed and damaged cells to the innate immune system. Hsp60 interacts with TLR2 and TLR4 and thereby induces a pro-inflammatory response. Hsp60 binds to LOX-1 and is endocytosed by immature bone marrow-derived DCs. Hsp60 most likely delivers antigens that are incorporated into MHC class I molecules, as measured by an increased IL-2 secretion. Both, uptake and cross-presentation in MHC class I complexes were inhibited by a LOX-1 antibody. Inhibition of LOX-1 also decreased the number of cytotoxic T lymphocytes in response to Hsp65, a protein that elicits protective immunity against tumors (Xie et al. 2010). As described above, LOX-1 is expressed on mature DCs and is upregulated during the maturation process from monocytes to DCs. The best characterized LOX-1 ligand, oxLDL, is taken up by DCs. This process is prevented by a LOX-1-targeting antibody (Nickel et al. 2009; Delneste et al. 2002). OxLDL induces the maturation and differentiation of DC and increases secretion of pro-inflammatory IL-6, decreases anti-inflammatory IL-10, and enhances NF-κB activity in DCs (Nickel et al. 2009). Like oxLDL, angiotensin II (Ang II) is able to stimulate dendritic cell maturation, as measured by an increased CD83 and HLA-DR expression. Cytokines IL-12 and IFN-γ were increased as well. Treatment of DCs with oxLDL or Ang II increased proliferation of T cells. It seems likely that oxLDL increases Ang II secretion in DCs in an autocrine manner. Ang II is one player in oxLDL-induced maturation of DCs. OxLDL and Ang II induced the LOX-1 expression in DCs. AT_1 receptor blocker losartan could restore this. The direct inhibition of LOX-1 by an antibody suppressed oxLDL- and Ang II-induced maturation of DCs. An additional protective effect of losartan as standard therapy in hypertension and heart failure could be the inhibition of LOX-1 (Huang et al. 2012). A recently published study indicates that LOX-1 is also part of the humoral immune response. DCs treated with an anti-LOX-1 antibody (αLOX-1) promoted naive B cell proliferation and differentiation into plasma blasts (PBs) that secrete class-switched immune globulins (Igs). B cells co-cultured with αLOX-1-treated DCs increased their production of IgM, IgG, and IgA, indicating an increased Ig secretion. αLOX-1-treated DCs induced activated B cells to express CCR10. This promotes the migration toward the ligands CCL28 and CCL27. This upregulation in combination with the downregulation of CXCR5 enables PBs to migrate out of the germinal centers toward the mucosal site or to the skin (Joo et al. 2014). To simulate the differentiation of monocytes to DCs in vitro, cells were typically treated with interferon-α (IFN-α). During the differentiation process, LOX-1 levels are high, whereas further treatment with LPS to induce ultimate maturation decreases LOX-1. IFN-α-differentiated DCs internalize apoptotic allogenic lymphocytes and stimulate production of autologous $CD8^+$ T cells. Uptake and proliferative response of $CD8^+$ T cell were prevented by an anti-LOX-1 antibody, suggesting an important role of LOX-1 in both processes. $CD4^+$ T cells were additionally activated, however, to a lesser extent. A greater activation of CD8+ T cells is often seen in HSP-induced activation, also mediated by LOX-1 in IFN-α-differentiated DCs. $CD4^+$ T cells might act as helper cells to induce priming, expansion, and survival of cytotoxic $CD8^+$ T cells. IFN-α-induced LOX-1 expression in DCs and subsequent uptake of apoptotic cells lead to a simultaneous rearrangement of MHC class I and II molecules. An intracellular co-localization of apoptotic cells with MHC class I and II molecules was blocked by an LOX-1 antibody. LOX-1 seems to be an important player in mediating endocytosis of apoptotic cells into IFN-α-shaped DCs and also in inducing molecules of MHC class I and II complexes. It is still under debate whether this pathway plays mainly beneficial or deleterious effects in cancer or autoimmune diseases (Parlato et al. 2010).

Earlier studies in endothelial cells revealed that LOX-1 mediates binding and phagocytosis of aged red blood cells and apoptotic cells. Phosphatidylserine (PS) is externalized during early steps of apoptosis and is recognized by LOX-1. LOX-1-mediated phagocytosis was inhibited by PS and oxLDL. Under physiological conditions, it might be possible that LOX-1 promotes removal of PS-positive cells to maintain the anticoagulant state of endothelial cells. However, under atherosclerotic conditions with increased oxLDL formation, aged and apoptotic cells remain in the blood stream and can promote thrombotic events (Oka et al. 1998).

First studies revealed that LOX-1 supports the adhesion of Gram-positive (*Staphylococcus aureus*) and Gram-negative (*Escherichia coli*) bacteria to endothelial cells. A monoclonal LOX-1 antibody blocked the binding of both types of bacteria. TNF-α treatment further enhanced the adhesion of *Escherichia coli* to endothelial cells, whereas *Staphylococcus aureus* did not while requiring additional molecules in this setting (Shimaoka et al. 2001). LOX-1 is also part of the recognition of outer membrane protein A (OmpA), a major compound of the outer membrane of Gram-negative *Enterobacteriaceae*. OmpA is a highly conserved moiety for recognition by innate immune receptors. Next to LOX-1 as an endocytotic receptor, toll-like receptor (TLR)-2 is required in the OmpA-induced innate immune response. LOX-1 and TLR2 co-localize and cooperate to trigger subsequent TLR2-mediated cellular responses (Jeannin et al. 2005). Attachment of bacteria to endothelial cells as well as mucosal or epithelial cells is an initial step in bacterial infections. LOX-1 seems to be involved in this process as a cell surface receptor (Jeannin et al. 2005; Shimaoka et al. 2001).

Recently published data underlined the role of LOX-1 in the adhesion of bacterial proteins to macrophages. GroEL is expressed on the surface of common bacteria and is involved in attachment and immune modulation. GroEL in *E. coli* is induced by heat-shock stress and increases phagocytosis by macrophages. LOX-1 facilitates the uptake of *E. coli* in macrophages, mainly by interaction with GroEL. This effect was inhibited by an anti-LOX-1 antibody. Furthermore, GroEL increases LOX-1 expression in macrophages, an effect also seen for other HSPs. Targeting LOX-1 by using a specific antibody increased survival of *E. coli*-induced peritonitis in mice (Zhu et al. 2013).

LOX-1 seems to be involved in the process of endotoxin-induced inflammation as well. Inhibition of LOX-1 could rescue rats from death after LPS treatment. Additionally, leukopenia was reduced by anti-LOX-1 treatment. In a rat model of low-dose endotoxin-induced uveitis, blocking of LOX-1 reduced the number of infiltrated leukocytes and protein exudation. In vivo experiments in retinal blood vessels confirmed that inhibition of LOX-1 decreases the number of rolling leukocytes and increases their velocity. LOX-1 seems to be an adhesion molecule promoting leukocyte recruitment and rolling during endotoxin-induced inflammation (Honjo et al. 2003). An important role of LOX-1 during sepsis was shown in mice lacking the functional active LOX-1 (LOX-1$^{-/-}$). These animals have substantial higher survival rates after inducing sepsis by cecal ligation and puncture, suggesting that LOX-1 is able to modulate host response. Systemic and lung-specific TNF-α and IL-6 levels were lower in septic LOX-1$^{-/-}$ mice and bacterial clearance as represented by reduced colony forming units (CFU) in peritoneal cavity.

Neutrophils are a subpopulation of leucocytes and play a pivotal role in host defense and during sepsis. Neutrophil migration is impaired during sepsis. LOX-1$^{-/-}$ mice revealed an increased number of neutrophils, less affected chemotaxis, and an improved CXCR2 surface expression which might help to maintain a better neutrophil response. Neutrophils also express higher levels of LOX-1. LPS treatment affects rather LOX-1 surface than total protein expression. The increased neutrophil surface expression during sepsis is modulated by TLR2 and TLR4 (Wu et al. 2011). An association between LOX-1 and TLR2 and TLR4 was shown in monocyte-derived DCs. Activators of TLR2 and TLR4 increased LOX-1 expression. Diesel exhaust particles (DEP) as major air pollutants from diesel cars further enhanced LOX-1 expression. It seems likely that DEP modulate signaling pathways activated by TLR and thereby control SR expression. These data suggest that LOX-1 and other SRs play also a role in DEP-induced lung diseases (Taront et al. 2009).

10.4 Conclusion

In conclusion, the LOX-1 receptor mediates a variety of physiological and pathophysiological processes. LOX-1 could be a therapeutic target in the pathogenesis of atherosclerosis, rheumatic, and bone diseases. Increasing evidence supports a novel role of LOX-1 in dendritic cells and immunity.

Acknowledgement This study has been supported by the European Section of the Aldosterone Council (ESAC) Deutschland (aldosterone, LOX-1, and adipose tissue to AH, HM), the *Else Kröner-Fresenius-Stiftung* (2010_A105 to HM), the Doktor Robert Pfleger Foundation, Bamberg, Germany (adiposity and endothelial function; LOX-1, aldosterone, mineralocorticoid receptors, and vascular function to HM), the German Federal Ministry of Education and Research (BMBF) (Professorship of Vascular Endothelium and Microcirculation to HM), and the *Deutsche Forschungsgemeinschaft* (DFG) (MO 1695/4-1, MO 1695/5-1, Excellence Initiative by the German Federal and State Governments (Institutional Strategy, measure "support the best" to HM).

References

Ait-Oufella H, Sage AP, Mallat Z, Tedgui A (2014) Adaptive (T and B cells) immunity and control by dendritic cells in atherosclerosis. Circ Res 114:1640–1660

Alanne-Kinnunen M, Lappalainen J, Oorni K, Kovanen PT (2014) Activated human mast cells induce LOX-1-specific scavenger receptor expression in human monocyte-derived macrophages. PLoS One 9:e108352

Aoyama T, Sawamura T, Furutani Y, Matsuoka R, Yoshida MC, Fujiwara H, Masaki T (1999) Structure and chromosomal assignment of the human lectin-like oxidized low-density-lipoprotein receptor-1 (LOX-1) gene. Biochem J 339(Pt 1):177–184

Bull C, Sobanov Y, Rohrdanz B, O'Brien J, Lehrach H, Hofer E (2000) The centromeric part of the human NK gene complex: linkage of LOX-1 and LY49L with the CD94/NKG2 region. Genes Immun 1:280–287

Chavez-Sanchez L, Espinosa-Luna JE, Chavez-Rueda K, Legorreta-Haquet MV, Montoya-Diaz E, Blanco-Favela F (2014) Innate immune system cells in atherosclerosis. Arch Med Res 45:1–14

Chen M, Masaki T, Sawamura T (2002) LOX-1, the receptor for oxidized low-density lipoprotein identified from endothelial cells: implications in endothelial dysfunction and atherosclerosis. Pharmacol Ther 95:89–100

Chen X, Zhang T, Du G (2008) Advanced glycation end products serve as ligands for lectin-like oxidized low-density lipoprotein receptor-1(LOX-1): biochemical and binding characterizations assay. Cell Biochem Funct 26:760–770

Delneste Y (2004) Scavenger receptors and heat-shock protein-mediated antigen cross-presentation. Biochem Soc Trans 32:633–635

Delneste Y, Magistrelli G, Gauchat J, Haeuw J, Aubry J, Nakamura K, Kawakami-Honda N, Goetsch L, Sawamura T, Bonnefoy J, Jeannin P (2002) Involvement of LOX-1 in dendritic cell-mediated antigen cross-presentation. Immunity 17:353–362

Draude G, Lorenz RL (2000) TGF-beta1 downregulates CD36 and scavenger receptor A but upregulates LOX-1 in human macrophages. Am J Physiol Heart Circ Physiol 278:H1042–H1048

English FA, McCarthy FP, Mcsweeney CL, Quon AL, Morton JS, Sawamura T, Davidge ST, Kenny LC (2013) Inhibition of lectin-like oxidized low-density lipoprotein-1 receptor protects against plasma-mediated vascular dysfunction associated with pre-eclampsia. Am J Hypertens 26:279–286

Fujita Y, Yamaguchi S, Kakino A, Iwamoto S, Yoshimoto R, Sawamura T (2011) Lectin-like oxidized LDL receptor 1 is involved in CRP-mediated complement activation. Clin Chem 57:1398–1405

Hashizume M, Mihara M (2012) Atherogenic effects of TNF-alpha and IL-6 via up-regulation of scavenger receptors. Cytokine 58:424–430

Higuchi S, Tanimoto A, Arima N, Xu H, Murata Y, Amada T, Makishima K, Sasaguri Y (2001) Effects of histamine and interleukin-4 synthesized in arterial intima on phagocytosis by monocytes/macrophages in relation to atherosclerosis. FEBS Lett 505:217–222

Hofnagel O, Luechtenborg B, Stolle K, Lorkowski S, Eschert H, Plenz G, Robenek H (2004) Proinflammatory cytokines regulate LOX-1 expression in vascular smooth muscle cells. Arterioscler Thromb Vasc Biol 24:1789–1795

Honjo M, Nakamura K, Yamashiro K, Kiryu J, Tanihara H, McEvoy LM, Honda Y, Butcher EC, Masaki T, Sawamura T (2003) Lectin-like oxidized LDL receptor-1 is a cell-adhesion molecule involved in endotoxin-induced inflammation. Proc Natl Acad Sci U S A 100:1274–1279

Hossain E, Ota A, Karnan S, Takahashi M, Mannan SB, Konishi H, Hosokawa Y (2015) Lipopolysaccharide augments the uptake of oxidized LDL by up-regulating lectin-like oxidized LDL receptor-1 in macrophages. Mol Cell Biochem 400:29–40

Huang D, Lu H, Liu H, Yao K, Sun A, Zou Y, Ge J (2012) Losartan attenuates human monocyte-derived dendritic cell immune maturation via downregulation of lectin-like oxidized low-density lipoprotein receptor-1. J Cardiovasc Pharmacol 60:133–139

Huysamen C, Brown GD (2009) The fungal pattern recognition receptor, Dectin-1, and the associated cluster of C-type lectin-like receptors. FEMS Microbiol Lett 290:121–128

Ishikawa M, Ito H, Akiyoshi M, Kume N, Yoshitomi H, Mitsuoka H, Tanida S, Murata K, Shibuya H, Kasahara T, Kakino A, Fujita Y, Sawamura T, Yasuda T, Nakamura T (2012) Lectin-like oxidized low-density lipoprotein receptor 1 signal is a potent biomarker and therapeutic target for human rheumatoid arthritis. Arthritis Rheum 64:1024–1034

Jeannin P, Bottazzi B, Sironi M, Doni A, Rusnati M, Presta M, Maina V, Magistrelli G, Haeuw JF, Hoeffel G, Thieblemont N, Corvaia N, Garlanda C, Delneste Y, Mantovani A (2005) Complexity and complementarity of outer membrane protein A recognition by cellular and humoral innate immunity receptors. Immunity 22:551–560

Joo H, Li D, Dullaers M, Kim TW, Duluc D, Upchurch K, Xue Y, Zurawski S, Le Grand R, Liu YJ, Kuroda M, Zurawski G, Oh S (2014) C-type lectin-like receptor LOX-1 promotes dendritic cell-mediated class-switched B cell responses. Immunity 41:592–604

Kita T (1999) LOX-1, a possible clue to the missing link between hypertension and atherogenesis. Circ Res 84:1113–1115

Kume N, Murase T, Moriwaki H, Aoyama T, Sawamura T, Masaki T, Kita T (1998) Inducible expression of lectin-like oxidized LDL receptor-1 in vascular endothelial cells. Circ Res 83:322–327

Lee JG, Lim EJ, Park DW, Lee SH, Kim JR, Baek SH (2008) A combination of Lox-1 and Nox1 regulates TLR9-mediated foam cell formation. Cell Signal 20:2266–2275

Li L, Roumeliotis N, Sawamura T, Renier G (2004) C-reactive protein enhances LOX-1 expression in human aortic endothelial cells: relevance of LOX-1 to C-reactive protein-induced endothelial dysfunction. Circ Res 95:877–883

Lintermans LL, Stegeman CA, Heeringa P, Abdulahad WH (2014) T cells in vascular inflammatory diseases. Front Immunol 5:504

Lubrano V, Balzan S (2013) LOX-1, a new marker of risk and prognosis in coronary artery disease? Mol Cell Biochem 383:223–230

Mehta JL, Rasouli N, Sinha AK, Molavi B (2006) Oxidative stress in diabetes: a mechanistic overview of its effects on atherogenesis and myocardial dysfunction. Int J Biochem Cell Biol 38:794–803

Mehta JL, Khaidakov M, Hermonat PL, Mitra S, Wang X, Novelli G, Sawamura T (2011) LOX-1: a new target for therapy for cardiovascular diseases. Cardiovasc Drugs Ther 25:495–500

Minami M, Kume N, Kataoka H, Morimoto M, Hayashida K, Sawamura T, Masaki T, Kita T (2000) Transforming growth factor-beta(1) increases the expression of lectin-like oxidized low-density lipoprotein receptor-1. Biochem Biophys Res Commun 272:357–361

Mitsuoka H, Kume N, Hayashida K, Inui-Hayashiada A, Aramaki Y, Toyohara M, Jinnai T, Nishi E, Kita T (2009) Interleukin 18 stimulates release of soluble lectin-like oxidized LDL receptor-1 (sLOX-1). Atherosclerosis 202:176–182

Morawietz H (2010) LOX-1 receptor as a novel target in endothelial dysfunction and atherosclerosis. Dtsch Med Wochenschr 135:308–312

Morawietz H, Rueckschloss U, Niemann B, Duerrschmidt N, Galle J, Hakim K, Zerkowski HR, Sawamura T, Holtz J (1999) Angiotensin II induces LOX-1, the human endothelial receptor for oxidized low-density lipoprotein. Circulation 100:899–902

Moriwaki H, Kume N, Kataoka H, Murase T, Nishi E, Sawamura T, Masaki T, Kita T (1998) Expression of lectin-like oxidized low density lipoprotein receptor-1 in human and murine macrophages: upregulated expression by TNF-alpha. FEBS Lett 440:29–32

Morton JS, Abdalvand A, Jiang Y, Sawamura T, Uwiera RR, Davidge ST (2012) Lectin-like oxidized low-density lipoprotein 1 receptor in a reduced uteroplacental perfusion pressure rat model of preeclampsia. Hypertension 59:1014–1020

Muller G, Morawietz H (2009) Nitric oxide, NAD(P)H oxidase, and atherosclerosis. Antioxid Redox Signal 11:1711–1731

Murase T, Kume N, Kataoka H, Minami M, Sawamura T, Masaki T, Kita T (2000) Identification of soluble forms of lectin-like oxidized LDL receptor-1. Arterioscler Thromb Vasc Biol 20:715–720

Nakayachi M, Ito J, Hayashida C, Ohyama Y, Kakino A, Okayasu M, Sato T, Ogasawara T, Kaneda T, Suda N, Sawamura T, Hakeda Y (2015) Lectin-like oxidized low-density lipoprotein receptor-1 abrogation causes resistance to inflammatory bone destruction in mice, despite promoting osteoclastogenesis in the steady state. Bone 75:170–182

Nickel T, Schmauss D, Hanssen H, Sicic Z, Krebs B, Jankl S, Summo C, Fraunberger P, Walli AK, Pfeiler S, Weis M (2009) oxLDL uptake by dendritic cells induces upregulation of scavenger-receptors, maturation and differentiation. Atherosclerosis 205:442–450

Oka K, Sawamura T, Kikuta K, Itokawa S, Kume N, Kita T, Masaki T (1998) Lectin-like oxidized low-density lipoprotein receptor 1 mediates phagocytosis of aged/apoptotic cells in endothelial cells. Proc Natl Acad Sci U S A 95:9535–9540

Parlato S, Romagnoli G, Spadaro F, Canini I, Sirabella P, Borghi P, Ramoni C, Filesi I, Biocca S, Gabriele L, Belardelli F (2010) LOX-1 as a natural IFN-alpha-mediated signal for apoptotic cell uptake and antigen presentation in dendritic cells. Blood 115:1554–1563

Pirillo A, Norata GD, Catapano AL (2013) LOX-1, OxLDL, and atherosclerosis. Mediators Inflamm 2013:152786

Sattler S, Ghadially H, Hofer E (2012) Evolution of the C-type lectin-like receptor genes of the DECTIN-1 cluster in the NK gene complex. Sci World J 2012:931386

Sawamura T, Kume N, Aoyama T, Moriwaki H, Hoshikawa H, Aiba Y, Tanaka T, Miwa S, Katsura Y, Kita T, Masaki T (1997) An endothelial receptor for oxidized low-density lipoprotein. Nature 386:73–77

Sawamura T, Kakino A, Fujita Y (2012) LOX-1: a multiligand receptor at the crossroads of response to danger signals. Curr Opin Lipidol 23:439–445

Sawamura T, Wakabayashi I, Okamura T (2015) LOX-1 in atherosclerotic disease. Clin Chim Acta 440:157–163

Shashkin P, Dragulev B, Ley K (2005) Macrophage differentiation to foam cells. Curr Pharm Des 11:3061–3072

Shimaoka T, Kume N, Minami M, Hayashida K, Sawamura T, Kita T, Yonehara S (2001) LOX-1 supports adhesion of gram-positive and gram-negative bacteria. J Immunol 166:5108–5114

Sobanov Y, Bernreiter A, Derdak S, Mechtcheriakova D, Schweighofer B, Duchler M, Kalthoff F, Hofer E (2001) A novel cluster of lectin-like receptor genes expressed in monocytic, dendritic and endothelial cells maps close to the NK receptor genes in the human NK gene complex. Eur J Immunol 31:3493–3503

Tanigawa H, Miura S, Zhang B, Uehara Y, Matsuo Y, Fujino M, Sawamura T, Saku K (2006) Low-density lipoprotein oxidized to various degrees activates ERK1/2 through Lox-1. Atherosclerosis 188:245–250

Tanimoto A, Murata Y, Nomaguchi M, Kimura S, Arima N, Xu H, Hamada T, Sasaguri Y (2001) Histamine increases the expression of LOX-1 via H2 receptor in human monocytic THP-1 cells. FEBS Lett 508:345–349

Taront S, Dieudonne A, Blanchard S, Jeannin P, Lassalle P, Delneste Y, Gosset P (2009) Implication of scavenger receptors in the interactions between diesel exhaust particles and immature or mature dendritic cells. Part Fibre Toxicol 6:9

Wu Z, Sawamura T, Kurdowska AK, Ji HL, Idell S, Fu J (2011) LOX-1 deletion improves neutrophil responses, enhances bacterial clearance, and reduces lung injury in a murine polymicrobial sepsis model. Infect Immun 79:2865–2870

Xie Q, Matsunaga S, Niimi S, Ogawa S, Tokuyasu K, Sakakibara Y, Machida S (2004) Human lectin-like oxidized low-density lipoprotein receptor-1 functions as a dimer in living cells. DNA Cell Biol 23:111–117

Xie J, Zhu H, Guo L, Ruan Y, Wang L, Sun L, Zhou L, Wu W, Yun X, Shen A, Gu J (2010) Lectin-like oxidized low-density lipoprotein receptor-1 delivers heat shock protein 60-fused antigen into the MHC class I presentation pathway. J Immunol 185:2306–2313

Xu S, Ogura S, Chen J, Little PJ, Moss J, Liu P (2013) LOX-1 in atherosclerosis: biological functions and pharmacological modifiers. Cell Mol Life Sci 70:2859–2872

Yamanaka S, Zhang XY, Miura K, Kim S, Iwao H (1998) The human gene encoding the lectin-type oxidized LDL receptor (OLR1) is a novel member of the natural killer gene complex with a unique expression profile. Genomics 54:191–199

Yoshida H, Kondratenko N, Green S, Steinberg D, Quehenberger O (1998) Identification of the lectin-like receptor for oxidized low-density lipoprotein in human macrophages and its potential role as a scavenger receptor. Biochem J 334(Pt 1):9–13

Zhao XQ, Zhang MW, Wang F, Zhao YX, Li JJ, Wang XP, Bu PL, Yang JM, Liu XL, Zhang MX, Gao F, Zhang C, Zhang Y (2011) CRP enhances soluble LOX-1 release from macrophages by activating TNF-alpha converting enzyme. J Lipid Res 52:923–933

Zhu H, Lee C, Zhang D, Wu W, Wang L, Fang X, Xu X, Song D, Xie J, Ren S, Gu J (2013) Surface-associated GroEL facilitates the adhesion of Escherichia coli to macrophages through lectin-like oxidized low-density lipoprotein receptor-1. Microbes Infect 15:172–180

Chapter 11
MGL/CD301 as a Unique C-Type Lectin Expressed on Dendritic Cells and Macrophages

Kaori Denda-Nagai and Tatsuro Irimura

Abstract Macrophage (MØ) galactose-type calcium-type lectin, MGL, also termed CD301 or Clec10A, is a type 2 transmembrane glycoprotein having a calcium-dependent carbohydrate recognition domain. Orthologous genes seem to be present throughout the vertebrates, and the lectin binds galactose and *N*-acetylgalactosamine among monosaccharides. Although mammalian MGL resembles the hepatic asialoglycoprotein receptor in its structure and its carbohydrate specificity, the cellular distribution of MGL is limited to specific subsets of dendritic cells (DCs) and MØs. Mouse MGL2, apparently the direct counterpart of human MGL, strongly binds clusters of carbohydrate chains with terminal *N*-acetylgalactosamine residues, a molecular feature characteristic to mucins. MGL was also shown to recognize carbohydrate chains from exogenous organisms. Comparisons of various oligosaccharides for their binding and identification of ligands/counter receptors by biochemical means strongly suggest that the important structural features recognized by MGL depend on the biological context. It is also likely that MGL functions on DCs to recognize and distinguish the self, altered self, pathogens, and commensal organisms, and that the regulatory outcomes depend on the unique nature of the DC/MØ subsets. Mice lacking the *Mgl1* or *Mgl2* gene do not show any abnormality as long as they are maintained under a controlled environment. However, upon encountering a pathogenic insult, they show a variety of outcomes depending on the organ sites and the type of the insult. Further investigations are necessary to explore how endogenous ligand/counter receptors are involved in the regulation of immunological and inflammatory responses by MGL.

Keywords MGL (CD301) • Calcium-type lectin • Dendritic cells • Macrophages • Tissue inflammation • Carbohydrate ligands

K. Denda-Nagai • T. Irimura, Ph.D. (✉)
Juntendo University School of Medicine, 2-1-1 Hongo, Bunkyo-ku, Tokyo 113-8421, Japan
e-mail: k-denda@juntendo.ac.jp; t-irimura@juntendo.ac.jp

S. Yamasaki (ed.), *C-Type Lectin Receptors in Immunity*,
DOI 10.1007/978-4-431-56015-9_11

11.1 Introduction: History and Facts on MGL/CD301

First reports on the presence of a C-type lectin on macrophages (MØs) that has specificity toward galactose and related monosaccharides appeared in the early 1980s (Imamura et al. 1984). Purification of the protein and subsequent molecular cloning from rats (Ii et al. 1988, 1990), mice (Sato et al. 1992; Oda et al. 1988), and humans (Suzuki et al. 1996) in the 1980s and 1990s revealed a lectin having a very similar structure to that of the hepatic C-type lectin, asialoglycoprotein receptor. Despite the subsequent discovery of a variety of C-type lectins in the immune system, its galactose-type specificity and the presence of a corresponding recognition motif at its carbohydrate recognition domain make this lectin distinct from others. The name MGL represents MØ galactose-type lectin.

MGL2, which shows strong homology to MGL1, was later found in mice (Tsuiji et al. 2002). Humans have a single MGL which apparently is more similar in its sequence, carbohydrate specificity, and expression profile to MGL2 than to MGL1, posing an unsolved question as to what kind of molecules cover the role of MGL1 in humans. MGL orthologues appear to be present in all vertebrates, suggesting that the molecule is evolutionally linked to the acquired immune system. MGL genes are located near the hepatic asialoglycoprotein receptor gene on chromosome 17 in humans and 6 in mice.

MGL in humans, MGL 1 and 2 in mice, and MGL in rats are type 2 transmembrane proteins with a ~M_r of 42 kDa. They have a *C*-terminal extracellular carbohydrate recognition domain, a stem domain responsible for multimer formation, a transmembrane domain, and a cytoplasmic *N*-terminal domain. They appear to form multimers, most likely trimers, and were shown to be present on the surfaces of activated or inflammatory MØs and on bone marrow-derived immature dendritic cells (DCs). A human MGL-specific antibody was used to define this molecule as CD301 and to show that it serves as a marker for myeloid DCs (Zola et al. 2005; Sano et al. 2007). CD301 expression is almost exclusive to MØs and DCs, with the only apparent exception being an expression on chondrocytes in embryonic mice as shown by the binding of a monoclonal antibody cross-reactive to MGL1 and MGL2 (Mizuochi et al. 1997, 1998). Antibodies specific for human MGL, mouse MGL1, or mouse MGL2 were generated and characterized for their specificity and other properties (Kimura et al. 1995; Denda-Nagai et al. 2010). In addition, knockout mice were prepared and backcrossed to C57BL/6 and BALB/c. Although they did not show any abnormality in their embryonic development, growth, behavior, or aging under controlled environmental conditions, they responded differently from wild-type mice upon the placement of a pathological insult. These observations are briefly described in the following chapters. For a summary of the milestones in the studies of MGL/CD301, refer to Table 11.1.

Table 11.1 Milestones in the studies on MGL/CD301

Year	Milestone	Reference
1984	Osawa's laboratory found Ca^{2+}-dependent carbohydrate-binding activity on activated MØs	Imamura et al. (1984)
1988	Galactose-binding lectin was purified from MØs of mice and rats	Oda 1988, Ii et al. (1988)
1990	cDNA cloning of the lectin later called MGL was done	Ii et al. (1990), Sato et al. (1992)
1995	Specific mAbs for mouse and human MGL were prepared	Kimura et al. (1995), Sano et al. (2007)
1996	Human MGL was identified and cloned	Suzuki et al. (1996)
1997	Tissue distribution of MGL^{+} cells was determined in mice	Mizuochi et al. (1997, 1998)
2002	The second MGL, MGL2, was found in mice	Tsuiji et al. (2002)
2002	*Mgl1−/−* mice were generated	Onami et al. (2002)
2002	MGL was found highly expressed on DCs	Denda-Nagai et al. (2002), Higashi et al. (2002)
2005	Human MGL was assigned as CD301 and became a marker for myeloid DCs	Zola et al. (2005)
2010	MGL2-specific mAb and *Mgl2−/−* mice were prepared	Denda-Nagai et al. (2010)
2013	$MGL2^{+}$ DC subset was found to promote Th2 response	Murakami et al. (2013); Kumamoto et al. (2013); Gao et al. (2013)

11.2 Carbohydrate Specificity and Ligands/Counter Receptors for MGL/CD301

It is important to keep in mind that the structural characteristics of ligands and counter receptors for MGL depend on the biological context. Considering that MGL orthologues are membrane-associated lectins exclusively expressed on MØs and DCs, i.e., cells specialized in surveying altered self and exogenous substances and organisms, MGL seemed to have evolved to recognize endogenous glycans and glycans from the outer milieu. This type of cellular distribution presents a strong contrast to the distribution of galectins (S-type lectins), which are expressed intracellularly and extracellularly by a variety of cells. Even though the monosaccharide and oligosaccharide specificity of some galectins and MGL may be similar, MGL is likely to perform functions associated with the initiation of immune and inflammatory processes. MGL is a family member of C-type lectins, which recognize saccharides in a Ca^{2+}-dependent manner. Most other C-type lectins are categorized as being either "mannose-specific" or "a protein having a C-type lectin fold but no carbohydrate ligands identified." Therefore, MGL and hepatic asialoglycoprotein receptors, which recognize galactose and *N*-acetylgalactosamine as a monosaccharide, were categorized into a unique and distinct category (Drickamer 1993).

Oligosaccharide specificity of recombinant mouse MGL1 and MGL2 as well as human MGL have been investigated by several methods employing defined collections of purified oligosaccharides. The most comprehensive comparison of K_D values was performed by frontal affinity chromatography, and the results corresponded well with fragmentary data from previous work (Suzuki et al. 1996; Tsuiji et al. 2002; Oo-Puthinan et al. 2008; Yamamoto et al. 1994). MGL1 has high affinity with oligosaccharides containing a terminal Gal-(Fuc-)GlcNAc, i.e., Lewis^X (Le^X) structure. MGL2 preferentially binds oligosaccharides with a terminal GalNAc residue (Fig. 11.1a). Our previous reports also indicated that MGL2 and human MGL preferentially bind clusters of *O*-linked GalNAc, i.e., Tn antigen and the cluster effect, is more prominent with the oligomeric form of intact MGL than with monomeric carbohydrate recognition domain of MGL (Fig. 11.1b) (Iida et al. 1999). Human MGL seems to have an affinity for sialylated Tn antigen, but MGL2 does not, making a slight distinction.

Important amino acid residues responsible for the differential specificity between MGL1 and MGL2 were investigated using pair-wise site-directed mutagenesis in the recombinant carbohydrate recognition domain (CRD) of MGL1 and MGL2 followed by the use of soluble polyacrylamide conjugated with either Le^X oligosaccharides or GalNAc (Oo-Puthinan et al. 2008). Mutation of Ala-89 to Arg and Thr -111 to Ser of MGL1 CRD caused significant reductions in Le^X binding. Significant increase in GalNAc binding was observed by single mutation of Ala-89 to Arg of MGL1. Likewise, mutation of Arg-89 to Ala and Ser-111 to Thr of MGL2 CRD resulted in the loss of β-GalNAc binding. MGL2 assumed Le^X binding by mutation of Arg-89 to Ala (Fig. 11.1c). Molecular modeling illustrated the potential for direct molecular interactions of Leu-61, Arg-89, and His-109 in MGL2 CRD with β-GalNAc (Fig. 11.1d).

To identify the Le^X-binding site on MGL1, a saturation transfer experiment for the MGL1-Le^X complex was carried out (Sakakura et al. 2008). To obtain sugar moiety-specific information on the interface between MGL1 and the Le^X trisaccharide, saturation transfer experiments where each of galactose-H5-, fucose-CH3-, and *N*-acetylglucosamine-CH3-selective saturations was applied to the MGL1-Le^X complex were performed. Based on the results, we proposed a Le^X-binding mode on MGL1 where the galactose moiety was bound to the primary sugar-binding site, including Asp-94, Trp-96, and Asp-118, and the fucose moiety interacted with the secondary sugar binding site, including Ala-89 and Thr-111. Ala-89 and Thr-111 in MGL1 are replaced with arginine and serine in MGL2, respectively. The hydrophobic environment formed by the small side chain of Ala-89 and the methyl group of Thr-111 seemed to be a requisite for the accommodation of the fucose moiety of the Le^X trisaccharide within the sugar-binding site of MGL1.

Cell surface lectins could potentially function as an uptake receptor (known for asialoglycoprotein receptor), as a regulator of cellular trafficking (known for selectins), or as a signaling molecule (according to recent reports). MGL could serve these functions for MØs and DCs. As described in the following sections, we have identified the cells expressing MGL1 and MGL2 in the skin and found that they migrate to regional lymph nodes (LNs) upon antigenic and other stimuli (Chun et al. 2000). The route of migration of these cells within the LNs, i.e., from subcapsular sinus to T cell areas, appeared to be unique. Thus, we assumed that

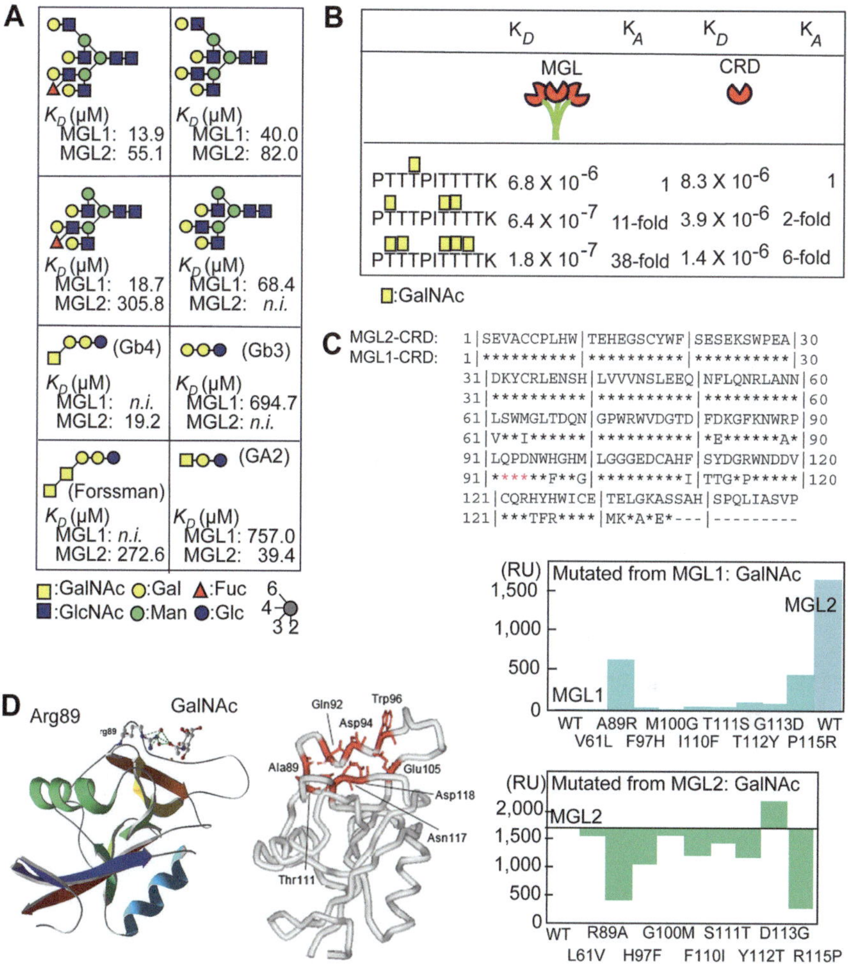

Fig. 11.1 Structural model and carbohydrate specificity of MGL1 and MGL2. (**a**) Oligosaccharide specificities of MGL1 and MGL2 analyzed by frontal affinity chromatography. A smaller K_D value indicates a higher affinity. (**b**) Cluster effects of GalNAc observed in human MGL. Human MGL interacting with Tn-bearing glycopeptides representing a sequence within the tandem repeats of MUC2. The cluster effect was more prominent with the oligomeric form of MGL. (**c**) *Upper panel*: Amino acid sequence of MGL1 and MGL2 CRD. *Two lower panels*: The binding of chimeric MGL1 and MGL2 CRD to GalNAc-polyacrylamide polymers based on the surface plasmon resonance data. Significant increase in GalNAc binding was observed by single mutation of Ala-89 to Arg of MGL1. Likewise, mutation of Arg-89 to Ala of MGL2 CRD resulted in the loss of GalNAc binding. (**d**) Predicted molecular interactions between MGL2 and GalNAc by the molecular modeling technique. These figures have been adapted from a previously published version (Oo-Puthinan et al. 2008)

counter receptors for MGL1 or MGL2 existed within LNs. By the use of affinity chromatography with immobilized MGL1, a few predominant proteins were purified from the lysates of LNs (Kumamoto et al. 2004). Mass spectroscopic analysis

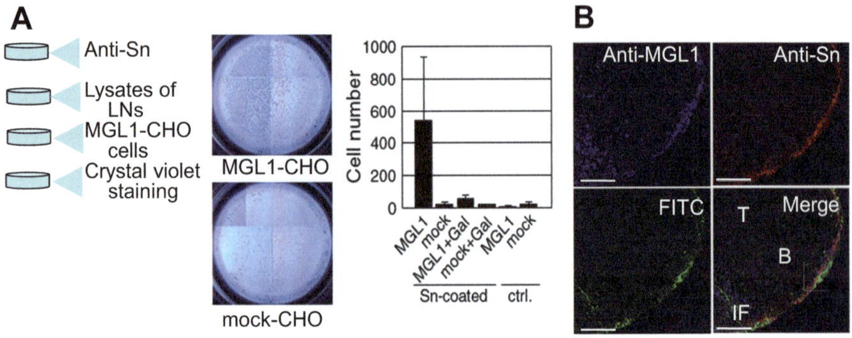

Fig. 11.2 Counter receptor in LNs for MGL1. (**a**) (*Left*) Schematic protocol to investigate the adhesion of cells expressing MGL1 (CHO-MGL1) to surfaces coated with sialoadhesin (Sn) from LNs. (*Middle*) Cell binding was monitored using a phase contrast microscope. (*Right*) Cell binding was shown as the number of adherent cells per well divided by 10. (**b**) Distribution of MGL1+ cells, Sn+ cells, and FITC in FITC-sensitized LNs. The merged image shows partial colocalization of MGL1 and Sn. Interfollicular regions (*IF*), and the border between T-cell areas (*T*) and B-cell areas (*B*) is shown. *Bars* indicate 100 μm. Figure 11.2a, b has been adapted from a previously published version (Kumamoto et al. 2004)

of one of these proteins indicated that the band corresponded to CD169, Siglec-1/sialoadhesin (Sn), which was known to be expressed on subcapsular sinus MØs. Sn purified from lysates of LNs was immobilized on microtiter plates precoated with anti-Sn monoclonal antibody, and binding of recombinant MGL1 as well as adhesion of cells expressing MGL1 was tested. The binding of recombinant MGL1 to Sn was shown to be dependent on Ca^{2+} and *N*-glycans present on Sn. MGL1-transfected Chinese hamster ovary cells adhered to the Sn-coated plates, whereas mock transfectants did not (Fig. 11.2a). Immunohistochemical localization of anti-Sn monoclonal antibody in LNs coincided with the subcapsular sinus area to which recombinant MGL1 was bound. Furthermore, after sensitization with fluorescein isothiocyanate (FITC), the distribution of MGL1-positive cells was demonstrated to overlap with that of Sn within the subcapsular sinus of draining LNs (Kumamoto et al. 2004) (Fig. 11.2b). These results suggest that Sn acts as an endogenous counter receptor for MGL1. Because the trafficking of cells expressing MGL1 was thought to be involved with the initiation phase of contact hypersensitivity, we investigated the distribution of MGL1-positive cells within the regional LNs with Sn knockout mice developed by Crocker and coworkers and the importance of the interaction on the development of hypersensitivity. The results revealed that despite the strong affinity of Sn with MGL1 and MGL1-positive cells, Sn was not involved with the regulation of trafficking of these cells to and within the regional LNs (unpublished data).

Apart from skin, MGL1 is also responsible for the signaling in intestinal MØs. Intestinal MØs express MGL1 to recognize commensal bacteria and down-modulate inflammatory responses by inducing IL-10 production as we previously reported (Saba et al. 2009). Although the counter receptor molecules in this scenario remain to be identified, it seems highly likely that they are bacterial components with a structure different from those previously investigated.

Infectious enveloped viruses are known to use MGL to attach to and/or enter mammalian cells. Well-known examples for human diseases are Ebola and Influenza (Takada et al. 2004; Upham et al. 2010). The counter receptors for MGL on these viruses are combination products of viral genes and glycosyltransferases of human cells. Thus, variations in the oligosaccharide portions of MGL counter receptors responsible for the viral infection should be limited to human glycans. From these results, it is still a mystery how MGL and its counter receptors evolved as in the examples seen in the evolution of infectious parasites and the immune system.

11.3 Cells Expressing MGL/CD301

As stated above, the cells expressing MGL are MØs and DCs. Although this statement sounds simple, it may not be the case due to the fact that both MØs and DCs comprise highly heterogeneous populations of cells potentially having diverse origins and tissue-specific differentiation. In mice, the situation is even more complex because of the presence of MGL1 and MGL2. We were successful in preparing monoclonal antibodies specific for mouse MGL1, MGL2, and human MGL, respectively. When anti-mouse MGL1 antibody was initially generated, MGL2 was not yet discovered. Among several antibodies prepared against mouse MGL purified from a cell line (Kimura et al. 1995), we later found that mAb LOM-14 was cross-reactive between MGL1 and 2, whereas mAb LOM-8.7 strictly bound MGL1 (Tsuiji et al. 2002). We also generated an MGL2-specific mAb URA-1. The distribution of the binding sites for mAb LOM-8.7 and mAb URA-1 was compared by immunohistochemistry (Denda-Nagai et al. 2010; Kumamoto et al. 2009). Both mAbs bound to cells in the connective tissue of almost all organs except for the brain, and the binding profiles were similar to those previously shown using cross-reactive mAb LOM-14. The number of cells that reacted with mAb URA-1 seemed to be similar to the number of cells that reacted with mAb LOM-8.7 in most organs as far as the results of immunohistochemistry were concerned; however, clear differences between the cell populations expressing MGL1 and MGL2 were shown by flow cytometry as discussed below.

In LNs, MGL1^{+} cells were mainly observed in the medullary and subcapsular sinuses, whereas MGL2^{+} cells were restricted to the cortical areas (Kumamoto et al. 2009). In spleens, MGL1^{+} cells appeared to have a more widespread distribution than MGL2^{+} cells, and the binding of these antibodies was limited to the white pulp (Denda-Nagai et al. 2010). In lung alveolar spaces, the number of MGL2^{+} cells appeared to be slightly greater than the number of MGL1^{+} cells, which was different from any other site (Denda-Nagai et al. 2010). Surface expression of MGL1 and MGL2 on cells from bone marrow (BM), LNs, spleen, and lung was analyzed by flow cytometry using mAb LOM-8.7 and mAb URA-1 in a cell suspension after exclusion of dead cells and $CD3^+$ or $CD19^+$ cells (Denda-Nagai et al. 2010; Kumamoto et al. 2009). These mAbs were found to bind subsets of $CD11c^+$ cells. BM, LNs, and spleen were populated with more $MGL1^+$ cells than $MGL2^+$ cells. All $MGL2^+$ cells appeared to co-express MGL1. MGL1 and MGL2 double-positive cells were $CD11c^+CD8^{-/low}CD11b^+MHCII^+F4/80^{low}$ conventional DCs. In BM, LNs, and spleen, a significant portion of MGL1 single-positive cells was observed and contained $CD11c^+B220^+$ plasmacytoid DCs.

It was found recently that $MGL2^+$ cells in the skin represent a unique subset of dermal DCs that skews the immune response toward Th2 (Murakami et al. 2013). Cells expressing MGL2 were found to be distinct from migratory Langerhans cells or $CD103^+$ dermal DCs (dDCs). Transfer of $MGL2^+$ dDCs but not $CD103^+$ dDCs from FITC-sensitized mice induced a Th2-type immune response in vivo in a model of contact hypersensitivity. Targeting $MGL2^+$ dDCs with a rat monoclonal antibody against MGL2 efficiently induced a humoral immune response with Th2-type properties, as determined by antibody subclass analysis.

Colonic lamina propria mononuclear cells were prepared from mice and analyzed by flow cytometry for the expression of cell surface markers using mAb LOM-8.7 and other antibodies (Saba et al. 2009). Cells expressing MGL1 were shown to express CD11b, CD11c, a high level of F4/80, and MHC class II. When we sorted cells with CD11b and F4/80, they also expressed *Mgl1* and *Mgl2* mRNA by RT-PCR. *Mgl1* expression was detected in $CD11b^+F4/80^{high}$ cells but not in $CD11b^+F4/80^{intermediate}$ cells. Sorted cells that expressed a combination of high levels of CD11b and F4/80 were shown to be $MGL1^+$ and incorporated FITC-labeled latex beads in vitro. These cells were also positively stained for nonspecific esterase. Therefore, the predominant populations of cells expressing MGL1 in the colonic wall were considered to be MØs, although they expressed CD11c.

From the above findings, $MGL2^+$ cells appeared to be a portion of $MGL1^+$ cells, mostly comprising conventional DCs. $MGL1^+$ cells were more diverse than $MGL2^+$ cells and included MØs and pDCs. Although several mouse myeloma cell lines, such as RAW264.7 cells and J774 cells, express MGL1, no cell line with surface expression of MGL2 or human MGL was identified so far (unpublished data). There is some reason to speculate that MGL1 and MGL2 form hetero-oligomers because $MGL2^+$ cells appear to also express MGL1, and because a homologous C-type lectin expressed on hepatocytes, hepatic asialoglycoprotein receptor, is known to form a hetero-oligomer.

11.4 MGL Knockout Mice and the Molecular Function of MGL/CD301

In the early days of MGL studies, we expected that the biological phenotypes of *Mgl1−/−* mice or *Mgl2−/−* mice should be easily interpretable because the expression profiles of these molecules are rather simple and limited to DCs and

MØs. However, contrary to our expectation, the molecular properties of MGL and the function of cells expressing MGL are not a straightforward matter.

Mgl1 −/− mice were generated by Hedrick's group and by our group around the same time, when the presence of MGL2 was not yet published. Because there was no obvious phenotype observed with *Mgl1* −/− mice, it was estimated by Hedrick's group that the biological role of this molecule might be redundantly carried out by multiple homologous gene products (Onami et al. 2002). Although this hypothesis remains to be confirmed, MGL2 is obviously not the answer, since, as partially explained above, it shows different expression profiles and different molecular properties. MGL1 has a so-called hemITAM motif in its cytoplasmic tail, suggesting signaling functions.

The first tangible observations regarding the phenotypic effects of *Mgl1* deficiency were the delay in the repair of damage in embryos caused by the irradiation of pregnant mice (Yuita et al. 2005) (Fig. 11.3a) and the lack of inflammatory tissue formation in an air pouch model (Sato et al. 2005). In the second model, the presence of granulation tissue induced by immunization with a chemically modified foreign protein antigen observed in wild-type mice was diminished in *Mgl1* −/− mice (Fig. 11.3b). Administration of anti-MGL1 antibody similarly suppressed granulation tissue formation in wild-type mice. A large number of cells, present only in the pouch of *Mgl1* −/− mice, were not myeloid or lymphoid lineage cells and the number significantly declined after administration of interleukin 1-alpha (IL-1α) into the pouch of *Mgl1* −/− mice. Furthermore, granulation tissue was restored by this treatment, and the cells obtained from the pouch of *Mgl1* −/− mice were incorporated into the granulation tissue after injection of IL-1α. Therefore, MGL1 expressed on a distinct subpopulation of MØs or DCs that secrete IL-1α seemed to regulate specific cellular interactions crucial to granulation tissue formation.

Almost opposite effects on inflammatory tissue formation were observed in an experimental colitis model (Saba et al. 2009). When dextran sulfate sodium salt (DSS) was orally administrated to *Mgl1* −/− mice and their wild-type littermates, *Mgl1* −/− mice showed significantly more severe inflammation than wild-type mice. $MGL1^{+}$ cells in colonic lamina propria corresponded to MØ-like cells with $F4/80^{high}$, $CD11b^{+}$, and $CD11c^{intermediate}$ cells. In *Mgl1* −/− mice, these cells produced a lower level of IL-10 mRNA compared with wild-type mice after administration of DSS for 2 days. Recombinant MGL1 was found to bind both *Streptococcus sp.* and *Lactobacillus sp.* among commensal bacteria isolated from mesenteric LNs of DSS-treated mice. Heat-killed *Streptococcus sp.* induced an increase in IL-10 secretion by $MGL1^{+}$ colonic lamina propria MØs in wild-type mice but not by MØs from *Mgl1* −/− mice (Fig. 11.3c). These results strongly suggest that MGL1/CD301a plays a protective role against colitis by effectively inducing IL-10 production by colonic lamina propria MØs in response to invading commensal bacteria. It appears that the signal generated through MGL1 leads to the production of cytokines that play a crucial positive or negative role in the process of inflammatory tissue formation in these two examples.

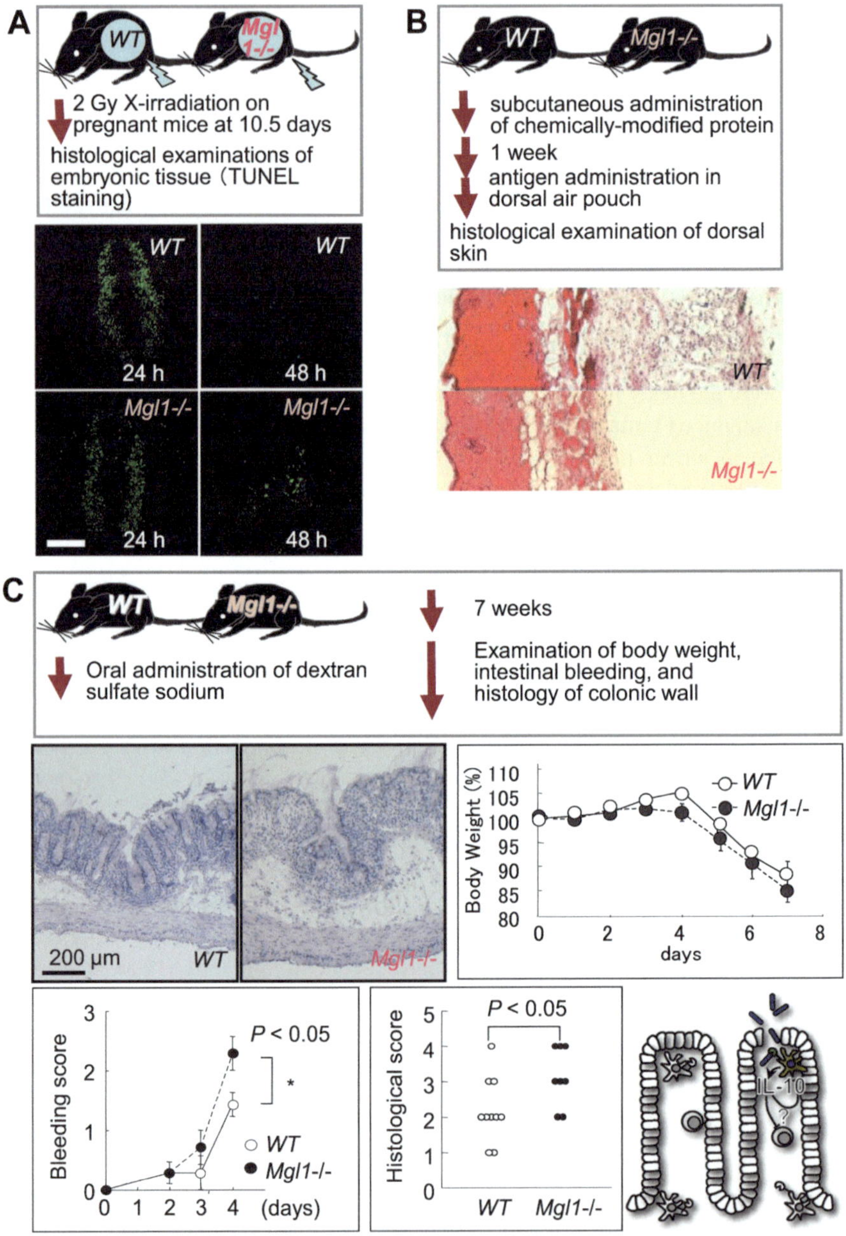

Fig. 11.3 Phenotypes observed with MGL1/CD301a knockout mice. (**a**) Histochemical localization of TUNEL-positive apoptotic cells (*green*) in neural tubes of *Mgl1+/+* mice and *Mgl1−/−* mice in histological sections 24 or 48 h after X-ray irradiation at 1 Gy. The bar represents 200 μm. The images in Fig. 11.3a are adapted from a previous paper (Yuita et al. 2005). (**b**) Lack of granulation tissue formation after re-challenge with chemically modified protein (azobenzene arsonate-conjugated acetylated bovine serum albumin) in pre-immunized *Mgl1−/−* mice. The images in Fig. 11.3b are adapted from a previous paper (Sato et al. 2005). (**c**) MGL1/CD301a plays

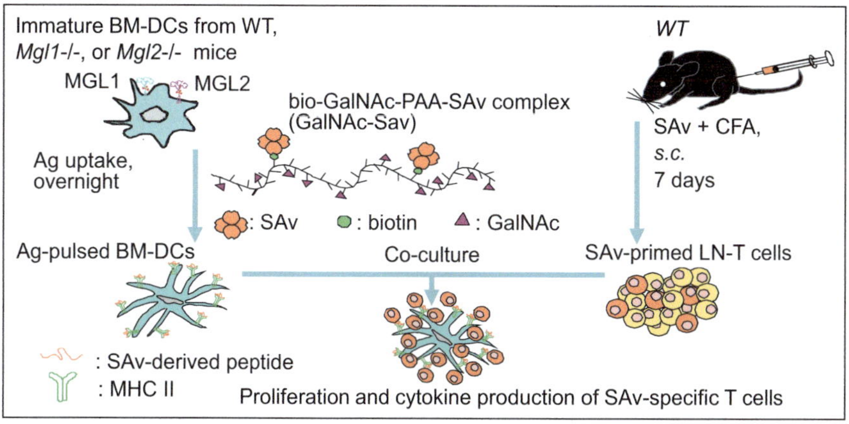

Fig. 11.4 Involvement of MGL1 and MGL2 in uptake of antigens with GalNAc residues by BM-DCs and presentation to T cells. The effects on T cell activation are examined according to this scheme as explained in the text

The function of MGL1 and MGL2 expressed on DCs in antigen uptake and presentation was examined using BM-DCs generated from BM cells of *Mgl1−/−* and *Mgl2−/−* mice. Wild-type BM-DCs expressed both MGL1 and MGL2. Streptavidin conjugated with biotinylated GalNAc-polyacrylamide polymer (GalNAc-SAv) was used as a model of glycosylated antigens (Fig. 11.4). GalNAc-SAv was more efficiently internalized and presented by wild-type BM-DCs to SAv-primed CD4^{+} T cells than GlcNAc-SAv or SAv alone. MGL2 was suggested to be responsible for efficient binding and uptake of GalNAc-SAv by a comparison of *Mgl1−/−* and *Mgl2−/−* BM-DCs with wild-type BM-DCs (Denda-Nagai et al. 2010).

11.5 Concluding Remarks

MGL/CD301 is a unique C-type lectin expressed by DCs and by MØs for MGL1 in mice. Studies conducted in the past revealed that the roles of MGL are diverse depending on the DC and MØ subsets. The origins and diversity of these subsets are not yet fully understood. MGL will contribute to understand the diversity and

Fig. 11.3 (continued) an anti-inflammatory role in murine dextran sulfate sodium salt (DSS)-induced experimental colitis. Body weights (*Middle-right panel*), hematoxylin/eosin-stained sections of colon (*Middle-left panel*), bleeding score in stools (*Lower-left panel*), and histological scores for the damage of colonic walls (*Lower-middle panel*) are shown. Conceptual representation of MGL1's anti-inflammatory role (*Lower-right panel*). Upon bacterial invasion into the lamina propria, MGL1-expressing MØs initiate IL-10 secretion to combat inflammation. Figure 11.3c has been adapted from a previous paper (Saba et al. 2009)

diverse regulatory roles of DCs and MØs in different organs, which may require distinct types of immunological defense.

Acknowledgments The authors would like to acknowledge those who are engaged with the studies introduced in this chapter and collaborators, particularly Dr. Steve Hedrick for *Mgl1–/–* mice and Dr. Paul Crocker, for contributing unpublished data. We would also like to acknowledge Dr. Katrin Ishii-Schrade and Ms. Miki Noji in the preparation of this manuscript. The work is supported by JSPS KAKENHI Grants to KDN and TI.

References

Chun KH, Imai Y, Higashi N, Irimura T (2000) Migration of dermal cells expressing a macrophage C-type lectin during the sensitization phase of delayed-type hypersensitivity. J Leukoc Biol 68:471–478

Denda-Nagai K, Aida S, Saba K, Suzuki K, Moriyama S, Oo-Puthinan S, Tsuiji M, Morikawa A, Kumamoto Y, Sugiura D, Kudo A, Akimoto Y, Kawakami H, Bovin NV, Irimura T (2010) Distribution and function of macrophage galactose-type C-type lectin 2 (MGL2/CD301b): efficient uptake and presentation of glycosylated antigens by dendritic cells. J Biol Chem 285:19193–19204

Denda-Nagai K, Kubota N, Tsuiji M, Kamata M, Irimura T (2002) Macrophage C-type lectin on bone marrow-derived immature dendritic cells is involved in the internalization of glycosylated antigens. Glycobiology 12(7):443–450

Drickamer K (1993) Recognition of complex carbohydrates by Ca(2+)-dependent animal lectins. Biochem Soc Trans 21(2):456–459

Gao Y, Nish SA, Jiang R, Hou L, Licona-Limón P, Weinstein JS, Zhao H, Medzhitov R (2013) Control of T helper 2 responses by transcription factor IRF4-dependent dendritic cells. Immunity 39(4):722–732

Higashi N, Fujioka K, Denda-Nagai K, Hashimoto S, Nagai S, Sato T, Fujita Y, Morikawa A, Tsuiji M, Miyata-Takeuchi M, Sano Y, Suzuki N, Yamamoto K, Matsushima K, Irimura T (2002) The macrophage C-type lectin specific for galactose/N-acetylgalactosamine is an endocytic receptor expressed on monocyte-derived immature dendritic cells. J Biol Chem 277(23):20686–20693

Ii M, Kawasaki T, Yamashina I (1988) Structural similarity between the macrophage lectin specific for galactose/N-acetylgalactosamine and the hepatic asialoglycoprotein binding protein. Biochem Biophys Res Commun 155(2):720–725

Ii M, Kurata H, Itoh N, Yamashina I, Kawasaki T (1990) Molecular cloning and sequence analysis of cDNA encoding the macrophage lectin specific for galactose and N-acetylgalactosamine. J Biol Chem 265(19):11295–11298

Iida S, Yamamoto K, Irimura T (1999) Interaction of human macrophage C-type lectin with O-linked N-acetylgalactosamine residues on mucin glycopeptides. J Biol Chem 274 (16):10697–10705

Imamura T, Toyoshima S, Osawa T (1984) Lectin-like molecules on the murine macrophage cell surface. Biochim Biophys Acta 805:235–244

Kimura T, Imai Y, Irimura T (1995) Calcium-dependent conformation of a mouse macrophage calcium-type lectin. Carbohydrate binding activity is stabilized by an antibody specific for a calcium-dependent epitope. J Biol Chem 270:16056–16062

Kumamoto Y, Higashi N, Denda-Nagai K, Tsuiji M, Sato K, Crocker PR, Irimura T (2004) Identification of sialoadhesin as a dominant lymph node counter-receptor for mouse macrophage galactose-type C-type lectin 1. J Biol Chem 279:49274–49280

Kumamoto Y, Denda-Nagai K, Aida S, Higashi N, Irimura T (2009) MGL2 dermal dendritic cells are sufficient to initiate contact hypersensitivity in vivo. PLoS One 4:e5619

Kumamoto Y, Linehan M, Weinstein JS, Laidlaw BJ, Craft JE, Iwasaki A (2013) CD301b(+) dermal dendritic cells drive T helper 2 cell-mediated immunity. Immunity 39(4):733–743

Mizuochi S, Akimoto Y, Imai Y, Hirano H, Irimura T (1997) Unique tissue distribution of a mouse macrophage C-type lectin. Glycobiology 7(1):137–146

Mizuochi S, Akimoto Y, Imai Y, Hirano H, Irimura T (1998) Immunohistochemical study on a macrophage calcium-type lectin in mouse embryos: transient expression in chondroblasts during endochondral ossification. Glycoconj J 15(4):397–404

Murakami R, Denda-Nagai K, Hashimoto S, Nagai S, Hattori M, Irimura T (2013) A unique dermal dendritic cell subset that skews the immune response toward Th2. PLoS One 8(9): e73270

Oda S, Sato M, Toyoshima S, Osawa T (1988) Purification and characterization of a lectin-like molecule specific for galactose/N-acetyl-galactosamine from tumoricidal macrophages. J Biochem 104(4):600–605

Onami TM, Lin MY, Page DM, Reynolds SA, Katayama CD, Marth JD, Irimura T, Varki A, Varki N, Hedrick SM (2002) Generation of mice deficient for macrophage galactose- and N-acetylgalactosamine-specific lectin: limited role in lymphoid and erythroid homeostasis and evidence for multiple lectins. Mol Cell Biol 22:5173–5181

Oo-Puthinan S, Maenuma K, Sakakura M, Denda-Nagai K, Tsuiji M, Shimada I, Nakamura-Tsuruta S, Hirabayashi J, Bovin NV, Irimura T (2008) The amino acids involved in the distinct carbohydrate specificities between macrophage galactose-type C-type lectins 1 and 2 (CD301a and b) of mice. Biochim Biophys Acta 1780:89–100

Saba K, Denda-Nagai K, Irimura T (2009) A C-type lectin MGL1/CD301a plays an anti-inflammatory role in murine experimental colitis. Am J Pathol 174:144–152

Sakakura M, Oo-Puthinan S, Moriyama C, Kimura T, Moriya J, Irimura T, Shimada I (2008) Carbohydrate binding mechanism of the macrophage galactose-type C-type lectin 1 revealed by saturation transfer experiments. J Biol Chem 283:33665–33673

Sano Y, Usami K, Izawa R, Denda-Nagai K, Higashi N, Kimura T, Suzuki N, Irimura T (2007) Properties of blocking and non-blocking monoclonal antibodies specific for human macrophage galactose-type C-type lectin (MGL/ClecSF10A/CD301). J Biochem 141:127–136

Sato M, Kawakami K, Osawa T, Toyoshima S (1992) Molecular cloning and expression of cDNA encoding a galactose/*N*-acetylgalactosamine-specific lectin on mouse tumoricidal macrophages. J Biochem 111:331–336

Sato K, Imai Y, Higashi N, Kumamoto Y, Onami TM, Hedrick SM, Irimura T (2005) Lack of antigen-specific tissue remodeling in mice deficient in the macrophage galactose-type calcium-type lectin 1/CD301a. Blood 106:207–215

Suzuki N, Yamamoto K, Toyoshima S, Osawa T, Irimura T (1996) Molecular cloning and expression of cDNA encoding human macrophage C-type lectin. Its unique carbohydrate binding specificity for Tn antigen. J Immunol 156:128–135

Takada A, Fujioka K, Tsuiji M, Morikawa A, Higashi N, Ebihara H, Kobasa D, Feldmann H, Irimura T, Kawaoka Y (2004) Human macrophage C-type lectin specific for galactose and N-acetylgalactosamine promotes filovirus entry. J Virol 78(6):2943–2947

Tsuiji M, Fujimori M, Ohashi Y, Higashi N, Onami TM, Hedrick SM, Irimura T (2002) Molecular cloning and characterization of a novel mouse macrophage C-type lectin, mMGL2, which has a distinct carbohydrate specificity from mMGL1. J Biol Chem 277:28892–28901

Upham JP, Pickett D, Irimura T, Anders EM, Reading PC (2010) Macrophage receptors for influenza A virus: role of the macrophage galactose-type lectin and mannose receptor in viral entry. J Virol 84:3730–3737

Yamamoto K, Ishida C, Shinohara Y, Hasegawa Y, Konami Y, Osawa T, Irimura T (1994) Interaction of immobilized recombinant mouse C-type macrophage lectin with glycopeptides and oligosaccharides. Biochemistry 33(26):8159–8166

Yuita H, Tsuiji M, Tajika Y, Matsumoto Y, Hirano K, Suzuki N, Irimura T (2005) Retardation of removal of radiation-induced apoptotic cells in developing neural tubes in macrophage galactose-type C-type lectin-1-deficient mouse embryos. Glycobiology 15(12):1368–1375
Zola H, Swart B, Nicholson I, Aasted B, Bensussan A, Boumsell L, Buckley C, Clark G, Drbal K, Engel P, Hart D, Horejsi V, Isacke C, Macardle P, Malavasi F, Mason D, Olive D, Saalmueller A, Schlossman SF, Schwartz-Albiez R, Simmons P, Tedder TF, Uguccioni M, Warren H (2005) CD molecules 2005: human cell differentiation molecules. Blood 106:3123–3126

Chapter 12
Structural Aspects of C-Type Lectin Receptors

Atsushi Furukawa, Shunsuke Kita, Takashi Tadokoro, Hideo Fukuhara, and Katsumi Maenaka

Abstract Numerous structural analyses (X-ray crystallography and NMR) of C-type lectin receptors (CLRs) have been performed, because CLRs are not only attractive as important molecules in immunity and infectious diseases but also as drug targets. In CLRs, high amino acid sequence similarity exists in the extracellular carbohydrate recognition domains (CRDs), which are responsible for ligand binding. However, recent functional analyses of CLRs implied that these molecules recognize a wide variety of ligands in addition to saccharides, including glycopeptides, glycolipids, and proteins. In this chapter, we focus on structural studies of CLRs. We first summarize the structural features conserved among the CRDs and then describe how each C-type lectin receptor elegantly achieves its distinct ligand specificity, by illustrating the structural aspects of several representative CLRs.

Keywords C-type lectin receptor • Protein structure • X-ray crystallography • Innate immunity • Carbohydrate recognition domain • Protein–protein interaction • Glycolipids • Glycoprotein

12.1 Conserved Structure of Carbohydrate Recognition Domains

In mammals, C-type lectin-(like) receptors (CLRs) play a crucial role in the immune response to pathogen invasions and physiological homeostasis. The CLRs are type II membrane proteins comprising an N-terminal intracellular domain, a transmembrane region, and extracellular neck and lectin domains (Fig. 12.1a). The CLRs are classified into 17 groups (groups I–XVII) on the basis of their domain organization and phylogeny (Zelensky and Gready 2005). Most of the CLRs involved in immune responses are categorized into either Group II or V, in which the extracellular domain consists of two parts, a neck domain (N-terminal)

A. Furukawa • S. Kita • T. Tadokoro • H. Fukuhara • K. Maenaka (✉)
Laboratory of Biomolecular Science, and Center for Research and Education on Drug Discovery, Faculty of Pharmaceutical Sciences, Hokkaido University, Sapporo, Japan
e-mail: maenaka@pharm.hokudai.ac.jp

S. Yamasaki (ed.), *C-Type Lectin Receptors in Immunity*,
DOI 10.1007/978-4-431-56015-9_12

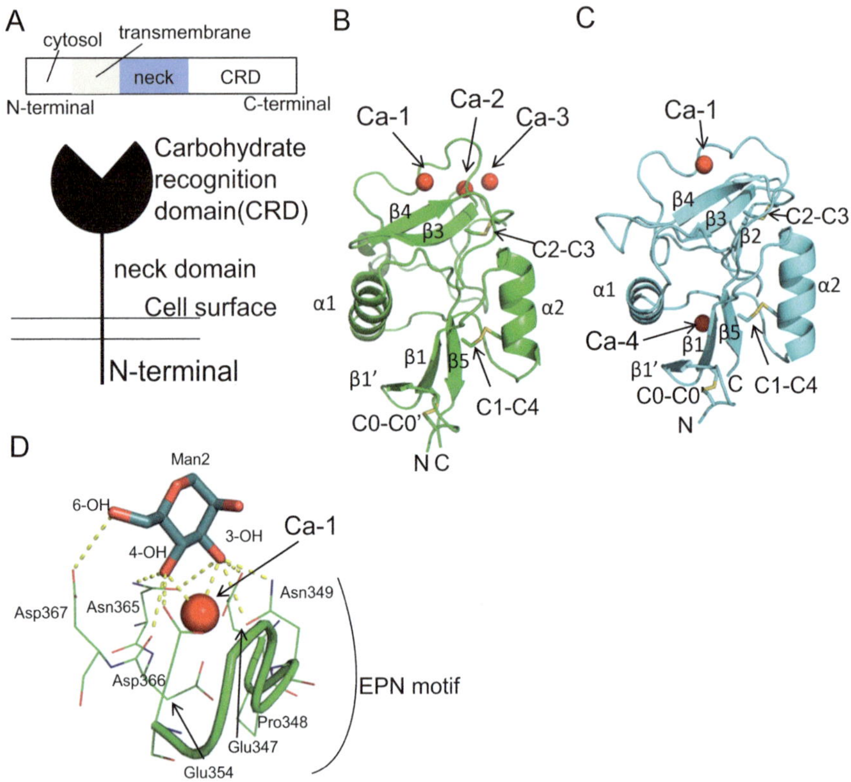

Fig. 12.1 (**a**) Schematic representations of the domains and surface expression of CLRs are shown. The crystal structure of a representative CLR, DC-SIGN (PDB ID: 19KI) (**b**), and another example of a CLR structure, Mincle (PDB ID: 3WH2) (**c**), are shown. Intramolecular disulfide bonds (C0′–C0, C1–C4, and C2–C3) are depicted by sticks. (**d**) Mechanism of mannose recognition by DC-SIGN. The DC-SIGN residues are represented by *lines*, and the complex sugar is depicted by a stick model. The calcium ion (Ca-1) bound to the sugar is shown as a *sphere*. *Dotted lines* indicate interactions between the protein and the sugar

and a carbohydrate recognition domain (CRD) (110–130 amino acid residues at the C-terminus) responsible for ligand binding. Various structural studies (X-ray crystallography and NMR analyses) of the CRD domains have been performed. The structure of DC-SIGN, as a representative CRD, is shown in Fig. 12.1b. The main body of the CRD is composed of two α-helices and a three-stranded antiparallel β-sheet. Six cysteines (C0–C0′ and C1–C4), which are the most conserved CRD residues, form disulfide bonds between the loops in DC-SIGN. The C1–C4 disulfide bond links the β5-sheet and the α1-helix, and the C2–C3 bond links β3 and β5. The C0–C0′ disulfide bond is conserved in almost all of the CLRs described in this chapter. The N-terminus and the C-terminus of the CRD approach each other, due to the formation of the antiparallel β-sheet pair (β1 and β5). Since the N-terminus is connected to the neck domain located near the cell surface, the membrane-distal

Table 12.1 Summary of the features of the CLRs described in this chapter

Ligand category	Ligand	Name of protein	Number of calcium ions	Calcium ion	Binding site	PDB ID
Glycan						
	High-mannose sugar, fucosylated oligosaccharides	DC-SIGN	3	Dependent	Top	1K9I, 1SL5
	High-mannose sugar	DC-SIGNR	3	Dependent	Top	1K9J
	Sialic acid/dextran sulfate	SIGN-R1	1	Dependent	Top	4CAJ, 4C9F
	High-mannose sugar, β-glucan	Langerin	1	Dependent	Top	3P5D, 3P5E, 3P5F
	Bisecting N-acetylglucosamine	DCIR2	1	Dependent	Top	3VYK
Glycolipid						
	TDM	Mincle, MCL	2, 1	Dependent	Top	3WH2 and 3WHD
Lipopeptide						
	Oxydized low-density lipoprotein	Lox-1	0	Independent	Top	1YPO, 1YPQ. . ..
Glycopeptides and proteinaceous ligands						
	Sulfated sugar	P-selectin	1	Dependent	Top	1G1S
	Rhodocytin (snake venom toxin)	CLEC-2	0	Independent	Side	3WWK
	Podoplanin					3WSR
	KACL	NKp65	0	Independent	Top	4IOP
	HLA-E	CD94/NKG2A complex	0	Independent	Top	3CII
	E-cadherin	KLRG1	0	Independent	Top	3FF7 and 3FF8
	SAP130	Mincle	2	Independent (?)	Top(?)	3WH2
	Actin filament	CLEC9A	1	Independent (?)	Top(?)	3VPP

face of the CRD is reasonably exposed to the outside for ligand binding. Four Ca^{2+} ion-binding sites (sites 1–4) are often found in CRD domains (Fig. 12.1b, c). Site 1, at the membrane-distal face of the CRD, is the most conserved and essentially responsible for Ca^{2+}-mediated sugar-binding. Sites 2 and 3 mainly contribute to additional recognition of sugar moieties (details in Sect. 12.3). In contrast, the Ca^{2+}

ion at site 4, which is coordinated by the α2-helix and the β1/β5-sheet, is not involved in ligand binding, but probably contributes to protein stability. Some CLRs lack Ca^{2+} ion-binding sites and instead utilize the "top" face to bind lipids or protein ligands, rather than sugars. Interestingly, CLRs often form homodimers, as observed in crystal structures and biochemical analyses (gel filtration, ultracentrifugation, etc.). Some CLRs further form heterodimers, to achieve efficient ligand binding using both membrane-distal surfaces. Moreover, various CLRs form trimers via the neck domain. Accordingly, CLRs adopt monomer/dimer/trimer/oligomer states to achieve appropriate signaling, but their mechanisms are not fully understood yet.

In the following sections, we take a closer look into the detailed structural characteristics of individual CLRs and discuss their substrate specificities mediated by their amino acid sequences and structural differences. The structural properties of the CLRs described in this chapter are summarized in Table 12.1.

12.2 Mono-/Oligosaccharide Recognition

The CRD of rat mannose-binding protein-A (MBP-A) complexed with Man_6-$GalNAc_2$-Asn was the first crystal structure determined of the complex between a CLR and a carbohydrate (Weis et al. 1992). As mentioned above, CRDs have a conserved Ca^{2+} ion (site 1), and in the case of the MBP-A complex, it electrostatically interacts with the 3- and 4-position hydroxyl groups of the mannose. CLRs also have the"Glu-Pro-Asn (EPN)" or "Gln-Pro-Asp (QPD)" sugar-binding motifs. Like MBP-A, CRDs with EPN motifs engage the equatorial/equatorial arrangement of the 3-OH/4-OH of the bound sugar residue, such as in glucose, N-acetylglucosamine, and mannose (Fig. 12.1d). In contrast, CLRs with QPD motifs engage the equatorial/axial configuration of 3-OH/4-OH, such as in galactose and N-acetylgalactosamine. The "Trp-Asn-Asp (WND)" motif is also conserved in CLRs and is involved in the Ca^{2+} ion coordination to site 1.

DC-SIGN and DC-SIGNR are closely related CLRs expressed on dendritic cell (DC) and endothelial cell surfaces, respectively. Both receptors recognize high-mannose N-linked glycans on viral glycoproteins, such as HIV, but only DC-SIGN can bind blood fucosylated oligosaccharides (Appelmelk et al. 2003; Feinberg et al. 2001, 2007; Guo et al. 2004). In the complex structure of human DC-SIGN with $GlcNAc_2$-Man_3, a secondary mannose binds DC-SIGN via the Ca^{2+} ion on site 1 (PDB ID: 1K9I) (Figs. 12.1d and 12.2a). The adjacent saccharides of the secondary mannose also interact along the groove of the CRD surface. In spite of the axial orientation of the 4-OH group in fucose, the α1-3-linked fucose of Lewis^x trisaccharide also bound to Ca^{2+} on DC-SIGN, with separated galactose recognition (PDB ID: 1SL5 and Fig. 12.2b). Recently, the complex structures of the macrophage receptor SIGN-R1/CD209b, one of the mouse homologs of human DC-SIGN, with α2-6 sialic acid or microbial dextran sulfate (DexS) were reported by Silva-Martín et al. (2014). SIGN-R1 bound to the glucose of DexS at the

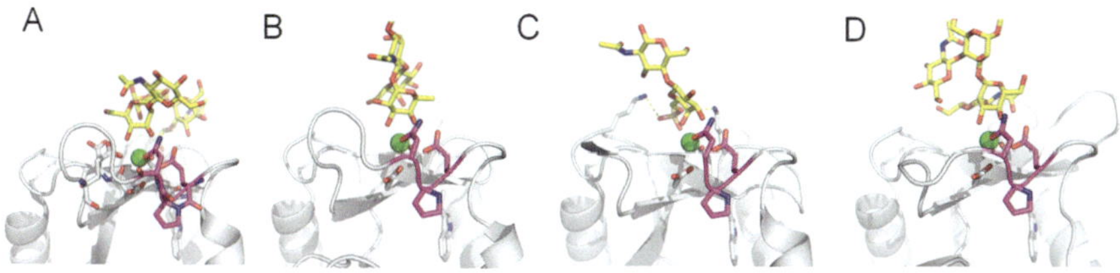

Fig. 12.2 CLRs with the "EPN motif" have versatile oligosaccharide recognition mechanisms. The human DC-SIGN with GlcNAc$_2$–Man$_3$ (1K9I) (**a**), DC-SIGN with Lewis^x trisaccharide (1SL5) (**b**), Langerin with 6SO$_4$–Galβ1–4GlcNAc (3P5I) (**c**), and mouse DCIR2 with bisecting N-acetylglucosamine (3VYK) (**d**) are shown. The *sphere* indicates the Ca-1 ion. The *dotted lines* in (**c**) represent the unique interactions between the lysines of DC-SIGN and the negatively charged moiety, mainly the sulfate group, from 6SO$_4$–Galβ1–4GlcNAc. The "EPN motif" in each CLR is colored differently and depicted by sticks

conventional calcium-binding site (site 1), regardless of the presence or absence of sulfate (PDB ID: 4C9F). Interestingly, sialic acid binding to the CRD of SIGN-R1 was observed on the outside of the EPN loop surrounding site 1 (PDB ID: 4CAJ). These characteristics suggest that SIGN-R1 may simultaneously bind to sialic acids on antibodies and immunoglycoproteins such as C1q complement factor and microbial polysaccharides, in the innate immune response to pathogen invasion.

Langerin (CD207/CLEC4K) is expressed on Langerhans cells (LCs), as well as some dermal and splenic DCs, to mediate immune responses (Valladeau et al. 2000). Langerin has a neck region for trimerization and a CRD that binds various mono- and oligosaccharides of endogenous and pathogenic glycans, in a calcium-dependent manner (Chatwell et al. 2008; Stambach and Taylor 2003). Feinberg et al. determined the crystal structures of the human Langerin CRD complexed with oligomannose, the blood group B antigen, and a β-glucan representative (Feinberg et al. 2011). In these complex structures, the Langerin CRD recognizes only a single sugar via the site 1 Ca^{2+} ion (PDB IDs: 3P5D, 3P5E, 3P5F, and 3P5I). The complex structure of Langerin with the 6SO$_4$–Galβ1–4GlcNAc unit from keratan sulfate revealed a unique interaction. Although Langerin has the EPN motif, which is suitable for mannose binding, the 3- and axial 4-OH groups of galactose in this unit were recognized via the site 1 Ca^{2+} ion. The salt bridges among Lys299, Lys313, and the sulfate group are essential to hold this ligand (Fig. 12.2c). These structures revealed the diversity of the interactions as well as the difficulty in predicting CRD ligands.

In 2013, Nagae et al. reported the complex structure of mouse dendritic cell immunoreceptor 2 (DCIR2)/CLEC4a CRD with a biantennary complex-type glycan containing bisecting N-acetylglucosamine (GlcNAc) (Nagae et al. 2013). DCIR2 is one of the mouse homologs of human DCIR. Both human DCIR and mouse DCIR2 are expressed on DCs as inhibitory receptors, with an intracellular immunoreceptor tyrosine-based inhibitory motif (ITIM). While human DCIR was shown to bind fucose and mannose preferably (Lee et al. 2011), Nagae et al. found that the mouse DCIR2 recognizes bisecting GlcNAc specifically. In the co-crystallized structure of mouse DCIR2 with a hexasaccharide bearing a bisecting GlcNAc, the calcium ion in site 1 was coordinated by the acidic side chains of the EPN motif residues (PDB ID: 3VYK and Fig. 12.2d). The hydroxyl groups of the primary-binding mannose and the branched GlcNAc interact with the DCIR2 residues, either directly or in a water-mediated fashion. The position of the primary-binding mannose overlapped well with the mannose on Langerin (PDB ID: 3P5D), but it faced the opposite direction to mannose on the DC-SIGN complex (PDB ID: 1K9I). While the precise ligand for human DCIR is still unknown, a different ligand recognition pattern is predicted because it has the EPS motif, instead of EPN, with a set of longer α3–β3 and β3–β4 loops than DCIR2.

12.3 Glycolipid Recognition (Mincle and MCL)

Macrophage inducible C-type lectin (Mincle) and macrophage C-type lectin (MCL) reportedly bind a glycolipid, trehalose dimycolate (TDM), from a mycobacterium. Mincle can also bind to other glycolipids, such as gentiobiosyl diacyl glycerides from *M. pachydermatis* and glycerol monomycolate. To understand the substrate specificity, structural analyses of Mincle were performed by two groups independently. Feinberg et al. reported the crystal structures of bovine Mincle (PDB ID: 4KZW), whereas we solved the crystal structures of human MCL (PDB ID: 3WHD) and human Mincle (PDB ID: 3WH3) (Feinberg et al. 2013; Furukawa et al. 2013). The crystals of bovine and human Mincle were grown under similar low-pH conditions with citrate buffer. In both structures, citrate binds the Ca^{2+} ion at site 1, which was predicted to be the primary sugar-binding site, but the binding modes of the citrate molecules in the two complexes are slightly different. We also reported the citrate-unbound structure (PDB ID: 3WH3). These crystals grew at neutral pH, and the structure superimposed well on the citrate-bound form (RMSD = 0.12 Å). These results indicated that citrate might be a (weak) ligand of Mincle, and its binding does not cause a large structural change.

Feinberg et al. successfully crystallized bovine Mincle complexed with trehalose (PDB ID: 4KZV) (Fig. 12.3c) under virtually identical conditions to those used for the citrate-binding form of bovine Mincle. In the trehalose-bound Mincle, a loop composed of Leu172 to Asp177 is located closer to the site 1 Ca^{2+} ion. This structural change is caused by the formation of an electrostatic network among Asp193, Glu176, the site 1 Ca^{2+} ion, and the site 3 Na^{+} ion. They found that the surface of bovine Mincle near the primary sugar-binding site possesses a hydrophobic channel, running between Phe197/Phe198 on one side and Leu172/Val173 on the other side. The entrance of this hydrophobic channel is located adjacent to the 6-OH group of the glucose residue at the primary binding site. The acyl chain is extended from the 6-OH group in TDM and would interact with the hydrophobic channel. Structural modeling revealed the binding mode of the octanoic acid attached to this 6-OH group of the primary glucose. The decreased affinity for the

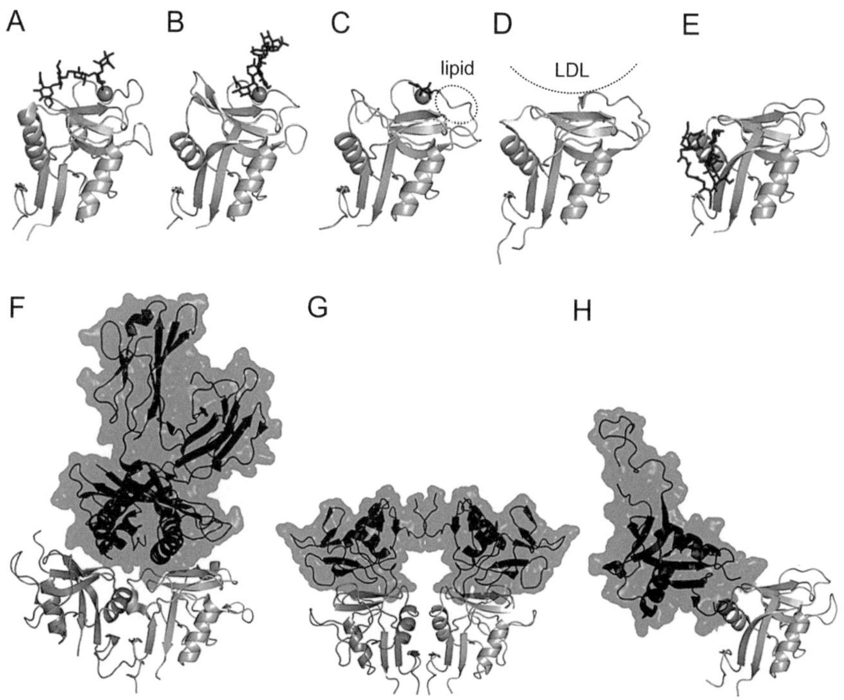

Fig. 12.3 Comparison of the ligand recognition patterns of CLRs. (**a**–**c**) Complex structures of DC-SIGN with oligomannose (**a**), DCIR2 with bisecting GlcNAc (**b**), and Mincle with trehalose (**c**). (**d**) Crystal structure of Lox-1. Predicted lipid-binding positions are shown by *dotted lines* in (**c**) and (**d**). (**e**, **h**) CLEC-2 in complex with podoplanin (**e**) or rhodocytin (**h**). (**f**) The NKG2A heterodimer with CD94 interacting with HLA-E. (**g**) KACL recognizes NKp65 as a homodimer. The CLRs are shown in the same orientation, as gray cartoon models. Sugar and protein ligands are represented by *black* sticks and cartoons with surfaces, respectively. Calcium ions bound for ligand interactions are shown as *spheres*

acyl trehalose, by either the Val173 or Phe197/Phe198 mutation of Mincle, proved the validity of this model. In addition, we also provided a glycolipid-binding model of human Mincle, by the superimposition of the human Mincle–citrate complex on the DC-SIGNR–mannose complex. The model of the trehalose moiety from TDM was placed on the sugar-binding site of Mincle, in a similar manner to the mode of mannose binding to DC-SIGNR. This suggested that a hydrophobic region located at the sugar-binding site could interact with the mycolic acid attached to the glucose 6-OH of TDM. The mutagenesis within this region of Mincle reduced the activity in reporter cell assays, proving the importance of the hydrophobic region for the interaction with TDM. The predicted lipid-binding sites of the two models described above overlapped well. Taken together, these structures suggested that the hydrophobic residues in the vicinity of the sugar-binding domain are important for the recognition of the lipid part of TDM.

The crystal structure of another glycolipid-binding CLR, MCL, a close relative of Mincle, was also reported (Furukawa et al. 2013). The structure indicated that MCL also has a shallow hydrophobic region near the canonical sugar-binding site (site 1), but the region is not obvious as compared to that in Mincle. This obscureness may account for the weaker binding affinity of MCL with TDM than Mincle.

12.4 Lipopeptide Recognition

Lectin-like, oxidized low-density lipoprotein (LDL) receptor 1, Lox-1, is the major receptor for oxidized LDL (OxLDL) in endothelial cells. The crystal structures of human Lox-1 were reported independently by two different groups in 2005 (Ohki et al. 2005; Park et al. 2005). As predicted from the amino acid sequence, the overall structure of Lox-1 forms an atypical-type CRD fold, such as in Ly49A, consisting of two antiparallel β-sheets and two α-helices (Fig. 12.3d). Lox-1 has three intramolecular disulfide bonds, Cys144–Cys155, Cys172–Cys264, and Cys243–Cys256, among which the latter two are commonly observed in all CRDs, while the former one stabilizes the short antiparallel β-sheet between β0 and β1. The Cys140 residue, which is unique in human Lox-1, is involved in the disulfide-linked homodimer formation, as often observed in the C-type lectin-like NK receptors (Natarajan et al. 2002), and is necessary for the proper dimerization of the Lox-1 CRDs (Ohki et al. 2005). The Lox-1 ligand-binding surface, which corresponds to the Ca^{2+}-binding site of the sugar-binding CLRs, exhibits a unique distribution of charged amino acids. The basic residues Arg229 and Arg231, in the long-loop region, and Arg248 are well aligned linearly across the homodimer surface, forming the so-called basic spine (Ohki et al. 2005). Lox-1 preferentially associates with negatively charged ligand molecules, including OxLDL. The mutations of all three residues abolished the ligand recognition, while the single mutation of the basic residues, Arg229 or Arg248, did not. Furthermore, a remarkable reduction of the LDL binding activity was observed when the Trp150 at the dimer interface was mutated. Taken together, proper homodimer formation by Lox-1 is quite crucial for the ligand recognition, to allow the key elements at the ligand-binding surface to align linearly (basic spine alignment) and provide a sufficient surface width for ligand association.

12.5 Glycopeptide and Protein Recognition

NKG2 family members are expressed on NK cells and T cells. The NKG2 family includes five isoforms (NKG2A, C, D, E, and F), and NKG2B and NKG2H are splice variants derived from NKG2A and NKG2E, respectively. Five NKG2 molecules (NKG2A, B, C, E, H) form disulfide-linked heterodimers with another CLR, CD94 (Borrego et al. 2006), which is encoded by a single gene and has low polymorphism. The NKG2A/CD94 heterodimer recognizes HLA-E, an MHC (major histocompatibility complex antigen) class I molecule. HLA-E presents the relatively conserved leader sequence of the MHC proteins. The structure of NKG2A/CD94 in complex with HLA-E (loaded with an HLA-G-derived peptide) was reported (PDB IDs: 3CII and 3CDG) (Fig. 12.3f) (Kaiser et al. 2008; Petrie et al. 2008). The antigen-presenting regions of HLA-E, the α1 and α2 domains, with the loaded peptide were recognized by the top CRD faces of both NKG2A and CD94. The Arg(P5) residue of the peptide is recognized by both NKG2A and CD94, whereas Phe(P8) is only recognized by CD94. These residues are basically conserved in the leader sequence and are dispensable for the interaction with the NKG2A/CD94 receptor. The structure clearly explained the significance of these residues at P5 and P8 in the HLA-E-loaded peptide.

Keratinocyte-specific C-type lectin-like receptor (KACL) is expressed in the skin and modulates the activities of natural killer (NK) cells through its receptor, NKp65 (Spreu et al. 2010). Recently, the crystal structure of the KACL–NKp65 complex was reported, as the first structure of a CLR–CLR complex (PDB ID: 4IOP) (Fig. 12.3g) (Li et al. 2013). KACL and NKp65 recognize each other by utilizing the top faces of their CRDs. The KACL dimer binds two NKp65 monomers independently. KACL forms a dimer in solution and crystals, while NKp65 exists as a monomer in the crystal. Thus, considering the fact that the normal state of NKp65 is the monomer, the KACL–NKp65 complex structure suggests that the dimerization of NKp65 upon KACL binding facilitates signal transduction.

CLEC9A is expressed on the surface of dendritic cells. CLEC9A recognizes a filamentous form of actin (F-actin), which dead cells and virally infected cells commonly display (Zhang et al. 2012). The crystal structure of CLEC9A was reported in the apo form (PDB ID: 3VPP) (Zhang et al. 2012). The putative ligand recognition site of CLEC9A is similar to those of other CLRs. Two mutations located at this face, W131A and W227A, disrupt the ability to bind F-actin, supporting the idea that this face is a ligand recognition site.

KLRG1 is expressed on NK cells, CD4+ T cells and CD8+ T cells, and recognizes the cell–cell adhesion-mediating molecule, cadherin. Cadherin is composed of five Ig-fold domains (EC1-5). The structures of human KLRG1 (hKLRG1) with human cadherin EC1 (hEC1) and mouse KLRG1 (mKLRG1) with or without hEC1 were reported (Li et al. 2009), as the hKLRG1–hEC1 complex (PDB ID: 3FF8), mKLRG1 (PDB ID: 3FF9), and the mKLRG1–hEC1 complex (PDB ID: 3FF7). Cadherin (EC1) is recognized by the top of the CRD in KLRG1. The KLRG1 monomer binds to the cadherin monomer. The region of cadherin recognized upon KLRG1 binding overlaps with the homodimer interface of cadherin, suggesting that KLRG1 detects the cadherin monomer exposed by the destruction of cell–cell adhesion elements.

CLEC-2 is expressed in the liver and blood cells. In platelets, CLEC-2 activation blocks lymphaticovenous connections, leading to separate blood and lymphatic vascular systems. CLEC-2 is unique in that it binds two different molecules, rhodocytin and podoplanin (Suzuki-Inoue et al. 2006, 2007). Rhodocytin, a snake venom toxin, is a heterodimeric protein composed of one α subunit and one β subunit. In contrast, podoplanin is a type I sialomucin-like glycoprotein. The binding of *O*-glycosylated podoplanin with CLEC-2 triggers the activation of cell spreading, via the downregulation of RhoA activity and myosin light-chain phosphorylation. The structures of CLEC-2 in complex with rhodocytin (PDB ID: 3WWK) (Fig. 12.3h) or podoplanin (PDB ID: 3WSR) (Fig. 12.3e) revealed that CLEC-2 utilizes a unique side face for ligand binding, which is distinct from the "top" conventional ligand recognition site of other CLRs (Nagae et al. 2014). The binding face of CLEC-2 has positively charged patches composed of basic residues (Arg107, Arg118, Arg152, and Arg157), which recognize the acidic residues of either rhodocytin or podoplanin. The sugar chain of podoplanin is recognized by Asn105, Arg118, and Tyr129, which are adjacent to the positive patches. The unique binding manner of CLEC-2 suggests that CLRs may utilize all surfaces as potential sites for the recognition of various ligands, including proteins, lipids, and sugars.

References

Appelmelk BJ, van Die I, van Vliet SJ, Vandenbroucke-Grauls CM, Geijtenbeek TB, van Kooyk Y (2003) Cutting edge: carbohydrate profiling identifies new pathogens that interact with dendritic cell-specific ICAM-3-grabbing nonintegrin on dendritic cells. J Immunol 170 (4):1635–1639

Borrego F, Masilamani M, Marusina AI, Tang X, Coligan JE (2006) The CD94/NKG2 family of receptors: from molecules and cells to clinical relevance. Immunol Res 35(3):263–278. doi:10.1385/IR:35:3:263

Chatwell L, Holla A, Kaufer BB, Skerra A (2008) The carbohydrate recognition domain of Langerin reveals high structural similarity with the one of DC-SIGN but an additional, calcium-independent sugar-binding site. Mol Immunol 45(7):1981–1994. doi:10.1016/j.molimm.2007.10.030

Feinberg H, Mitchell DA, Drickamer K, Weis WI (2001) Structural basis for selective recognition of oligosaccharides by DC-SIGN and DC-SIGNR. Science 294(5549):2163–2166. doi:10.1126/science.1066371

Feinberg H, Taylor ME, Weis WI (2007) Scavenger receptor C-type lectin binds to the leukocyte cell surface glycan Lewis(x) by a novel mechanism. J Biol Chem 282(23):17250–17258. doi:10.1074/jbc.M701624200

Feinberg H, Taylor ME, Razi N, McBride R, Knirel YA, Graham SA, Drickamer K, Weis WI (2011) Structural basis for langerin recognition of diverse pathogen and mammalian glycans through a single binding site. J Mol Biol 405(4):1027–1039. doi:10.1016/j.jmb.2010.11.039

Feinberg H, Jégouzo SA, Rowntree TJ, Guan Y, Brash MA, Taylor ME, Weis WI, Drickamer K (2013) Mechanism for recognition of an unusual mycobacterial glycolipid by the macrophage receptor mincle. J Biol Chem. doi:10.1074/jbc.M113.497149

Furukawa A, Kamishikiryo J, Mori D, Toyonaga K, Okabe Y, Toji A, Kanda R, Miyake Y, Ose T, Yamasaki S, Maenaka K (2013) Structural analysis for glycolipid recognition by the C-type lectins Mincle and MCL. Proc Natl Acad Sci U S A 110(43):17438–17443. doi:10.1073/pnas.1312649110

Guo Y, Feinberg H, Conroy E, Mitchell DA, Alvarez R, Blixt O, Taylor ME, Weis WI, Drickamer K (2004) Structural basis for distinct ligand-binding and targeting properties of the receptors DC-SIGN and DC-SIGNR. Nat Struct Mol Biol 11(7):591–598. doi:10.1038/nsmb784

Kaiser BK, Pizarro JC, Kerns J, Strong RK (2008) Structural basis for NKG2A/CD94 recognition of HLA-E. Proc Natl Acad Sci U S A 105(18):6696–6701. doi:10.1073/pnas.0802736105

Lee RT, Hsu TL, Huang SK, Hsieh SL, Wong CH, Lee YC (2011) Survey of immune-related, mannose/fucose-binding C-type lectin receptors reveals widely divergent sugar-binding specificities. Glycobiology 21(4):512–520. doi:10.1093/glycob/cwq193

Li Y, Hofmann M, Wang Q, Teng L, Chlewicki LK, Pircher H, Mariuzza RA (2009) Structure of natural killer cell receptor KLRG1 bound to E-cadherin reveals basis for MHC-independent missing self recognition. Immunity 31(1):35–46. doi:10.1016/j.immuni.2009.04.019

Li Y, Wang Q, Chen S, Brown PH, Mariuzza RA (2013) Structure of NKp65 bound to its keratinocyte ligand reveals basis for genetically linked recognition in natural killer gene complex. Proc Natl Acad Sci U S A 110(28):11505–11510. doi:10.1073/pnas.1303300110

Nagae M, Yamanaka K, Hanashima S, Ikeda A, Morita-Matsumoto K, Satoh T, Matsumoto N, Yamamoto K, Yamaguchi Y (2013) Recognition of bisecting N-acetylglucosamine: structural basis for asymmetric interaction with the mouse lectin dendritic cell inhibitory receptor 2. J Biol Chem 288(47):33598–33610. doi:10.1074/jbc.M113.513572

Nagae M, Morita-Matsumoto K, Kato M, Kaneko MK, Kato Y, Yamaguchi Y (2014) A platform of C-type lectin-like receptor CLEC-2 for binding O-glycosylated podoplanin and nonglycosylated rhodocytin. Structure 22(12):1711–1721. doi:10.1016/j.str.2014.09.009

Natarajan K, Dimasi N, Wang J, Mariuzza RA, Margulies DH (2002) Structure and function of natural killer cell receptors: multiple molecular solutions to self, nonself discrimination. Annu Rev Immunol 20:853–885. doi:10.1146/annurev.immunol.20.100301.064812

Ohki I, Ishigaki T, Oyama T, Matsunaga S, Xie Q, Ohnishi-Kameyama M, Murata T, Tsuchiya D, Machida S, Morikawa K, Tate S (2005) Crystal structure of human lectin-like, oxidized low-density lipoprotein receptor 1 ligand binding domain and its ligand recognition mode to OxLDL. Structure 13(6):905–917. doi:10.1016/j.str.2005.03.016

Park H, Adsit FG, Boyington JC (2005) The 1.4 angstrom crystal structure of the human oxidized low density lipoprotein receptor lox-1. J Biol Chem 280(14):13593–13599. doi:10.1074/jbc.M500768200

Petrie EJ, Clements CS, Lin J, Sullivan LC, Johnson D, Huyton T, Heroux A, Hoare HL, Beddoe T, Reid HH, Wilce MC, Brooks AG, Rossjohn J (2008) CD94-NKG2A recognition of human leukocyte antigen (HLA)-E bound to an HLA class I leader sequence. J Exp Med 205(3):725–735. doi:10.1084/jem.20072525

Silva-Martín N, Bartual SG, Ramírez-Aportela E, Chacón P, Park CG, Hermoso JA (2014) Structural basis for selective recognition of endogenous and microbial polysaccharides by macrophage receptor SIGN-R1. Structure 22(11):1595–1606. doi:10.1016/j.str.2014.09.001

Spreu J, Kuttruff S, Stejfova V, Dennehy KM, Schittek B, Steinle A (2010) Interaction of C-type lectin-like receptors NKp65 and KACL facilitates dedicated immune recognition of human keratinocytes. Proc Natl Acad Sci U S A 107(11):5100–5105. doi:10.1073/pnas.0913108107

Stambach NS, Taylor ME (2003) Characterization of carbohydrate recognition by langerin, a C-type lectin of Langerhans cells. Glycobiology 13(5):401–410. doi:10.1093/glycob/cwg045

Suzuki-Inoue K, Fuller GL, García A, Eble JA, Pöhlmann S, Inoue O, Gartner TK, Hughan SC, Pearce AC, Laing GD, Theakston RD, Schweighoffer E, Zitzmann N, Morita T, Tybulewicz VL, Ozaki Y, Watson SP (2006) A novel Syk-dependent mechanism of platelet activation by the C-type lectin receptor CLEC-2. Blood 107(2):542–549. doi:10.1182/blood-2005-05-1994

Suzuki-Inoue K, Kato Y, Inoue O, Kaneko MK, Mishima K, Yatomi Y, Yamazaki Y, Narimatsu H, Ozaki Y (2007) Involvement of the snake toxin receptor CLEC-2, in podoplanin-mediated platelet activation, by cancer cells. J Biol Chem 282(36):25993–26001. doi:10.1074/jbc.M702327200

Valladeau J, Ravel O, Dezutter-Dambuyant C, Moore K, Kleijmeer M, Liu Y, Duvert-Frances V, Vincent C, Schmitt D, Davoust J, Caux C, Lebecque S, Saeland S (2000) Langerin, a novel C-type lectin specific to Langerhans cells, is an endocytic receptor that induces the formation of Birbeck granules. Immunity 12(1):71–81

Weis WI, Drickamer K, Hendrickson WA (1992) Structure of a C-type mannose-binding protein complexed with an oligosaccharide. Nature 360(6400):127–134. doi:10.1038/360127a0

Zelensky AN, Gready JE (2005) The C-type lectin-like domain superfamily. FEBS J 272 (24):6179–6217. doi:10.1111/j.1742-4658.2005.05031.x

Zhang JG, Czabotar PE, Policheni AN, Caminschi I, Wan SS, Kitsoulis S, Tullett KM, Robin AY, Brammananth R, van Delft MF, Lu J, O'Reilly LA, Josefsson EC, Kile BT, Chin WJ, Mintern JD, Olshina MA, Wong W, Baum J, Wright MD, Huang DC, Mohandas N, Coppel RL, Colman PM, Nicola NA, Shortman K, Lahoud MH (2012) The dendritic cell receptor Clec9A binds damaged cells via exposed actin filaments. Immunity 36(4):646–657. doi:10.1016/j.immuni.2012.03.009

Chapter 13
The Ligands of C-Type Lectins

Amy J. Foster, Jessie H. Bird, Mattie S.M. Timmer, and Bridget L. Stocker

Abstract In this chapter, a comprehensive overview of the known ligands for the C-type lectins (CTLs) is provided. Emphasis has been placed on the chemical structure of the glycans that bind to the different CTLs and the amount of structural variation (or overlap) that each CTL can tolerate. In this way, both the synthetic carbohydrate chemist and the immunologist can more readily gain insight into the existing structure-activity space for the CTL ligands and, ideally, see areas of synergy that will help identify and refine the ligands for these receptors.

Keywords C-type lectin • Receptor • Ligand • Pathogen • Immunity • Carbohydrates • Glycolipids

13.1 Introduction

There has been much interest in identifying the ligands that bind to and activate C-type lectins (CTLs) and in determining how this ligand-receptor binding modulates the immune response. As illustrated in Table 13.1, most CTL ligands (CTLLs) are from exogenous sources, although several examples of endogenous CTLLs can also be found. Most CTLLs contain a glycan motif; however, the breath of CTLLs is diverse and also includes proteins and oligonucleotides, as well as molecules whose structure is still to be determined. There are also several CTLs, such as CLEC12B, CLEC1-1, DCAL-1, DCAR and mDCAR1, for which there are no known ligands. As previous chapters in this book have focussed on the biochemical pathways and immunomodulatory activities of the CTLs, our focus herein has been to showcase the CTLLs and, where relevant, to highlight the degree of structural variation each CTL can tolerate. By doing so, we hope to equip the reader with a more readily digestible data set on the vast array of CTLLs and to encourage natural products or synthetic chemists to identify new, or further refine, ligands for these receptors.

A.J. Foster • J.H. Bird • M.S.M. Timmer • B.L. Stocker (✉)
School of Chemical and Physical Sciences, Victoria University of Wellington, Wellington, New Zealand
e-mail: mattie.timmer@vuw.ac.nz; bridget.stocker@vuw.ac.nz

S. Yamasaki (ed.), *C-Type Lectin Receptors in Immunity*,
DOI 10.1007/978-4-431-56015-9_13

Table 13.1 C-type lectins and their ligands

Lectin	Structure	Origin	Lectin	Structure	Origin
Dectin-2	$Man_{7\text{-}9}GlcNAc_2$	Glycan array	SIGN-R1	Man, Fuc, Le^x, Le^y, SLe^x	Glycan array
	O-linked mannobiose-rich glycoproteins	*Malassezia*		Lacto-N-fucopentaose III (LNFP III)	Milk
	Mannose-capped LAM	*Mycobacterium*		Dextran	*Leuconostoc* spp.
MDL-1	Unidentified	Endogenous	LSECtin	GlcNAc, Fuc	Glycan array
BDCA-2	Non-sialylated complex-type glycans	Glycan array	Langerin	Man, Fuc, GlcNAc, $Man_9GlcNAc_2$	Glycan array
	Asialo-galactosyl-oligosaccharides	Chemical synthesis		Le^y, $6'\text{-}SO_4\text{-}Le^x$	Glycan array
Mincle	Trehalose dimycolate (TDM)	*Mycobacterium*		Dextran sulphate	*Leuconostoc* spp.
	Trehalose diesters (TDE) and trehalose monoesters (TME)	Chemical synthesis		$6\text{-}SO_4$-Gal, $6\text{-}SO_4$-GlcNAc, $6'\text{-}SO_4$-LacNAc	Polyacrylamide (PAA)-conjugates
	α-Mannose and mannitol derivatives	*Malassezia*		Heparan sulphate (HS) and chondroitin sulphate (CS)	Porcine
MCL	TDM	*Mycobacterium*		Heparin (HEP)	Chemical synthesis
Dectin-1	T-cell ligand	Endogenous		Laminarin (β-1,6- and β-1,3-glucan)	*Laminaria digitata* (brown alga)
	Unidentified N-glycans	Tumour cells		Fucoidan (α-1,3/4-(2/3-SO_4)-fucan)	*Fucus vesiculosus*
	Linear and branched β-1,3-glucans	Plant and fungal cell walls		Galactan (β-1,4-)	Plant
CLEC-2	Rhodocytin	Snake venom		Mannan (β-1,4-)	*Saccharomyces cerevisiae*
	NeuAcα2-3Galα1-3(NeuAcα2-6) GalNAcα	Endogenous podoplanin		Dextran (α-1,3-branched α-1,6-glucans)	*Leuconostoc* spp.
	Fucoidan (α-1,3/4-(2/3-SO_4)-fucan)	*Fucus vesiculosus*		Zymosan (β-1,3-glucan)	*Saccharomyces cerevisiae*
DNGR-1	Filamentous actin (F-actin)	Necrotic cells		Mannose and β-glucans	Fungi

SIGN-R3	Dextran (α-1,3-branched α-1,6-glucans)	*Leuconostoc* spp.	MGL	α- and β-GalNAc (Tn antigen), LacdiNAc, Sialyl-Tn	Endogenous
	Zymosan (β-1,3-glucan)	*Saccharomyces cerevisiae*		MUC-1 and Muc-2	Endogenous
	High mannose	Glycan array	MGL-1	Gal, GalNAc, Le^x, Le^a	Glycan array
	Fucosylated glycans (Le^a and Le^b)	Glycan array		Gb5	Chemical synthesis
	ManLAM and LM	*Mycobacterium*	MGL-2	α- and β-GalNAc, Gal, Tn, TF, core 2,	Glycan array
	Unidentified	*Leishmania*, commensal fungi and bacteria		GalNAc	Endogenous
DCIR	HIV-1 glycoprotein 140 (gp140)	HIV	LOX-1	Modified lipoprotein	Chemically modified endogenous
	Sulfo-Le^a, Le^a, Le^b	Synthetic PAA conjugates		Advanced glycation end product (AGE)	Synthetic BSA Conjugates
	Manα1-3(Manα1-6)Man	Synthetic BSA conjugates		Polyinosinic acid (Poly I)	Synthetic
	Gal, GalNAc, Glc and GlcNAc	Synthetic BSA conjugates		Carrageenan (type III Kappa)	Red algae
DCIR-2	N-glycan with bisecting GlcNAc	Chemical synthesis		Phosphatidylserine, phosphatidylinositol	Liposomes
MICL	Mono sodium urate (MSU) crystals	Synthetic		Phosphatidic acid, cardiolipin, phosphatidylglycerol	Liposomes
LY49Q	MHC-I	Endogenous	MR	ManLAM	*Mycobacterium*
DC-SIGN	Le^x, Le^y, LDNF	*Schistosoma mansoni*		Mannan	*Saccharomyces cerevisiae*
	ManLAM, PIM, α-glucan	*Mycobacterium*		Fuc, Man	Neoglycoproteins
	$Man_9GlcNAc_2$	Glycan array		3- and 4-SO_4-GalNAc	Chemical synthesis
	Mannan	*Saccharomyces cerevisiae*		Chondroitin sulphates A and B	Bovine and porcine
	TAG-72, semen clusterin	Endogenous		SO_4-Le^x, SO_4-Le^a	Chemical synthesis
	HIV-1 glycoprotein 120 (gp120)	HIV		High-mannose-containing glycoproteins	Endogenous
L-SIGN	$Man_9GlcNAc_2$	Chemical synthesis	DEC-205	CpG oligonucleotides	Chemical synthesis

13.2 C-Type Lectins Containing ITAM-Like Signalling Motifs

13.2.1 Dectin-2

Dectin-2, otherwise known as CLEC6A, is an FcRγ-coupled receptor found on macrophages (Mφs), monocytes and several subsets of dendritic cells (DCs) (Sancho and Reis e Sousa 2012). Binding assays have demonstrated that the extracellular portion of Dectin-2 can recognise the hyphal portions of *Candida albicans*, *Microsporum audouinii* and *Trichophyton rubrum* (Sato et al. 2006). In addition to fungal species, Dectin-2 has been shown to recognise *Schistosoma mansoni* egg antigens (Ritter et al. 2010). Although the exact ligand structure for Dectin-2 is not well defined, carbohydrate binding studies have revealed that Dectin-2 recognises high-mannose structures such as $Man_9GlcNAc_2$ (Fig. 13.1) (McGreal et al. 2006). Here, Dectin-2 was screened against 109 synthetic carbohydrates, and although Dectin-2 displayed the highest specificity for $Man_9GlcNAc_2$, it also recognised $Man_8GlcNAc_2$ and $Man_7GlcNAc_2$. Dectin-2 has also been shown to bind α-mannans of fungal cell walls (Saijo et al. 2010).

Dectin-2, in cooperation with Mincle, has been shown to recognise the pathogenic fungus *Malassezia*. Using solvent-based fractionation, it was determined that Dectin-2 recognised the hydrophilic components of *Malassezia*, and with the aid of mass spectrometry and NMR analysis, an *O*-linked mannobiose-rich glycoprotein was determined to be the Dectin-2 ligand (Ishikawa et al. 2013). The mannose-capped lipoarabinomannan LAM from mycobacterial species has also been identified as a ligand for Dectin-2 (Yonekawa et al. 2014). Moreover, Dectin-2 recognises a ligand on $CD4^+CD25^+$ T cells; however, the exact ligand structure is unknown (Aragane et al. 2003).

13.2.2 MDL-1

MDL-1 (myeloid DAP12-associating lectin-1 or CLEC5A) is a CTL expressed on the cell surface of monocytes, Mφs and osteoclasts (Sancho and Reis e Sousa 2012). MDL-1 has been shown to bind the dengue virion (DV), resulting in DAP12 phosphorylation and cytokine production (Chen et al. 2008). If, on the other hand, the DV-MDL-1 interaction is blocked with monoclonal antibodies, symptoms associated with DV infection, such as plasma leakage and vital organ haemorrhaging, were reduced in a murine model. In addition to the dengue virion, MDL-1 also directly interacts with the Japanese encephalitis virus and induces cytokine production by Mφs (Chen et al. 2012). MDL-1 knockout or blocking with an MDL-1 antibody was shown to reduce the symptoms of arthritis, which indicates that there may be an unidentified self-ligand for MDL-1 (Joyce-Shaikh et al. 2010). The same authors later proposed that galectin-9 was a ligand for MDL-1 and

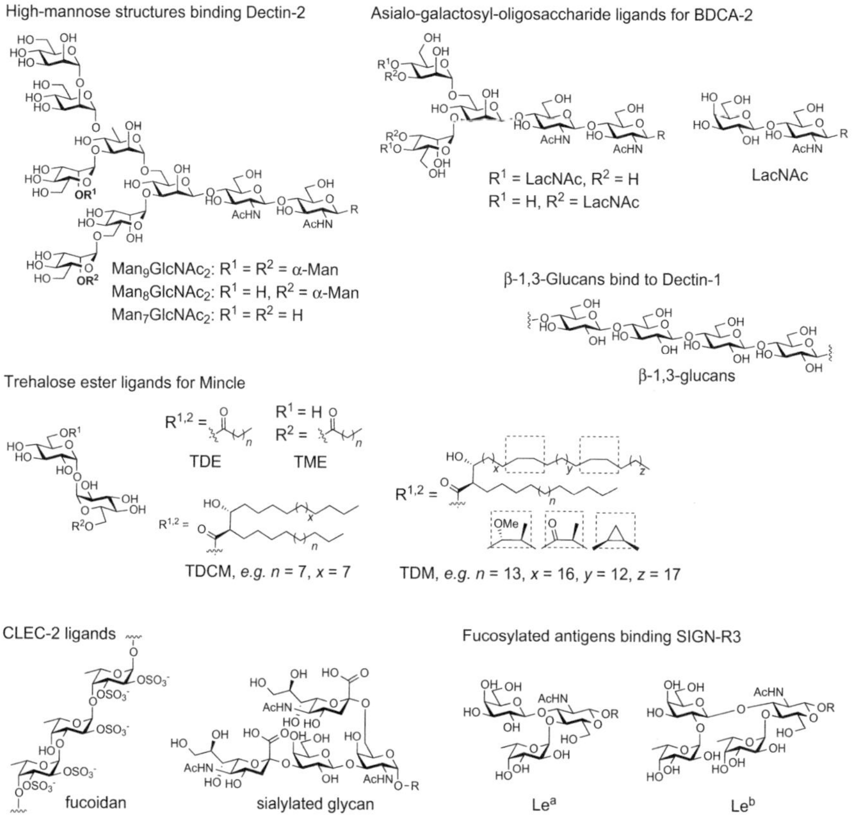

Fig. 13.1 Representative ligands for ITAM-like and Hem-ITAM-like receptors

demonstrated that treatment with galectin-9 intensified disease in a murine model of arthritis (Joyce-Shaikh et al. 2014).

13.2.3 BDCA-2

Blood dendritic cell antigen 2 (BDCA-2, also known as CLEC4C or CD303) belongs to the group II ITAM-coupled family of myeloid CLRs. BDCA-2 expression is restricted to human plasmacytoid DCs (Dzionek et al. 2001), and the targeting of antigens to this receptor has been suggested to be a promising strategy for inducing antigen-specific tolerance. BDCA-2 was first shown to bind the HIV protein, gp120, leading to the inhibition of toll-like receptor (TLR)-9-mediated activation and interferon (IFN)-γ secretion in plasmacytoid DCs (Martinelli et al. 2007), while more recently it was determined that BDCA-2 recognises complex-type sugars that have lost their terminal sialic residues (Fig. 13.1), such as

Galβ1-4GlcNAcβ1-2Manα1-3(Galβ1-4GlcNAcβ1-2Manα1-6)-Manβ1-4GlcNAcβ1-4GlcNAcβ and Galβ1-3GlcNAcβ1-2Manα1-3(Galβ1-3GlcNAcβ1-2Manα1-6)-Manβ1-4GlcNAcβ1-4GlcNAcβ (Riboldi et al. 2011). The recognition of complex galactose-terminated glycans by BDCA-2 has been suggested to be a mechanism that tumour cells or invading pathogens use to downregulate IFN-γ production and immune surveillance.

13.2.4 Mincle

Macrophage-inducible C-type lectin (Mincle, CLEC4e or CLECf9) is a group II FcRγ-coupled receptor that is expressed in low levels on Mφs and DCs (Sancho and e Sousa 2012). In a ligand-binding study that employed an NFAT-GFP reporter cell line and Mincle-deficient mice, it was demonstrated that trehalose dimycolate (TDM), the most abundant glycolipid in the mycobacterial cell wall, is a ligand for Mincle (Ishikawa et al. 2009). TDMs consist of a trehalose disaccharide core bound to two mycolic acid chains, which can be of varying structural complexity (Fig. 13.1). Additionally, both human and murine Mincle have been shown to recognise the yeast species *Candida albicans* (Bugarcic et al. 2008; Wells et al. 2008) and the fungal species *Malassezia* (Yamasaki et al. 2009). In *Malassezia* fungal species, specific mannitol-containing ligands were identified as direct ligands for Mincle (Ishikawa et al. 2013). Mincle has also been shown to sense dead cells through a protein component of small nuclear ribonucleoprotein (SAP30) (Yamasaki et al. 2008).

Structure-activity relationship studies have since confirmed that many simple TDM synthetic analogues also bind Mincle. Specifically, long-chain trehalose diesters (TDEs), including trehalose dibehenate (TDB), have been found to lead to the robust activation of Mφs (Schoenen et al. 2010; Khan et al. 2011), while more recently it has been demonstrated that trehalose esters with only one lipid chain (trehalose monoesters, TMEs) can also bind and activate Mincle (Stocker et al. 2014). Moreover, functionalised trehalose glycolipids, including those containing fluorescent reporter groups or photoaffinity probes, have also been shown to bind and activate Mincle (Khan et al. 2013; Kodar et al. 2015). In addition to trehalose diesters, another mycobacterial immunostimulatory lipid, glycerol monomycolate, has been shown to activate human but not mouse Mincle (Hattori et al. 2014). Corynomycolic esters of trehalose (TDCM) were also found to be potent activators of both mouse and human Mincle, while 2-*S*-corynomycolic esters of glycerol were found to activate human but not mouse Mincle (van der Peet et al. 2015).

13.2.5 MCL

Macrophage C-type lectin (MCL or CLEC4d) is another FcRγ-coupled receptor that is constitutively expressed on myeloid cells (Sancho and Reis e Sousa 2012). MCL is thought to arise via a gene duplication of Mincle, and while MCL contains a calcium coordination site, it does not retain the exact EPN motif of Mincle (Sancho and Reis e Sousa 2012). The exact ligand structure of MCL is not well defined; however, MCL has been shown to bind TDM (Miyake et al. 2013; Furukawa et al. 2013).

13.3 C-Type Lectins Containing Hem-ITAM-Like Signalling Motifs

13.3.1 Dectin-1

Dectin-1 (also known as CLEC7a) was the first non-toll-like receptor shown to mediate its own intracellular signalling (Sancho and Reis e Sousa 2012). The receptor was originally found to recognise an endogenous T-cell ligand of unknown structure (Ariizumi et al. 2000) and has since been shown to bind other endogenous ligands including undefined N-glycans on the surface of tumour cells, which leads to tumour destruction by natural killer (NK) cells (Chiba et al. 2014), and vimentin (a type III intermediate filament protein) (Thiagarajan et al. 2013). An unknown ligand in mycobacteria also binds Dectin-1 (Yadav and Schorey 2006; Rothfuchs et al. 2007).

The most insight into the structure of the Dectin-1 ligands, however, concerns the ability of the receptor to bind glucans, particularly β-1,3-glucans (Fig. 13.1), which are found in the cell wall of plants and fungi (Brown and Gordon 2001; Sancho and Reis e Sousa 2012). Microarray studies suggest that ten to eleven β-1,3-linked glucose oligomers are required for optimal Dectin-1 binding, and while β-1,6-glucans of comparable length do not bind (Palma et al. 2006), branched β-1,3-glucans do bind to the receptor (Palma et al. 2006; Adams et al. 2008). More recent studies have demonstrated that smaller glucan mimetics can also bind Dectin-1, as evidenced by studies using synthetic β-glu6 containing an α-(1→3)-linked bond (Li et al. 2013) and di- and trimeric hydroxylamine-based mimetics (Ferry et al. 2014). In the latter study, the binding of the small oligosaccharide fragments was attributed to the increased hydrophobic interaction between the α-face of the di- or trisaccharide and the aromatic side chains of Trp 221 ad His 223 in the binding site of Dectin-1. A distinction has also been made between the binding of soluble and particulate glucans. Different downstream Dectin-1 signalling occurs when myeloid cells come into contract with particulate β-glucan, and in particular, that prolonged Dectin-1 signalling occurs when myeloid cells come into contact with large β-glucan particles (Sancho and Reis e Sousa 2012). A few polysaccharides can also interfere with the binding of β-glucans to Dectin-1, thereby thwarting the host immune response. These include chitin-like components found on sclerotic cells in murine models of chromoblastomycosis (Dong et al. 2014) and α-(1,3)-glucan, which is a cell wall constituent of most fungal respiratory pathogens (Rappleye et al. 2007).

13.3.2 CLEC-2

Like Dectin-1, CLEC-2 (also known as CLEC-1B or CLEC-2B) also belongs to Group V of the non-calcium-dependent CTLs. CLEC-2 was discovered with CLEC-1 in a bioinformatics screen for NK receptors, and while mRNA for CLEC-2 has been found in bone marrow cells, DCs, monocytes, granulocytes and some NK cell populations, most studies have focussed on the role of CLEC-2 on platelets (Plato et al. 2013). Several CLEC-2 ligands have been identified, including rhodocytin (aggretin), which is an exogenous multimeric protein found in snake venom that leads to platelet activation and aggregation and subsequent coagulation of the blood (Suzuki-Inoue et al. 2011). Soon thereafter, the endogenous protein podoplanin was found to be a ligand for CLEC-2 (Suzuki-Inoue et al. 2007). Podoplanin is found in multiple cell types such as lymphatic endothelial cells, type I lung alveolar cells and in some cancer cells and consists of the sialylated glycan NeuAcα2-3Galβ1-3(NeuAcα2-6)GalNAc (Fig. 13.1) conjugated via an α-linkage to Thr52 in the platelet aggregation-stimulating domain (Kaneko et al. 2007). Both the disialyl core and the stereostructure of the protein were found to be critical for CLEC-2 binding, as evidenced by the observation that CLEC-2-Fc terminal deletion mutants and human podoplanin glycopeptides containing truncated glycans were unable to bind the receptor (Kato et al. 2008). Fucoidan (Fig. 13.1), which is a sulphated polysaccharide from *Fucus vesiculosus*, is an agonist for CLEC-2 (Manne et al. 2013), and HIV is also thought to be recognised by CLEC-2, although it is proposed that this recognition is due to the incorporation of a host protein into HIV during budding (Suzuki-Inoue et al. 2011). Finally, a series of synthetically prepared nucleic acid CLEC-2 ligands (aptamers) were identified using the systematic evolution of ligands by exponential enrichment (SELEX) methodology (Layzer et al. 2010).

13.3.3 DNGR-1

The DC, NK lectin group receptor-1 (DNGR-1), also known as CLEC9A, is expressed on specific subsets of DCs (Sancho and Reis e Sousa 2012). The receptor recognises filamentous actin (F-actin), which is exposed on the surface of necrotic cells and thus serves as an evolutionarily conserved damage-associated molecular pattern (Ahrens et al. 2012; Zhang et al. 2012). The binding of F-actin to DNGR-1 does not lead to pro-inflammatory responses, but, rather, signalling from the receptor is required for antigen cross-presentation and effective immunity (Plato et al. 2013). The ability of DNGR-1 to promote antigen cross-presentation has seen interest in the development of a peptide-conjugate vaccine via use of an anti-DNGR-1 antibody conjugated to the tumour-associated glycoprotein antigen, MUC1 (Picco et al. 2014).

13.3.4 SIGN-R3

Mouse SIGN-R3 (CD209d) is a receptor with endocytic activity and is part of a cluster of mouse SIGN-R genes that are highly homologous to human DC-SIGN; although unlike human DC-SIGN, mouse SIGN-R3 signals via a Syk-dependent pathway (Sancho and Reis e Sousa 2012). Ligand studies using transfected non-macrophage cell lines demonstrated that SIGN-R3 endocytosed dextran of 40 kDa or greater and zymosan (Takahara et al. 2004). A comprehensive glycan array analysis further refined the SIGN-R3 ligands and demonstrated that SIGN-R3 preferentially binds to high-mannose glycans and fucosylated glycans, particularly Lewis^{a} (Le^{a}) and Le^{b} antigens (Fig. 13.1) (Galustian et al. 2004), with subsequent array studies supporting these findings (Powlesland et al. 2006). SIGN-R3 has been found to contribute to early host resistance to *M. tuberculosis* infection with mycobaterial ManLAM and LM, but not AraLAM, binding and activating the receptor (Tanne et al. 2009). More recently, SIGN-R3 has been shown to play a role in leishmaniasis infection (Lefèvre et al. 2013) and to recognise ligands in commensal fungi and bacteria thereby potentially mediating colitis (Eriksson et al. 2013; Lightfoot et al. 2015); however, in each study, the specific ligand was not identified.

13.4 C-Type Lectins Containing ITIM-Like Signalling Motifs

13.4.1 DCIR

The C-type lectin dendritic cell immunoreceptor (DCIR) is an ITIM-coupled receptor. Human DCIR (CLEC4a) is expressed in monocytes, Mφs, granulocytes, B cells and DCs (Sancho and Reis e Sousa 2012). It has been demonstrated that DCIR can bind HIV-1 (Lambert et al. 2008), and while the specific ligand was not identified during this study, later work demonstrated that DCIR interacts with Le^{b} and Le^{a} (Fig. 13.2), mannotriose and sulfo-Le^{a} (Fig. 13.2) as well as the HIV-1-type glycoprotein, gp140 (Bloem et al. 2014). In a competitive binding study that compared the binding characteristics of several CTLs, DCIR was found to bind

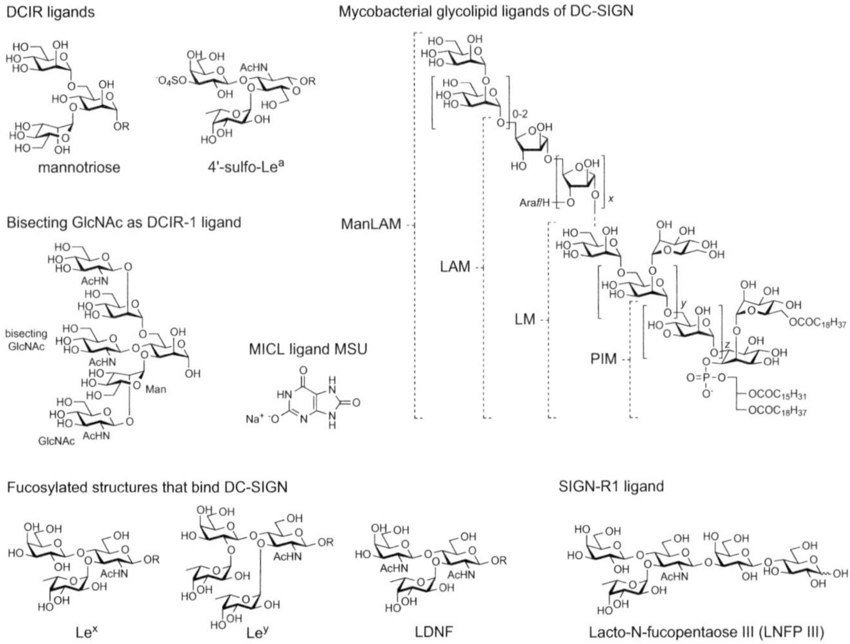

Fig. 13.2 Representative ligands for ITIM-like and ITAM-ITIM-independent CTLs

mannose- and fucose-based ligands as well as thio-linked Gal-, GalNAc-, Glc- and GlcNAc-BSA (Lee et al. 2011). In an additional study, it was demonstrated that purified DCIR could bind the glycan structures Le^b and Man_3; however, this binding was not detected when the DCIR was expressed on the cell surface (Bloem et al. 2013).

13.4.2 DCIR-1 and DCIR-2

Four DCIR homologues have been identified in mice (DCIR-1-4); however, only DCIR-1 and DCIR-2 contain an ITIM sequence. DCIR-1 is expressed in B cells, monocytes, Mφs and DCs; however, very little is known about its ligands (Sancho and Reis e Sousa 2012). DCIR-2, on the other hand, is expressed on DCs and has been found to specifically bind *N*-glycans that incorporate bisecting *N*-acetyl-glucosamine (a β-GlcNAc moiety attached to the N-glycan β-mannose 4-position) (Nagae et al. 2013). Here, the authors noted that DCIR-2 primarily recognises two residues including the GlcNAcβ1-2Manα1-3- and bisecting GlcNAc residues (Fig. 13.2).

13.4.3 MICL

Human myeloid inhibitory C-type lectin (MICL also known as DCAL-2, KLRL-1, CLL-1 and CLEC-12a) is an ITIM-coupled receptor expressed in granulocytes, monocytes, Mφs and DCs (Sancho and Reis e Sousa 2012). Murine MICL on the other hand is expressed in myeloid cells, B cells, $CD8^+$ T cells and bone marrow NK cells. Using flow cytometry and an Fc-mMICL fusion protein, MICL was found to bind several endogenous ligands from the heart, lung, liver, spleen and kidney (Pyz et al. 2008). More recently, MICL was determined to be a receptor for dead cells derived from 293T cells or thymocytes (Neumann et al. 2014). Here, ligand-binding studies demonstrated that both human and mouse MICL can recognise uric acid crystals (monosodium urate, Fig. 13.2), which are well-known cell death danger signals (Neumann et al. 2014).

13.4.4 LY49Q

Mouse Ly49Q (Klral7) is an inhibitory receptor that is expressed in $Ly6C/G^+$ myeloid precursors, immature monocytes and plasmacytoid DCs (Sancho and Reis e Sousa 2012). Reporter cell analysis was employed to demonstrate that $H\text{-}2^b$-derived tumour cells contain a high-affinity MHC-Ia-like ligand for LY49Q (Tai et al. 2007). In a subsequent study, LY49Q was identified as a direct receptor for MHC-I in mice (Scarpellino et al. 2007).

13.5 ITAM-ITIM-Independent C-Type Lectins

13.5.1 DC-SIGN

Dendritic cell-specific intercellular adhesion molecule-3-grabbing nonintegrin (DC-SIGN), also known as CD209, is a CTL involved in DC-T-cell contact. DC-SIGN has a single carbohydrate recognition domain (CRD) and binds branched D-mannose and L-fucose motifs common on pathogen surfaces (Sancho and Reis e Sousa 2012), with clustering of the lectins resulting in the formation of tetramers to enhance ligand binding (Mitchell et al. 2001). High-mannose oligosaccharide ligands include mannan, mannosylated lipoarabinomannan (ManLAM, Fig. 13.2) (Maeda et al. 2003), phosphatidyl-*myo*-inositol mannosides (PIM, Fig. 13.2) (Driessen et al. 2009) and $Man_9GlcNAc_2$ (Fig. 13.1), whereby the latter binds DC-SIGN with 130 times higher affinity than mannose (Mitchell et al. 2001). DC-SIGN has a higher affinity for L-fucose than mannose and recognises branched fucosylated structures with terminal galactose residues, such as the Lewis antigens (Coombs et al. 2005), in particular Le^x and possibly LDNF (Fig. 13.2) (van Die et al. 2003). DC-SIGN binds *Helicobacter pylori* (Miszczyk et al. 2012) and *S. mansoni* (Meyer et al. 2005) through the Le^x and Le^y antigens, while *Mycobacterium tuberculosis* is recognised via ManLAM (Maeda et al. 2003), PIM (Driessen et al. 2009) and α-glucan (Geurtsen et al. 2009), as well as further unidentified ligands (Ehlers 2010). Although this implies broad specificity for branched sugars, DC-SIGN discriminates between ligands through secondary binding sites and the α/β-linkage of adjacent saccharides (van Die et al. 2003). The signalling pathways induced by DC-SIGN are dependent on the nature of the ligand, leading to endocytosis or modulation of gene expression (Sancho and Reis e Sousa 2012). Endogenous ligands include tumour-associated glycoprotein-72 (TAG-72) (Laskarin et al. 2011) and semen clusterin (Sabatte et al. 2011), while DC-SIGN acts as a receptor for HIV through binding to the HIV-1 gp120 envelope protein (Curtis et al. 1992).

13.5.2 L-SIGN

L-SIGN or DC-SIGNR (also known as CD299, CD209L and CLEC4M) is a type II transmembrane C-type lectin receptor with 77 % sequence homology to DC-SIGN. In contrast to DC-SIGN which is expressed on DCs, L-SIGN is highly expressed on liver sinusoidal cells, endothelial vascular cells and in the lymph nodes. Like DC-SIGN, L-SIGN has high-affinity binding to a variety of ligands, including ICAM-3, HIV gp120-binding protein, simian immunodeficiency virus, Ebola virus, hepatitis C virus and respiratory syncytial virus. The CRD of L-SIGN binds the Man9GlcNAc2 oligosaccharide (Fig. 13.1) 17-fold more tightly than mannose, and its affinity for a glycopeptide bearing two $Man_9GlcNAc_2$ oligosaccharides is further increased by fivefold to 25-fold. These results indicate that the CRDs contain extended or secondary oligosaccharide binding sites. When the CRDs are clustered in the tetrameric extracellular domain, their arrangement provides a means of amplifying specificity for multiple glycans on host molecules targeted by DC-SIGN and L-SIGN (Mitchell et al. 2001).

13.5.3 SIGN-R1

The mouse CTL-specific ICAM-3 grabbing nonintegrin-related 1 (SIGN-R1, CD209b) is a homologue of hDC-SIGN and is expressed on a limited subset of Mφs and endothelial cells with a cell-specific expression similar to that of hL-SIGN. SIGN-R1 binds mannose- and fucose-containing ligands and Lewis blood antigens, thereby mirroring the specificity of hDC-SIGN and hL-SIGN, but in addition, SIGN-R1 also interacts with sialylated Le^x (Galustian et al. 2004; Koppel et al. 2005). Lacto-N-fucopentaose III (LNFPIII, Fig. 13.2) binds to the surface of cells transfected with SIGN-R1, and binding of LNFPIII-NGC to SIGN-R1 has been demonstrated by ELISA (Srivastava et al. 2014). SIGN-R1 binds zymosan, a glucan with repeating glucose units connected by β-1,3-glycosidic linkages, and the capsular polysaccharide of *Streptococcus pneumoniae*, and while SIGN-R1-Fc did not interact with dextran, which contains a combination of α-1,3- and α-1,6-glucose linkages, cellular-expressed SIGN-R1 does interact with dextran, as demonstrated by several groups (Geijtenbeek et al. 2002; Kang et al. 2003).

13.5.4 LSECtin

Liver and lymph node sinusoidal endothelial cell C-type lectin (LSECtin, CLEC4G) is a ~40 kDa type II integral membrane protein with a single C-type lectin-like domain, closest in homology to DC-SIGNR, DC-SIGN and CD23 (Liu et al. 2004). LSECtin functions as an attachment factor for Ebola virus and SARS, but it does not bind HIV or hepatitis C virus (Gramberg et al. 2005). LSECtin exhibits ligand-induced internalisation, and its sugar recognition specificity differs from that of DC-SIGN, as sugar-binding studies indicate that LSECtin specifically recognises *N*-acetyl-glucosamine (Dominguez-Soto et al. 2007) and L-fucose (Liu et al. 2004), whereas no LSECtin binding to mannan, *N*-acetyl-galactosamine and galactose were observed. The presence of LSECtin on myeloid cells should therefore contribute to expanding their antigen-capture and pathogen-recognition capabilities.

13.5.5 Langerin

Langerin (CLEC4K, CD207) is a type II transmembrane receptor with an extracellular region consisting of a neck and a C-terminal C-type CRD and is highly expressed on Langerhans cells, $CD103^+$ DCs and splenic $CD8^+$ DCs. Langerin recognises a wide array of carbohydrates including mannose, fucose and GlcNAc structures and especially $Man_9GlcNAc_2$ (Stambach et al. 2003). Langerin also recognises the difucosylated oligosaccharide Le^y (Holla et al. 2011) and Le^x-type sequences that are sulphated at the 6-position of the outer galactose (6′-sulfo-3′-SLe ^{x}, Fig. 13.3). This specificity is unique among the CTLs and contrasts markedly with the selectins which bind the analogous Le^x structures that are sulphated at the 3-position of the galactose. Of the sulphated saccharides, Langerin has also been shown to bind dextran-sulphate (Galustian et al. 2004) and 6-sulphated GlcNAc and especially 6′-sulfo-LacNAc (Tateno et al. 2010), whereas no binding was observed for either its positional isomer, 6-sulfo-LacNAc, or its unsulphated form. Langerin also binds to glioblastoma tissues via Gal-6-sulphated glycans (Tateno et al. 2010). Taken together, this suggests that the sulphate at C-6 of the non-reducing end sugar might be important for Langerin recognition.

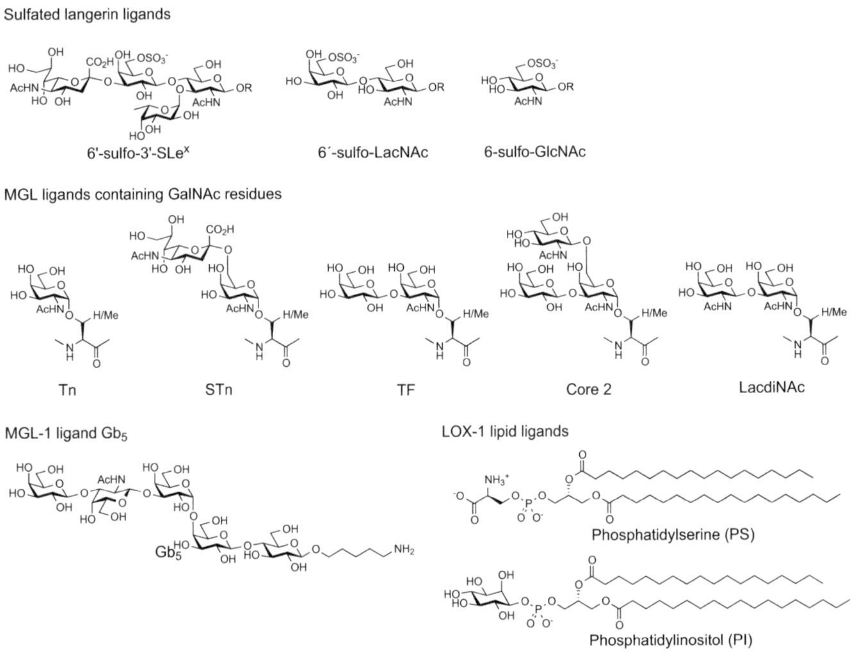

Fig. 13.3 Representative ligands for ITAM-ITIM-independent CTLs

Other studies have shown that glycosaminoglycans such as heparan sulphate (HS) and chondroitin sulphate (CS) (preferentially 4-O-sulphated) interact with Langerin through a Ca^{2+}-independent glycosaminoglycan (GAG)-specific binding site (Chabrol et al. 2012). Depending on ligand size, there are two binding modes for HS/heparin (HEP) oligosaccharides: a Ca^{2+}-dependent mode for small HEP trisaccharides and binding at a positively charged groove at the interface of two CRDs and the neck domain for large (6 kDa) HEP oligosaccharides (Muñoz-García et al. 2015). Further, polysaccharides found to bind Fc-Langerin include laminarin, fucoidan, galactan and α-mannan (Hsu et al. 2009), and langerin was shown to bind to high molecular weight dextran (250–2,000 kDa) and zymosan, although binding was inhibited by mannan (Takahara et al. 2004).

Langerin binds to a variety of microorganisms: langerin is a receptor for *Yersinia pestis* phagocytosis and promotes dissemination (Yang et al. 2015), and carbohydrate-dependent binding of langerin to a cell wall glycoprotein of *Mycobacterium leprae* has been observed (Kim et al. 2015). Langerin binds HIV-1, which prevents transmission (De Witte et al. 2007). Also *Candida* and *Saccharomyces* species and *Malassezia furfur* are recognised by langerin, but very weak binding was observed to *Cryptococcus gattii* and *Cryptococcus neoformans*. No binding was observed for the Gram-positive bacteria *Staphylococcus aureus* or Gram-negative *Escherichia coli* and *Salmonella typhimurium* (Takahara et al. 2004). Notably, Langerin has been identified as the primary fungal receptor on Langerhans cells (LCs), since the interaction of LCs with fungi was blocked by antibodies against Langerin. Langerin recognises both mannose and β-glucans present on fungal cell walls and appears to be an important fungal pathogen receptor on human LCs that recognises pathogenic and commensal fungi (De Jong et al. 2010). Interestingly, common polymorphisms in human langerin change the receptor specificity for glycan ligands (Feinberg et al. 2013).

13.5.6 MGL

Macrophage galactose-type lectin (MGL, CLEC-10A, CD301, DC-ASGPR), a type II transmembrane receptor, is expressed on immature and tolerogenic DCs, Mφs, dermal CD1a$^+$ DCs and blood CD1c$^+$ myeloid DCs. MGL specifically recognises α- and β-linked GalNAc residues (Suzuki et al. 1996), including the Tn antigen and LacdiNAc (Fig. 13.3) and 6-substituted GalNAc derivatives such as the sialyl-Tn antigen (Mortezai et al. 2013). In addition, MGL binds to CD45 on effector T cells and interacts with lymphatic endothelial cells through an unknown ligand. Of the self-antigens, MGL binds to the GalNAc moieties on the tumour-derived MUC1 and MUC2 glycoproteins (Iida et al. 1999). Pathogenic organisms that engage with MGL include *Neisseria gonorrhoeae*, *Campylobacter jejuni*, Ebola virus, *Schistosoma mansoni* (Van Vliet et al. 2005) and *Trichuris suis* (Van Kooyk et al. 2015).

13.5.7 MGL-1

The mouse CTL MGL-1 (CD301a, CLEC-10a) is one of the two hMGL orthologues with distinct carbohydrate recognition. Early studies demonstrated that mMGL-1 recognises galactose-related structures such as Le^x (Tsuiji et al. 2002). Using a glycan array, murine MGL-1 was found to be highly specific for Le^x and Le^a structures. The generation of MGL-1-Fc proteins allowed the identification of high endothelial venules as ligands in the lymph nodes (Singh et al. 2009), while the incubation of arrays with an MGL-1-hFc fusion protein showed up to tenfold increased binding to multiantennary N-glycans displaying Le^x structures compared to monovalent Le^x trisaccharide (Eriksson et al. 2014). In another glycan array, MGL-1 was found to bind to a range of terminal galactose and GalNAc glycans, which is consistent with the known galactose-binding motif QPD in the CRD of MGL-1. In these studies, MGL-1-Fc was found to bind the stage-specific tumour antigen Gb5 (Fig. 13.3), which suggests an association of this interaction with tumour progression (Maglinao et al. 2014).

MGL-1 also binds *Trypanosoma cruzi*, presumably through surface-expressed galactose moieties (Vázquez et al. 2014), and triggers phosphorylation of Raf-1 in response to products excreted/secreted by the helminth parasite *Taenia crassiceps*, thereby being involved in the induction of Th2 responses against the parasite (Terrazas et al. 2013). MGL-1 acts as an attachment and entry receptor for influenza virus, independent of sialic acid expression (Upham et al. 2010; Ng et al. 2014). Recombinant MGL-1 was found to bind both *Streptococcus* sp. and *Lactobacillus* sp. among commensal bacteria isolated from mesenteric lymph nodes of mice treated with dextran sulphate sodium salt (DSS) (Saba et al. 2009).

13.5.8 MGL-2

MGL-2 (CD301b) recognises carbohydrates containing GalNAc, which is similar to the carbohydrate specificity of human MGL (Tsuiji et al. 2002). Using a glycan array, MGL-2 was found to recognise GalNAc and galactose, including the O-linked Tn- and TF-antigens and core 2 O-GalNAc glycans (Fig. 13.3). Strikingly, MGL-2 interacted strongly with adenocarcinoma cells, suggesting a potential role in tumour immunity (Singh et al. 2009). MGL-2 specifically binds tumour-associated GalNAc, and modification of an antigen with GalNAc targeted the antigen specifically to the MGL-2 on bone marrow-derived (BM) DCs and splenic DCs and promoted antigen internalisation in DCs and presentation to CD4 T cells, as well as differentiation of IFN-γ producing CD4 T cells (Singh et al. 2011). Mice infected with the natural rodent hookworm pathogen *Nippostrongylus brasiliensis* required MGL-2^{+} DCs for efficient Th2 development, but these cells were dispensable for T follicular helper or B-cell responses, as MGL-2^{+} DC-depleted animals showed normal levels of IgG1 and IgE antibodies (Kumamoto et al. 2013).

13.5.9 LOX-1

Lectin-like oxidised low-density lipoprotein (ox-LDL) receptor-1 (LOX-1, CLEC8A) has been identified as the receptor for five diverse ligand classes (Chen and Du 2007). The first class is modified lipoproteins including ox-LDL (Sawamura et al. 1997), hypochlorite-modified high-density lipoprotein (HOCl-HDL) (Marsche et al. 2001), carbamylated LDL (Apostolov et al. 2009), electronegative LDL (Lu et al. 2009), apolipoprotein B (Gillotte et al. 2000; Okamura et al. 2013, and advanced glycation end product (AGE) proteins (Jono et al. 2002). Native LDL is not recognised by LOX-1 (Yoshimoto et al. 2011), while some studies suggest that acetylated LDL is recognised (Shi et al. 2001), and others indicate it is not (Moriwaki et al. 1998). The second group of ligands are polyanionic structures including polyinosinic acid and carrageenan (Moriwaki et al. 1998). Anionic phospholipids are also recognised, including the cellular ligands phosphatidylserine (PS) and phosphatidylinositol (PI) (Fig. 13.3) (Oka et al. 1998), which are expressed on apoptotic and aged cells. Cellular ligands comprise the fourth group; however, the exact molecules recognised by LOX-1 on these cells are less well defined. Apoptotic cells (Oka et al. 1998), platelets (Kakutani et al. 2000) and both Gram-negative bacteria such as *Escherichia coli* and Gram-positive bacteria such as *Staphylococcus aureus* are recognised (Shimaoka et al. 2001). Finally, other macromolecules that act as LOX-1 ligands include bile salt-dependent lipase (Chen and Du 2007), heats-hock proteins (Murshid et al. 2011) and C-reactive proteins (Shih et al. 2009).

13.5.10 Mannose Receptor

The mannose receptor (MR), also known as CD-206, is a 175-kDa transmembrane C-type lectin widely expressed on tissue Mφs and DCs (Martinez-Pomares 2012). MR binds branched high-mannose-containing motifs, such as ManLAM (Kang et al. 2005) and mannan (Taylor et al. 1992), and while MR preferentially binds branched α-linked oligomannoses, its specificity is not limited to D-mannose-containing glycoconjugates. Affinity binding competition studies demonstrate that MR binds to glycoconjugates with the following specificity: L-Fuc = D-Man > D-GlcNac ≈ D-Glc > D-Xyl >> D-Gal = L-Ara = D-Fuc, while galactose and GalNAc do not bind (Shepherd et al. 1981). Ligand binding is mediated by eight tandem CRDs, and while CRDs 1–3 have weak ligand binding, CRD-4 is able to elicit monosaccharide binding, and CRDs 4–8 are necessary for the binding of complex glycans (Mullin et al. 1994; Taylor et al. 1992). MR has a secondary lectin binding site located at the cysteine-rich N-terminus of the protein, and this mediates binding to sulphated sugar residues (Leteux et al. 2000), with special affinity for GalNAc residues sulphated at the 3- and 4-positions, including chondroitin sulphates A and B and sulphated Le^x and Le^a.

Given the breath of glycans to which MR binds, the receptor is thus able to recognise a variety of pathogens including viruses, fungi, bacteria and helminths, resulting in their phagocytosis, and many specific allergens including Ara h 1 (peanut), Bla g2 (cockroach), Can f 1 (dog), Der p 1 (mite), Der p 2 (mite) and Fel d 1 (cat) (Martinez-Pomares 2012). MR also acts as a molecular scavenger, and indeed the receptor was initially identified due to its ability to clear high-mannose-containing glycoproteins, such as lysosomal enzymes, from blood (Stahl et al. 1976; Martinez-Pomares 2012). Other endogenous ligands include salivary amylase, tissue plasminogen activator, thyroglobulin and serum secretory phospholipase A2-IIA (Martinez-Pomares 2012).

13.5.11 DEC-205

Mouse DEC-205 (CD205), a CTL that is highly expressed on $CD8\alpha^+$ DCs and to a lesser degree on macrophages, T cells, B cells, and granulocytes, recognises plasminogen activator (PLA) expressing bacteria such as Yersinia pestis and

Escherichia coli but not the PLA-negative controls (Sancho and Reis e Sousa 2012; Zhang et al. 2008). Both murine and human DEC-205 (which is widely expressed) act as receptors for dying cells (Shrimpton et al. 2009).

13.6 Summary

As evidenced above, the repertoire of CTLs and their associated ligands is immense. Some CTLs have been studied for many years, and, accordingly, their associated ligands are, by and large, well defined; however, for other CTLs, and especially those that appear to accommodate a vast array of ligands, much remains unknown about the specificity of ligand binding and how this influences the immune response. It is without a doubt that insight into the specific ligand structure for such CTLs will further assist in understanding how pathogens can either be recognised by the immune system or how they can thwart the immune response. Moreover, the association between CTLs and endogenous ligands can assist in understanding deleterious cellular process such as tumour growth and also regular cellular 'housekeeping' processes, such as debris clearance. Thus, it is imperative that immunologists and chemists continue to work closely together in order to determine how CTL-ligand interactions influence the many varied aspects of the immunology.

References

Adams EL, Rice PJ, Graves B et al (2008) Differential high-affinity interaction of Dectin-1 with natural or synthetic glucans is dependent upon primary structure and is influenced by polymer chain length and side-chain branching. J Pharmacol Exp Ther 325:115–123

Ahrens S, Zelenay S, Sancho D et al (2012) F-Actin is an evolutionarily conserved damage-associated molecular pattern recognized by DNGR-1, a receptor for dead cells. Immunity 36:635–645

Apostolov EO, Shah SV, Ray D et al (2009) Scavenger receptors of endothelial cells mediate the uptake and cellular proatherogenic effects of carbamylated LDL. Arterioscler Thromb Vasc Biol 29:1622–1630

Aragane Y, Maeda A, Schwarz A et al (2003) Involvement of Dectin-2 in ultraviolet radiation-induced tolerance. J Immunol 171:3801–3807

Ariizumi K, Shen GL, Shikano S et al (2000) Identification of a novel, dendritic cell-associated molecule, dectin-1, by subtractive cDNA cloning. J Biol Chem 275:20157–20167

Bloem K, Vuist IM, van der Plas AJ et al (2013) Ligand binding and signaling of dendritic cell immunoreceptor (DCIR) is modulated by the glycosylation of the carbohydrate recognition domain. PLoS ONE 8, e66266

Bloem K, Vuist IM, van den Berk M et al (2014) DCIR interacts with ligands from both endogenous and pathogenic origin. Immunol Lett 158:33–41

Brown GD, Gordon S (2001) Immune recognition: a new receptor for beta-glucans. Nature 413:36–37

Burgarcic A, Hitchens K, Beckhouse AG et al (2008) Human and mouse macrophage-inducible C-type lectin (Mincle) bind *Candida albicans*. Glycobiology 18:679–685

Chabrol E, Nurisso A, Daina A et al (2012) Glycosaminoglycans are interactants of Langerin: comparison with gp120 highlights an unexpected calcium-independent binding mode. PLoS ONE 7, e50722

Chen XP, Du GH (2007) Lectin-like oxidized low-density lipoprotein receptor-1: protein, ligands, expression and pathophysiological significance. Chin Med J 120:421–426

Chen ST, Lin YL, Huang MT (2008) CLEC5A is critical for dengue-virus-induced lethal disease. Nature 453:672–676

Chen S, Liu R, Wu M et al (2012) CLEC5A regulates Japanese encephalitis virus-induced neuroinflammation and lethality. PLoS Pathog 8, e1002655

Chiba S, Ikushima H, Ueki H et al (2014) Recognition of tumor cells by Dectin-1 orchestrates innate immune cells for anti-tumor responses. eLife 3, e04177

Coombs PJ, Graham SA, Drickamer K et al (2005) Selective binding of the scavenger receptor C-type lectin to Lewis(x) trisaccharide and related glycan ligands. J Biol Chem 280:22993–22999

Curtis BM, Scharnowske S, Watson AJ (1992) Sequence and expression of a membrane-associated C-type lectin that exhibits CD4-independent binding of human immunodeficiency virus envelope glycoprotein gp120. Proc Nat Acad Sci 89:8356–8360

De Jong MA, Vriend LE, Theelen B et al (2010) C-type lectin Langerin is a beta-glucan receptor on human Langerhans cells that recognizes opportunistic and pathogenic fungi. Mol Immunol 47:1216–1225

De Witte L, Nabatov A, Pion M et al (2007) Langerin is a natural barrier to HIV-1 transmission by Langerhans cells. Nat Med 13:367–371

Dominguez-Soto A, Aragoneses-Fenoll L, Martin-Gayo E et al (2007) The DC-SIGN-related lectin LSECtin mediates antigen capture and pathogen binding by human myeloid cells. Blood 109:5337–5345

Dong B, Li D, Li R et al (2014) A chitin-like component on sclerotic cells of Fonsecaea pedrosoi inhibits Dectin-1-mediated murine Th17 development by masking β-glucans. PLoS ONE 9, e114113

Driessen NN, Ummels R, Maaskant JJ et al (2009) Role of phosphatidylinositol mannosides in the interaction between mycobacteria and DC-SIGN. Infect Immun 77:4538–4547

Dzionek A, Sohma Y, Nagafune J et al (2001) BDCA-2, a novel plasmacytoid Dendritic Cell-specific Type II C-type Lectin, mediates antigen capture and is a potent inhibitor of interferon α/β Induction. J Exp Med 194:1823–1834

Ehlers S (2010) DC-SIGN and mannosylated surface structures of *Mycobacterium tuberculosis*: a deceptive liaison. Eur J Cell Biol 89:95–101

Eriksson M, Johannssen T, von Smolinski D et al (2013) The C-Type lectin receptor SIGNR3 binds to fungi present in commensal microbiota and influences immune regulation in experimental colitis. Front Immunol 4:196

Eriksson M, Serna S, Maglinao M et al (2014) Biological evaluation of multivalent LewisX-MGL-1 interactions. ChemBioChem 15:844–851

Feinberg H, Rowntree TJ, Tan SL et al (2013) Common polymorphisms in human langerin change specificity for glycan ligands. J Biol Chem 288:36762–36771

Ferry A, Malik G, Guinchard X et al (2014) Synthesis and evaluation of di- and trimeric hydroxylamine-based β-(1→3)-glucan mimetics. J Am Chem Soc 136:14852–14857

Furukawa A, Kamishikiryo J, Mori D et al (2013) Structural analysis for glycolipid recognition by the C-type lectins Mincle and MCL. Proc Natl Acad Sci U S A 110:17438–17443

Galustian C, Park CG, Chai W et al (2004) High and low affinity carbohydrate ligands revealed for murine SIGN-R1 by carbohydrate array and cell binding approaches, and differing specificities for SIGN-R3 and langerin. Int Immunol 16:853–866

Geijtenbeek TB, Groot PC, Nolte MA et al (2002) Marginal zone macrophages express a murine homologue of DC-SIGN that captures blood-borne antigens in vivo. Blood 100:2908–2916

Geurtsen J, Chedammi S, Mesters J et al (2009) Identification of mycobacterial α-glucan as a novel ligand for DC-SIGN: involvement of mycobacterial capsular polysaccharides in host immune modulation. J Immunol 183:5221–5231
Gillotte KL, Hörkkö S, Witztum JL et al (2000) Oxidized phospholipids, linked to apolipoprotein B of oxidized LDL, are ligands for macrophage scavenger receptors. J Lipid Res 41:824–833
Gramberg T, Hofmann H, Möller P et al (2005) LSECtin interacts with filovirus glycoproteins and the spike protein of SARS coronavirus. Virology 340:224–236
Hattori Y, Morita D, Fujiwara N et al (2014) Glycerol monomycolate is a novel ligand for the human, but not mouse macrophage inducible C-type lectin, Mincle. J Biol Chem 289:15405–15412
Holla A, Skerra A (2011) Comparative analysis reveals selective recognition of glycans by the dendritic cell receptors DC-SIGN and langerin. Protein Eng Des Sel 24:659–669
Hsu TL, Cheng SC, Yang WB et al (2009) Profiling carbohydrate-receptor interaction with recombinant innate immunity receptor-Fc fusion proteins. J Biol Chem 284:34479–34489
Iida S, Yamamoto K, Irimura T (1999) Interaction of human macrophage C-type lectin with O-linked N-acetylgalactosamine residues on mucin glycopeptides. J Biol Chem 274:10697–10705
Ishikawa E, Ishikawa T, Morita YS et al (2009) Direct recognition of the mycobacterial glycolipid, trehalose dimycolate, by C-type lectin Mincle. J Exp Med 206:2879–2888
Ishikawa T, Itoh F, Yoshida S et al (2013) Identification of distinct ligands for the C-type lectin receptors Mincle and Dectin-2 in the pathogenic fungus *Malassezia*. Cell Host Microbe 13:477–488
Jono T, Miyazaki A, Nagai R et al (2002) Lectin-like oxidized low density lipoprotein receptor-1 (LOX-1) serves as an endothelial receptor for advanced glycation end products (AGE). FEBS Lett 511:170–174
Joyce-Shaikh B, Bigler ME, Cao CC et al (2010) Myeloid DAP12-associating lectin (MDL)-1 regulates synovial inflammation and bone erosion associated with autoimmune arthritis. J Exp Med 207:579–589
Joyce-Shaikh B, Wilson DC, Cua DJ et al (2014) Mdl-1 Ligand. Patent WO20140227719
Kakutani M, Masaki T, Sawamura T (2000) A platelet-endothelium interaction mediated by lectin-like oxidized low-density lipoprotein receptor-1. Proc Natl Acad Sci U S A 97:360–364
Kaneko MK, Kato Y, Kameyama A et al (2007) Functional glycosylation of human podoplanin: glycan structure of platelet aggregation-inducing factor. FEBS Lett 581:331–336
Kang YS, Yamazaki S, Iyoda T et al (2003) SIGN-R1, a novel C-type lectin expressed by marginal zone macrophages in spleen, mediates uptake of the polysaccharide dextran. Int Immunol 15:177–186
Kang PB, Azad AK, Torrelles JB et al (2005) The human macrophage mannose receptor directs Mycobacterium tuberculosis lipoarabinomannan-mediated phagosome biogenesis. J Exp Med 202:987–999
Kato Y, Kaneko MK, Kunita A et al (2008) Molecular analysis of the pathophysiological binding of the platelet aggregation-inducing factor podoplanin to the C-type lectin-like receptor CLEC-2. Cancer Sci 99:54–61
Khan AA, Chee SH, McLaughlin RJ et al (2011) Long-chain lipids are required for the innate recognition of trehalose diesters by macrophages. ChemBioChem 12:2572–2576
Khan AA, Kamena F, Timmer MS et al (2013) Development of a benzophenone and alkyne functionalised trehalose probe to study trehalose dimycolate binding proteins. Org Biomol Chem 11:881–885
Kim HJ, Brennan PJ, Heaslip D et al (2015) Carbohydrate-dependent binding of langerin to SodC, a cell wall glycoprotein of Mycobacterium leprae. J Bacteriol 197:615–625
Kodar K, Eising S, Khan AA et al (2015) The uptake of Trehalose glycolipids by macrophages is independent of Mincle. ChemBioChem 16:683–693
Koppel EA, Ludwig IS, Appelmelk BJ et al (2005) Carbohydrate specificities of the murine DC-SIGN homologue mSIGNR1. Immunobiology 210:195–201

Kumamoto Y, Linehan M, Weinstein JS et al (2013) $CD301b^+$ dermal dendritic cells drive T helper 2 cell-mediated immunity. Immunity 39:733–743
Lahoud MH, Ahmet F, Zhang JG et al (2012) DEC-205 is a cell surface receptor for CpG oligonucleotides. Proc Natl Acad Sci U S A 109:16270–16275
Lambert AA, Gilbert C, Richard M et al (2008) The C-type lectin surface receptor DCIR acts as a new attachment factor for HIV-1 in dendritic cells and contributes to trans- and cis-infection pathways. Blood 112:1299–1307
Laskarin G, Redzovic A, Vlastelic I et al (2011) Tumor-associated glycoprotein (TAG-72) is a natural ligand for the C-type lectin-like domain that induces anti-inflammatory orientation of early pregnancy decidual $CD1a^+$ dendritic cells. Am J Reprod Immunol 88:12–23
Layzer JM, Mahanty SK, Redick CC et al (2010) Nucleic acid modulators of CLEC-2. Patent WO2012051571
Lee RT, Hsu TL, Huang SK et al (2011) Survey of immune-related, mannose/fucose-binding C-type lectin receptors reveals widely divergent sugar-binding specificities. Glycobiology 21:512–520
Lefèvre L, Lugo-Villarino G, Meunier E et al (2013) The C-type lectin receptors Dectin-1, MR, and SIGNR4 contribute both positively and negatively to the macrophage response to Leishmania infantum. Immunity 38:1038–1049
Leteux C, Chai W, Loveless RW et al (2000) The cysteine-rich domain of the macrophage mannose receptor is a multispecific lectin that recognizes chondroitin sulfates A and B and sulfated oligosaccharides of blood group $Lewis^a$ and $Lewis^x$ types in addition to the sulfated *N*-glycans of lutropin. J Exp Med 191:1117–1126
Li X, Wang J, Wang W et al (2013) Immunomodulatory activity of a novel, synthetic beta-glucan (β-glu6) in murine macrophages and human peripheral blood mononuclear cells. PLoS ONE 8, e80399
Lightfoot YL, Selle K, Yang T et al (2015) SIGNR3-dependent immune regulation by lactobacillus acidophilus surface layer protein A in colitis. EMBO J 34:881–895
Liu W, Tang L, Zhang G et al (2004) Characterization of a novel C-type lectin-like gene, LSECtin: demonstration of carbohydrate binding and expression in sinusoidal endothelial cells of liver and lymph node. J Biol Chem 279:18748–18758
Lu J, Yang JH, Burns AR et al (2009) Mediation of electronegative low-density lipoprotein signaling by LOX-1: a possible mechanism of endothelial apoptosis. Circ Res 104:619–627
Maeda N, Nigou J, Herrmann JL et al (2003) The cell surface receptor DC-SIGN discriminates between mycobacterium species through selective recognition of the mannose caps on lipoarabinomannan. J Biol Chem 278:5513–5516
Maglinao M, Eriksson M, Schlegel MK et al (2014) A platform to screen for C-type lectin receptor-binding carbohydrates and their potential for cell-specific targeting and immune modulation. J Control Release 175:36–42
Manne BK, Getz TM, Hughes CE et al (2013) Fucoidan is a novel platelet agonist for the C-type lectin-like receptor 2 (CLEC-2). J Biol Chem 288:7717–7726
Marsche G, Levak-Frank S, Quehenberger O et al (2001) Identification of the human analog of SR-BI and LOX-1 as receptors for hypochlorite-modified high-density lipoprotein on human umbilical venous endothelial cells. FASEB J 15:1095–1097
Martinelli E, Cicala C, Van Ryk D et al (2007) HIV-1 gp120 inhibits TLR9-mediated activation and IFN-γ secretion in plasmacytoid dendritic cells. Proc Natl Acad Sci U S A 104:3396–3401
Martinez-Pomares L (2012) The mannose receptor. J Leukoc Biol 92:1177–1186
McGreal EP, Rosas M, Brown GD et al (2006) The carbohydrate-recognition domain of Dectin-2 is a C-type lectin with specificity for high mannose. Glycobiology 16:422–430
Meyer S, Van Liempt E, Imberty A et al (2005) DC-SIGN mediates binding of dendritic cells to authentic pseudo-$Lewis^Y$ glycolipids of *Schistosoma mansoni* cercariae, the first parasite-specific ligand of DC-SIGN. J Biol Chem 280:37349–37359

Miszczyk E, Rudnicka K, Moran AP et al (2012) Interaction of Helicobacter pylori with C-type lectin dendritic cellspecific ICAM grabbing nonintegrin. J Biomed Biotechnol 2012, Article ID 206463, 10 pp

Mitchell DA, Fadden AJ, Drickamer K (2001) A novel mechanism of carbohydrate recognition by the C-type lectins DC-SIGN and DC-SIGNR: subunit organization and binding to multivalent ligands. J Biol Chem 276:28939–28945

Miyake Y, Toyonaga K, Mori D et al (2013) C-type lectin MCL is an FcRγ-coupled receptor that mediates the adjuvanticity of mycobacterial cord factor. Immunity 38:1050–1062

Moriwaki H, Kume N, Sawamura T et al (1998) Ligand specificity of LOX-1, a novel endothelial receptor for oxidized low density lipoprotein. Arterioscler Thromb Vasc Biol 18:1541–1547

Mortezai N, Behnken HN, Kurze AK et al (2013) Tumor-associated Neu5Ac-Tn and Neu5Gc-Tn antigens bind to C-type lectin CLEC10A (CD301, MGL). Glycobiology 23:844–852

Mullin NP, Hall KT, Taylor ME (1994) Characterization of ligand binding to a carbohydrate-recognition domain of the macrophage mannose receptor. J Biol Chem 269:28405–28413

Muñoz-García JC, Chabrol E, Vivès RR et al (2015) Langerin-heparin interaction: two binding sites for small and large ligands as revealed by a combination of NMR spectroscopy and cross-linking mapping experiments. J Am Chem Soc 137:4100–4110

Murshid A, Theriault J, Gong J et al (2011) Investigating receptors for extracellular heat shock proteins. Methods Mol Biol 787:289–302

Nagae M, Yamanaka K, Hanashima S et al (2013) Recognition of bisecting N-acetylglucosamine: structural basis for asymmetric interaction with the mouse lectin dendritic cell inhibitory receptor 2. J Biol Chem 288:33598–33610

Neumann K, Castiñeiras-Vilariño M, Höckendorf U et al (2014) CLEC12a is an inhibitory receptor for uric acid crystals that regulates inflammation in response to cell death. Immunology 40:389–399

Ng WC, Liong S, Tate MD et al (2014) The macrophage galactose-type lectin can function as an attachment and entry receptor for influenza virus. J Virol 88:1659–1672

Oka K, Sawamura T, Kikuta KI et al (1998) Lectin-like oxidized low-density lipoprotein receptor 1 mediates phagocytosis of aged/apoptotic cells in endothelial cells. Proc Natl Acad Sci U S A 95:9535–9540

Okamura T, Sekikawa A, Sawamura T et al (2013) LOX-1 ligands containing apolipoprotein B and carotid intima-media thickness in middle-aged community-dwelling US Caucasian and Japanese men. Atherosclerosis 229:240–245

Palma AS, Feizi T, Zhang Y et al (2006) Ligands for the β-glucan receptor, Dectin-1, assigned using "designer" microarrays of oligosaccharide probes (neoglycolipids) generated from glucan polysaccharides. J Biol Chem 281:5771–5779

Picco G, Beatson R, Taylor-Papadimitriou J et al (2014) Targeting DNGR-1 (CLEC9A) with antibody/MUC1 peptide conjugates as a vaccine for carcinomas. Eur J Immunol 44:1947–1955

Plato A, Willment JA, Brown GD (2013) C-Type lectin-like receptors of the Dectin-1 cluster: ligands and signalling pathways. Int Rev Immunol 32:134–156

Powlesland AS, Ward EM, Sadhu SK et al (2006) Widely divergent biochemical properties of the complete set of mouse DC-SIGN-related proteins. J Biol Chem 281:20440–20449

Pyż E, Huysamen C, Marshall ASJ et al (2008) Characterisation of murine MICL (CLEC12A) and evidence for an endogenous ligand. Eur J Immunol 38:1157–1163

Rappleye CA, Groppe Eissenberg L, Goldman WE (2007) Histoplasma capsulatum α-(1,3)-glucan blocks innate immune recognition by the β-glucan receptor. Proc Natl Acad Sci 104:1366–1370

Riboldi E, Daniele R, Parola C et al (2011) Human C-type lectin domain family 4, member C (CLEC4C/BDCA-2/CD303) is a receptor for asialo-galactosyl-oligosaccharides. J Biol Chem 286:35329–35333

Ritter M, Gross O, Kays S et al (2010) *Schistosoma mansoni* triggers Dectin-2, which activates the Nlrp3 inflammasome and alters adaptive immune responses. Proc Natl Acad Sci U S A 107:20459–20464

Rothfuchs AG, Bafica A, Feng CG et al (2007) Dectin-1 interaction with *Mycobacterium tuberculosis* leads to enhanced IL-12p40 production by splenic dendritic cells. J Immunol 179:3463–3471
Saba K, Denda-Nagai K, Irimura T (2009) A C-type lectin MGL1/CD301a plays an anti-inflammatory role in murine experimental colitis. Am J Pathol 174:144–152
Sabatte J, Faigle W, Ceballos A et al (2011) Semen clusterin is a novel DC-SIGN ligand. J Immunol 187:5299–5309
Saijo S, Ikeda S, Yamebe K et al (2010) Dectin-2 recognition of α-mannans and induction of Th17 cell differentiation is essential for host defense against *Candida albicans*. Immunity 32:681–691
Sancho D, Reis e Sousa C (2012) Signalling by myeloid C-type lectin receptors in immunity and homeostasis. Annu Rev Immunol 30:491–529
Sato K, Yang XI, Yudate T et al (2006) Dectin-2 is a pattern recognition receptor for fungi that couples with the Fc receptor γ chain to induce innate immune responses. J Biol Chem 281:38854–38866
Sawamura T, Kume N, Aoyama T et al (1997) An endothelial receptor for oxidized low-density lipoprotein. Nature 386:73–77
Scarpellino L, Oeschger F, Guillaume P (2007) Interactions of Ly49 family receptors with MHC class I ligands in trans and cis. J Immunol 178:1277–1284
Schoenen H, Bodendorfer B, Hitchens K et al (2010) Cutting edge: mincle is essential for recognition and adjuvanticity of the mycobacterial cord factor and its synthetic analog trehalose-dibehenate. J Immunol 184:2756–2760
Shepherd VL, Lee YC, Schlesinger PH et al (1981) L-Fucose-terminated glycoconjugates are recognized by pinocytosis receptors on macrophages. Proc Nat Acad Sci 78:1019–1022
Shi X, Niimi S, Ohtani T et al (2001) Characterization of residues and sequences of the carbohydrate recognition domain required for cell surface localization and ligand binding of human lectin-like oxidized LDL receptor. J Cell Sci 114:1273–1282
Shih HH, Zhang S, Cao W et al (2009) CRP is a novel ligand for the oxidized LDL receptor LOX-1. Am J Physiol Heart Circ Physiol 296:H1643–H1650
Shimaoka T, Kume N, Minami M et al (2001) LOX-1 supports adhesion of Gram-positive and Gram-negative bacteria. J Immunol 166:5108–5114
Shrimpton RE, Butler M, Morel AS et al (2009) CD205 (DEC-205): a recognition receptor for apoptotic and necrotic self. Mol Immunol 46:1229–1239
Singh SK, Streng-Ouwehand I, Litjens M et al (2009) Characterization of murine MGL1 and MGL2 C-type lectins: distinct glycan specificities and tumor binding properties. Mol Immunol 46:1240–1249
Singh SK, Streng-Ouwehand I, Litjens M et al (2011) Tumour-associated glycan modifications of antigen enhance MGL2 dependent uptake and MHC class I restricted CD8 T cell responses. Int J Cancer 128:1371–1383
Srivastava L, Tundup S, Choi BS et al (2014) Immunomodulatory glycan lacto-N-fucopentaose III requires clathrin-mediated endocytosis to induce alternative activation of antigen-presenting cells. Infect Immun 82:1891–1903
Stahl P, Schlesinger PH, Rodman JS et al (1976) Recognition of lysosomal glycosidases in vivo inhibited by modified glycoproteins. Nature 264:86–88
Stambach NS, Taylor ME (2003) Characterization of carbohydrate recognition by langerin, a C-type lectin of Langerhans cells. Glycobiology 13:401–410
Stocker BL, Khan AA, Chee SH et al (2014) On one leg: trehalose monoesters activate macrophages in a Mincle-dependent manner. ChemBioChem 15:382–388
Suzuki N, Yamamoto K, Toyoshima S et al (1996) Molecular cloning and expression of cDNA encoding human macrophage C-type lectin. Its unique carbohydrate binding specificity for Tn antigen. J Immunol 156:128–135
Suzuki-Inoue K, Kato Y, Inoue O et al (2007) Involvement of the snake toxin receptor CLEC-2, in podoplanin-mediated platelet activation, by cancer cells. J Biol Chem 282:25993–26001

Suzuki-Inoue K, Inoue O, Ozaki Y (2011) Novel platelet activation receptor CLEC-2: from discovery to prospects. J Thromb Haemost 9(Suppl 1):44–55

Tai LH, Goulet ML, Belanger S et al (2007) Recognition of H-2K^b by Ly49Q suggests a role for class Ia MHC regulation of plasmacytoid dendritic cell function. Mol Immunol 44:2638–2646

Takahara K, Yashima Y, Omatsu Y et al (2004) Functional comparison of the mouse DC-SIGN, SIGNR1, SIGNR3 and Langerin, C-type lectins. Int Immunol 16:819–829

Tanne A, Ma B, Boudou F et al (2009) A murine DC-SIGN homologue contributes to early host defense against Mycobacterium tuberculosis. J Exp Med 206:2205–2220

Tateno H, Ohnishi K, Yabe R et al (2010) Dual specificity of Langerin to sulfated and mannosylated glycans via a single C-type carbohydrate recognition domain. J Biol Chem 285:6390–6400

Taylor ME, Bezouska K, Drickamer K (1992) Contribution to ligand binding by multiple carbohydrate-recognition domains in the macrophage mannose receptor. J Biol Chem 267:1719–1726

Terrazas CA, Alcántara-Hernández M, Bonifaz L et al (2013) Helminth-excreted/secreted products are recognized by multiple receptors on DCs to block the TLR response and bias Th2 polarization in a cRAF dependent pathway. FASEB J 27:4547–4560

Thiagarajan PS, Yakubenko VP, Elsori DH et al (2013) Vimentin is an endogenous ligand for the pattern recognition receptor Dectin-1. Cardiovasc Res 99:494–504

Tsuiji M, Fujimori M, Ohashi Y et al (2002) Molecular cloning and characterization of a novel mouse macrophage C-type lectin, mMGL2, which has a distinct carbohydrate specificity from mMGL1. J Biol Chem 277:28892–28901

Upham JP, Pickett D, Irimura T et al (2010) Macrophage receptors for influenza A virus: role of the macrophage galactose-type lectin and mannose receptor in viral entry. J Virol 84:3730–3737

Van der Peet PL, Gunawan C, Torigoe S et al (2015) Corynomycolic acid-containing glycolipids signal through the pattern recognition receptor Mincle. Chem Commun 51:5100–5103

Van Die I, Van Vliet SJ, Nyame AK et al (2003) The dendritic cell-specific C-type lectin DC-SIGN is a receptor for *Schistosoma mansoni* egg antigens and recognizes the glycan antigen Lewis x. Glycobiology 13:471–478

Van Kooyk Y, Ilarregui JM, Van Vliet SJ (2015) Novel insights into the immunomodulatory role of the dendritic cell and macrophage-expressed C-type lectin MGL. Immunobiology 220:185–192

Van Vliet SJ, Van Liempt E, Saeland E et al (2005) Carbohydrate profiling reveals a distinctive role for the C-type lectin MGL in the recognition of helminth parasites and tumor antigens by dendritic cells. Int Immunol 17:661–669

Vázquez A, Ruiz-Rosado Jde D, Terrazas LI et al (2014) Mouse macrophage galactose-type lectin (mMGL) is critical for host resistance against *Trypanosoma cruzi* infection. Int J Biol Sci 10:909–920

Wells CA, Salvage-Jones JA, Li X et al (2008) The Macrophage-inducible C-type lectin, Mincle, is an essential component of the innate immune response to *Candida albicans*. J Immunol 180:7404–7413

Yadav M, Schorey JS (2006) The β-glucan receptor dectin-1 functions together with TLR2 to mediate macrophage activation by mycobacteria. Blood 108:3168–3175

Yamasaki S, Ishikawa E, Sakuma M et al (2008) Mincle is an ITAM-coupled activating receptor that senses damaged cells. Nat Immunol 9:1179–1188

Yamasaki S, Matsumoto M, Takeuchi O et al (2009) C-type lectin Mincle is an activating receptor for pathogenic fungus, *Malassezia*. Proc Natl Acad Sci U S A 106:1897–1902

Yang K, Park CG, Cheong C et al (2015) Host Langerin (CD207) is a receptor for Yersinia pestis phagocytosis and promotes dissemination. Immunol Cell Biol 93:815–24

Yonekawa A, Saijo S, Hoshino Y et al (2014) Dectin-2 is a direct receptor for mannose-capped lipoarabinomannan of Mycobacteria. Immunity 41:402–413

Yoshimoto R, Fujita Y, Kakino A et al (2011) The discovery of LOX-1, its ligands and clinical significance. Cardiovasc Drugs Ther 25:379–391

Zhang SS, Park C, Zhang P et al (2008) Plasminogen activator Pla of Yersinia pestis utilizes murine DEC-205 (CD205) as a receptor to promote dissemination. J Biol Chem 283:31511–31521

Zhang JG, Czabotar PE, Policheni AN et al (2012) The dendritic cell receptor CLEC9A binds damaged cells via exposed actin filaments. Immunity 36:646–657

MIX
Papier aus verantwortungsvollen Quellen
Paper from responsible sources
FSC® C105338

If you have any concerns about our products,
you can contact us on
ProductSafety@springernature.com

In case Publisher is established outside the EU,
the EU authorized representative is:
Springer Nature Customer Service Center GmbH
Europaplatz 3, 69115 Heidelberg, Germany

Printed by Libri Plureos GmbH
in Hamburg, Germany